Second Philosophy

Second Philosophy

A naturalistic method

Penelope Maddy

OXFORD
UNIVERSITY PRESS

Great Clarendon Street, Oxford OX2 6DP

Oxford University Press is a department of the University of Oxford.
It furthers the University's objective of excellence in research, scholarship,
and education by publishing worldwide in

Oxford New York

Auckland Cape Town Dar es Salaam Hong Kong Karachi
Kuala Lumpur Madrid Melbourne Mexico City Nairobi
New Delhi Shanghai Taipei Toronto

With offices in

Argentina Austria Brazil Chile Czech Republic France Greece
Guatemala Hungary Italy Japan Poland Portugal Singapore
South Korea Switzerland Thailand Turkey Ukraine Vietnam

Oxford is a registered trade mark of Oxford University Press
in the UK and in certain other countries

Published in the United States
by Oxford University Press Inc., New York

First published 2007

British Library Cataloguing in Publication Data

Data available

Library of Congress Cataloging in Publication Data

Data available

Typeset by Laserwords Private Limited, Chennai, India
Printed in Great Britain
on acid-free paper by
Biddles Ltd, King's Lynn, Norfolk

ISBN 978–0–19–927366–9

10 9 8 7 6 5 4 3 2 1

For David

Preface

The roots of this project trace to my previous book—*Naturalism in Mathematics*—in two ways, one foreseen and one not. The first is implicit in *Naturalism* itself, where typically 'philosophical' questions about mathematics are explicitly set aside in single-minded pursuit of purely methodological concerns (see [1997], p. 203). I intended from the start to return to these broader issues, an undertaking that seemed to me to require a preliminary foray into the philosophy of logic before circling back; the result is Part III and the three central sections on mathematics in Part IV. The second came as a surprise: when I set out on *Naturalism*, I assumed that everyone knew what it is to be a naturalist, that my job was to explain how to extend this idea to mathematics. What I discovered—from reactions to various talks and papers and to the book—was that everyone, naturalist or not, seemed to harbor his or her own firm notion of what 'naturalism' requires. To complete the story of *Naturalism*, I had to make my own firm notion explicit, which eventually led to the Second Philosophy described in Parts I, II, and the remainder of Part IV. Along the way, the nature of word–world relations presented itself as an opportune example for Part II because of the way debates over truth often intermingle with the topics of IV.4.

Several papers published since *Naturalism* contain embryonic versions of discussions found here. Much of 'Second philosophy' ([2003]) has made its way into I.1, I.2, I.6, and IV.1, and 'Mathematical existence' ([2005b]) makes up a goodly portion of IV.4. Smaller chunks of the survey paper 'Three forms of naturalism' ([2005a]) appear in I.6, and of 'Some naturalistic reflections on set theoretic method' ([2001a]) in IV.2.iii and IV.3. Traces of 'Three forms', 'Logic and the discursive intellect' ([1999]), 'Naturalism and the a priori' ([2000]), and 'Naturalism: friends and foes' ([2001b]) survive in III.1, III.1 and III.2, I.4, and I.7 and IV.1, respectively. My thanks to the Indian Council of Philosophical Research, the Association for Symbolic Logic, Oxford University Press, Springer Science and Business Media, the *Notre Dame Journal of Formal Logic*, and *Philosophical Perspectives* for permission to use this material. In addition,

'A naturalistic look at logic' ([2002]) constitutes a rough first pass at the position of Part III.

Finally, it's a great pleasure to acknowledge my many intellectual debts. Let me first extend special thanks to John Burgess and Mark Wilson, whose thinking has influenced far more than the specific sections (IV.4 and II.6) where their work is explicitly discussed. Many others have helped in a variety of ways, including Karl Americks, Jody Azzouni, Mark Balaguer, Jeff Barrett, Janet Broughton, Joshua Brown, Mark Colyvan, Christina Conroy, William Demopoulos, Lara Denis, Sean Ebels-Duggan, Hartry Field, Michael Friedman, Sam Hillier, Kent Johnson, Peter Koellner, Joe Lambert, Mary Leng, Cathay Liu, Colin McLarty, David Malament, Ruth Marcus, Patricia Marino, Teri Merrick, Alan Nelson, Charles Parsons, Brendan Purdy, John Rapalino, Brian Rogers, Waldemar Rohloff, Jeffrey Roland, Adina Roskies, Barbara Sarnecka, Sally Sedgwick, Stewart Shapiro, Rory Smead, Kyle Stanford, John Steel, Jamie Tappenden, Carla Valenzuela, Nick White, Crispin Wright, Steve Yablo, and Kevin Zollman. I'm particularly grateful to Hillier, Liu, Malament, Purdy, Rapalino, Rogers, Rohloff, Smead, Valenzuela, and Zollman for participating in an informal discussion group that read through the penultimate draft of the entire manuscript and made many good suggestions, and to Malament again, for keeping me honest on the physics and for encouraging words when they were needed most. Thanks finally to Peter Momtchiloff of Oxford University Press for supporting the project from the beginning.

Near the end of the *Tractatus* (in 6.53), Wittgenstein remarks, 'The right method of philosophy would be this. To say nothing except what can be said, *i.e.*, the propositions of natural science.' Of course, 'what can be said' and 'the propositions of natural science' are heavily weighted theoretical terms for Wittgenstein, but if the words 'to say nothing except ... natural science' were understood in their ordinary, rough and ready senses, this 'right method' would be Second Philosophy. Ever tempted to overstep, Wittgenstein adds 'and when someone else wished to say something metaphysical, to demonstrate to him that he had given no meaning to certain signs'. The Second Philosopher indulges in no such corrective project: her reaction to extra-scientific philosophy is puzzlement; she asks methodically after its standards and goals, and assesses these by her own lights. Wittgenstein concludes that 'this method would be unsatisfying to the other—he would not have the feeling that we were teaching him philosophy'. My hope here is to discredit this last claim, to suggest that the Second Philosopher can address traditionally philosophical questions—and that she can answer them.

P. M.

Irvine, California
July 2006

Contents

Introduction

These days, as more and more philosophers count themselves as naturalists, the term has come to mark little more than a vague science-friendliness. To qualify as unnaturalistic, a contemporary thinker has to insist, for example, that epistemology is an a priori discipline with nothing to learn from empirical psychology or that metaphysical intuitions show quantum mechanics to be false. There are those who take such positions, of course, but to lump everybody else under one rubric is clearly too crude a diagnostic. My goal in this book is to delineate and to practice a particularly austere form of naturalism. One minor difficulty is that the term 'naturalism' has acquired so many associations over the years that using it tends to invite indignant responses of the form, 'but that can't be naturalism! Naturalism has to be like this!' As my project is to spell out an approach that differs in subtle but fundamental ways from other 'naturalisms', it seems best to coin a new term, on the assumption that I will then be permitted to stipulate what I intend it to mean. Thus, 'Second Philosophy'.

A deeper difficulty springs from the lesson won through decades of study in the philosophy of science: there is no hard and fast specification of what 'science' must be, no determinate criterion of the form '*x* is science iff... '. It follows that there can be no straightforward definition of Second Philosophy along the lines 'trust only the methods of science'. Thus Second Philosophy, as I understand it, isn't a set of beliefs, a set of propositions to be affirmed; it has no theory. Since its contours can't be drawn by outright definition, I resort to the device of introducing a character, a particular sort of idealized inquirer called the Second Philosopher, and proceed by describing her thoughts and practices in a range of contexts; Second Philosophy is then to be understood as the product of her inquiries.

This Second Philosopher is equally at home in anthropology, astronomy, biology, botany, chemistry, linguistics, neuroscience, physics, physiology, psychology, sociology, ... and even mathematics, once she realizes how central it is to her ongoing effort to understand the world. Her interest in other subjects, at least as far as we see her here, is limited to her pursuit of their anthropology, psychology, sociology, and so on. She uses what we typically describe with our rough and ready term 'scientific methods', but again without any definitive way of characterizing exactly what that term entails. She simply begins from commonsense perception and proceeds from there to systematic observation, active experimentation, theory formation and testing, working all the while to assess, correct, and improve her methods as she goes.

Though the Second Philosopher's approach is what we would typically term 'scientific', I contend that she is fully capable of appreciating and addressing a wide range of questions we would just as typically regard as 'philosophical'. The central examples here are fundamental questions in the philosophies of logic and mathematics: what is the ground of logical and mathematical truth? how do we come to know such truths? what role do they play in our investigation of the world? These issues take center stage in Parts III and IV, but before I can describe the Second Philosopher's take on them, I need to explain who she is and how she operates. This is the main goal of Parts I and II. The Second Philosopher is introduced in I.1 in contrast with Descartes's First Philosopher and the rough outlines of her character emerge gradually, by a sustained exercise in compare and contrast, through the step-by-step historical review of Part I. Toward the end of Part I, she begins to advance positions of her own (in I.6 and I.7).

Part II employs a different technique to illuminate Second Philosophy. It takes up a well-known contemporary debate over the nature of truth and reference and of word–world relations more generally and asks to what extent it can be understood as a piece of naturally occurring Second Philosophy; the process of reconfiguring the question in the Second Philosopher's terms should provide further insight into her motivations and methods. In the end, she stakes out a tentative position of her own on the topics at issue (and her second-philosophical understanding of the available options eventually helps clarify the central ontological discussion of IV.4). The stage is then set for the sustained pursuit of Second Philosophy in Parts III and IV, primarily in the philosophy of logic and mathematics.

Though 'Second Philosophy' is never explicitly defined in all this, I hope that Parts I and II provide enough guidance for at least some sympathetic readers to get the hang of how to carry on. I should note that the delineation of Second Philosophy itself and the pursuit of particular questions by the Second Philosopher are independent: for example, one might adopt the account of logical truth in Part III or of mathematical ontology in Part IV without buying into the full austerity of the second-philosophical method in all things; conversely one might sign on as a Second Philosopher while thinking I've gone astray in my pursuit of the particulars. Finally of course none of this amounts to an argument that we should all strive to conduct ourselves as Second Philosophers. My hope is that the appeal of the approach will be obvious to the susceptible.

Those content to allow the book's line of thought to unfold in its own time are encouraged to skip from here directly to Part I, perhaps returning to the rest of this introduction for summaries as desired. For those who prefer to read the reviews before seeing the movie, even at the risk of spoilers, let me sketch the upcoming terrain in more detail.

In I.1 we see our first example of the practical consequences of the Second Philosopher's lack of a criterion for demarcating science from non-science: when Descartes proposes that she adopt his Method of Doubt, she doesn't reject it as 'unscientific'; impressed by the promised pay-off—a firmer foundation for her beliefs—she's quite willing to give his proposal a try; she eventually discards it only as it proves ineffective. When the contemporary skeptic issues his challenge in I.2—claiming that her commonsense methods, even as corrected by her more developed and self-conscious inquiries, lead to radical skepticism—the Second Philosopher is troubled; only the hard-won conviction that there is some sleight of hand in his arguments, that they don't actually proceed from common sense, sets her mind at ease. Once she understands the peculiarly philosophical way he wishes to pose the question of knowledge—an understanding apparently closed to Moore—she can sympathize with his desire that all our methods be justified in a way that doesn't presuppose any of them, but she doesn't regard the impossibility of gratifying that desire as undermining her reasonable beliefs about the world. I.3 raises the possibility that the naturalistic Hume, originator of the empirical Science of Man, may have uncovered a more viable route from common sense to radical skepticism. Here we see our first example of another recurring phenomenon: a noble attempt at naturalism that loses its way.

The discussion of Kant in I.4 introduces another perennial motif: the two-level philosophical theory. In his effort to account for a priori knowledge of the world, Kant undertakes a transcendental inquiry wholly distinct from ordinary science (in his terms, empirical inquiry). (Here it's Kant, not the Second Philosopher, who draws a science/non-science distinction. This is typical of two-level views.) What makes the view two-leveled in the intended sense is that Kant's empirical inquiry is methodologically independent of his transcendental considerations, that is, ordinary science is entirely in order for purposes of investigating the empirical world, it's just that Kant also has other purposes. Notice that in these terms, the Descartes of I.1 isn't proposing a two-level project: he doesn't regard science as methodologically independent, as entirely in order for its purposes; he's out to correct it, to improve its foundations. Still, as in her reaction to Descartes, the Second Philosopher doesn't reject transcendental inquiry as extra-scientific; open-minded as always, she asks Kant, as she did Descartes, why she should undertake his distinctive study, what purposes it's intended to serve, and simply comes away unpersuaded. (I.4 also lays the groundwork for the discussion of Kant's view of logic in III.2.) I.5 on Carnap's project of rational reconstruction presents another two-level position and an analogous second-philosophical response.

Consideration of Carnap leads inevitably to the celebrated Quine, father of contemporary naturalism and direct inspiration for Second Philosophy. Alas, the task of I.6 is to point out how the Second Philosopher differs from the Quinean naturalist in matters great and small, and this initial separation broadens in the second-philosophical account of logic in Part III and of mathematical ontology in Part IV. For now perhaps it's enough to note that the Second Philosopher is born native to her scientific (our term) worldview, she isn't driven to it, as Quine's naturalist seems to be, by despair over the failed Cartesian project of grounding science. More substantive disagreements concern radical skepticism and the nature of naturalized epistemology, and holism and the confirmation of theories (as illustrated by the case of atomic theory which returns at intervals throughout the book).

Part I concludes with a look at Putnam's Quine-inspired naturalism of the 1970s and especially his subsequent anti-naturalism of the 1980s. This later Putnam develops yet another two-level position, with predictable reactions from the Second Philosopher, but his critique also helps clarify her position on the status of inquirers whose evidential standards differ starkly from her own. Of course the Second Philosopher, using her methods, can justify her standards and expose the shortcomings of the astrologer's, but presumably the astrologer, using his methods, can likewise justify his standards and find fault with hers. A certain breed of naturalist might conclude that the astrologer's position is as good as her own—this is 'relativism'—but the Second Philosopher is unimpressed by the astrologer's efforts; she has every reason to trust her well-honed means of investigation and they show his to be misguided—this is 'imperialism'. In addition, Putnam's critique spotlights the theory of truth and the role it might play in an empirical explanation of how human language use functions in our dealings with each other and with the world—a question that helps shape the discussion of Part II.

In sum, then, Part I aims to zero in on the nature of second-philosophical inquiry by tracing the Second Philosopher's reaction to various skeptical challenges and two-level positions, and by comparing and contrasting her with such naturalistic thinkers as Hume and Quine. Part II takes a different approach. Given the current tendency toward 'naturalism' in its various forms, given that Second Philosophy isn't a theory but simply a way of conducting philosophical inquiry, we might expect to find Second Philosophy taking place somewhere, in practice if not in name. After a brief opening discussion (in II.1) to allay the worry that there's nothing left for the Second Philosopher to do, Part II explores a particularly promising case: the contemporary debate over the nature of word–world connections that arose after Hartry Field's 1972 criticism of Tarski's theory of truth. II.2 traces the discussion from Tarski to Field to Stephen Leeds; II.3 focuses on Field's valence analogy to show how the Second Philosopher is motivated by more concrete explanatory goals than the others. With the debate reconfigured second-philosophically, II.4 sketches a form of disquotationalism that descends from Field and Leeds, and II.5 clarifies its structure by contrasting it with the minimalism of Crispin Wright and Paul Horwich. Finally, in II.6, we meet in Mark Wilson a true Second Philosopher and help ourselves to a few of his many insights to deepen the account begun in II.4.

In Part II, then, I hope to have illustrated how an apparently naturalistic philosophical discussion is subtly reconfigured when regarded from the Second Philosopher's point of view, and to have demonstrated that there is at least one natural-born Second Philosopher at work today. Along the way, a tentative second-philosophical take on truth, reference, and word–world relations has emerged, but what carries forward to the ontological discussion of IV.4 is not this particular view, but the accompanying second-philosophical understanding of both correspondence and disquotational theories. (There I argue that the ontological distinctions at issue are independent of one's stand on truth.)

Parts I and II are designed primarily to illustrate the nature of Second Philosophy; assuming they've done their job, Parts III and IV attempt to practice it. My aim is to provide a philosophical backdrop for the methodological views of *Naturalism in Mathematics*, but this effort requires a prior investigation of the nature of logical truth, the subject of Part III. III.1 gives a brief survey of familiar naturalistic options. Building on the sketch of Kant's views in I.4, III.2 outlines a Kantian account of logic; III.3 converts it into a possibility open to the Second Philosopher. III.4 and III.5 examine the viability of the claims this position makes about the structure of the physical world and of human cognition, producing a modified account with considerable claim to empirical support. The status of the rudimentary logic it validates is examined in III.6: contingent, perhaps a priori in some sense, empirical though difficult to revise, not obviously analytic in any useful way. III.7 catalogs the restrictions and idealizations along the path from this rudimentary logic to full classical logic, pausing to note the various deviant logics that present themselves as alternatives. Finally, the empirical contingencies on which this view rests are reviewed in III.8.

At last the stage is set for a second-philosophical look at mathematics. The opening section of Part IV returns to the themes of Part I—skepticism and two-level positions—but this time in the context of current debates in the philosophy of science; this serves as a transition to the discussion of applied mathematics in IV.2. There I draw the consequences of the Second Philosopher's rejection of holism (in I.6) for the Quine/Putnam indispensability arguments for mathematics realism, explore the extent to which mathematical structures are physically realized, and attempt a mild debunking of the purported 'miracle of applied mathematics'. IV.3 returns to the central topic of *Naturalism*, the methodology of pure mathematics: having discerned that mathematics is an invaluable aid to her investigation of the world, the Second Philosopher undertakes to pursue it herself; her methodological decisions are then based on an analysis of the goals of the practice and of the effectiveness of available means for reaching them. This brings us finally to the classic ontological and epistemological questions about mathematics: what is the nature of mathematical truth?

and how can we come to know it? In IV.4, I present three very general styles of answer to these questions—Robust Realism, Thin Realism, and Arealism—and argue that despite appearances Thin Realism is closer to Arealism than to Robust Realism. For the Second Philosopher, Robust Realism is problematic (including, alas, the position of my [1990]). I suggest that Thin Realism is independent of the debate over truth (a pay-off from II.4), that Arealism can accommodate the application of mathematics, and indeed that Thin Realism and Arealism are superficial variants of the same underlying position. Part IV concludes with a look at the broader prospects for metaphysics in the second-philosophical spirit.

One last comment by way of orientation: the discussions of IV.3, IV.4, and IV.5 may, indeed should, strike the reader as especially open-ended. Surely more can be said about the search for new set theoretic axioms and possible solutions to the Continuum Problem, about the workings of Thin Realism and Arealism, about the case for the atomic hypothesis, and about Second Metaphysics more generally! Let me just say, that is my hope.

PART I
What is Second Philosophy?

I.1

Descartes's first philosophy

To explain what 'Second Philosophy' is supposed to be, I should begin with René Descartes and his *Meditations on First Philosophy* (1641a). The key to this work is Descartes's dramatic Method of Doubt.[1] It begins modestly enough, noting that our senses sometimes deceive us about objects that are very small or very distant, but quickly moves on to perceptual reports that seem beyond question, like my current belief that 'this is a hand' (as I hold up my hand and look at it). Still, the meditator wonders, might I not be mad, or asleep?

> Yet at the moment my eyes are certainly wide awake ... as I stretch out and feel my hand I do so deliberately, and I know what I am doing. All this would not happen with such distinctness to one asleep. Indeed! As if I did not remember other occasions when I have been tricked by exactly similar thoughts while asleep! As I think about this more carefully, I see plainly that there are never any sure signs by means of which being awake can be distinguished from being asleep. The result is that I begin to feel dazed ... Perhaps ... I do not even have ... hands ... at all. (Descartes [1641a], p. 13)

In his dizziness, the meditator anxiously grasps for a fixed point:

> ... whether I am awake or asleep, two and three added together are five, and a square has no more than four sides. It seems impossible that such transparent truths should incur any suspicion of being false. (Descartes [1641a], p. 14)

But the midnight fears cannot be stopped. What if God is a deceiver, or worse, what if there is no God, and I am as I am by mere chance? Mightn't I then be wrong in absolutely all my beliefs?

[1] The following account of Descartes's goals and strategies comes from the elegant and enlightening Broughton [2002].

I have no answer to these arguments, but am finally compelled to admit that there is not one of my former beliefs about which a doubt may not properly be raised. (Descartes [1641a], pp. 14–15)

And he concludes that

in future I must withhold my assent from these former beliefs just as carefully as I would from obvious falsehoods, if I want to discover any certainty.... I will suppose therefore that... some malicious demon of the utmost power and cunning has employed all his energies in order to deceive me. I shall think that the sky, the air, the earth, colours, shapes, sounds and all external things are merely the delusions of dreams which he has devised to ensnare my judgement. I shall consider myself as not having hands or eyes, or flesh, or blood or senses, but as falsely believing that I have all these things... this is an arduous undertaking... (Descartes [1641a], p. 15)

Arduous, indeed, for me to deny that I have hands, that I'm now typing these words, or for you to deny that you're reading them, following along as I rehearse the familiar Cartesian catechism. We might fairly ask, what is the point of this difficult exercise?

The point is not that I am somehow unjustified in believing these things. Despite the doubts that have just been raised, Descartes and his meditator continue to regard my ordinary beliefs as

highly probable... opinions, which, despite the fact that they are in a sense doubtful... it is still much more reasonable to believe than to deny. (Descartes [1641a], p. 15)

The very reasonableness of these beliefs is what makes it so difficult to suspend them. For this purpose, some exaggeration[2] is needed:

I think it will be a good plan to turn my will in completely the opposite direction and deceive myself, by pretending for a time that these former opinions are utterly false and imaginary. (Descartes [1641a], p. 15)

So, the Evil Demon Hypothesis is designed to help to unseat my otherwise reasonable beliefs, though the doubt raised thereby is 'a very slight, and, so to speak, metaphysical one' (Descartes [1641a], p. 25).

[2] In the 'Fourth Replies', Descartes refers to 'the exaggerated doubts which I put forward in the First Meditation', and in the 'Seventh Replies' he reminds us that 'I was dealing merely with the kind of extreme doubt which, as I frequently stressed, is metaphysical and exaggerated and in no way to be transferred to practical life' (Descartes [1642], pp. 159, 308). See Broughton [2002], p. 48.

But this just pushes the question one step back. We now wonder, why should I wish to unseat my otherwise reasonable beliefs? The meditator is explicit on this point. He is concerned about the status of natural science, and he holds that

> It [is] necessary, once in the course of my life, to demolish everything completely and start again right from the foundations if I [want] to establish anything at all in the sciences that [is] stable and likely to last. (Descartes [1641a], p. 12)

The Method of Doubt, the suspension of belief in anything in any way doubtful, is just that, a method—designed to lead us to a firm foundation for the sciences:

> I must withhold my assent from these former beliefs just as carefully as I would from obvious falsehoods, *if I want to discover any certainty in the sciences*. (Descartes [1641a], p. 15, emphasis mine, underlined phrase from the 1647 French edition)

The hope is that once we set aside all our ordinary beliefs, reasonable or not, some absolutely indubitable foundational beliefs will then emerge, on the basis of which science and common sense can then be given a firm foundation. The Method of Doubt is the one-time expedient that enables us to carry out this difficult task.

Janet Broughton, the scholar whose account of Descartes I've been following here, describes the mediator's situation like this:

> Of course, there is nothing about the strategy of this [Method of Doubt] that guarantees it will do what we want it to do. Perhaps we will find that all claims can be impugned by a reason for doubt. Perhaps we will find some that cannot, but then discover that they are very general or have few interesting implications. (Broughton [2002], p. 53)

Of course, this is not the fate of Descartes's meditator. In the second Meditation, he quickly establishes that he must exist—as he must exist even for the Evil Demon to be deceiving him!—and that he is a thinking thing. From there, he moves to the existence of a benevolent God, the dependability of 'clear and distinct ideas', and so on, returning at last to the reasonable beliefs of science and common sense.[3]

Alas, a sad philosophical history demonstrates that the path leading from the Evil Demon Hypothesis to hyperbolic doubt has always been

[3] Though not quite in their original form, as we'll see in a moment.

considerably more compelling than the route taken by the meditator back to belief in his hands. Still, the Cartesian hope of securing an unassailable foundation for science has persisted, down the centuries. So, for example, the good Bishop Berkeley (in his [1710]) suggested that our sense impressions are incontrovertible evidence for the existence of physical objects, because such objects simply *are* collections of impressions, but the price he paid—subjective idealism[4]—was one nearly all but Berkeley have found entirely too high. More recently, Bertrand Russell (in his [1914]) and the young Rudolf Carnap (in his [1928][5]) applied the full scope and power of modern mathematical logic to the project of construing physical objects as more robust logical constructions from sensory experiences, but both efforts ultimately failed, even in the opinions of their authors.[6] There is surely much in this historical record—both in the detail of each attempt and in the simple fact of this string of failures—to lead us to despair of founding science and common sense on some more trustworthy emanations of First Philosophy. Thus, Willard van Orman Quine speaks of a 'forlorn hope' and a 'lost cause' (Quine [1969a], p. 74).

But perhaps the situation is not as tragic as it is sometimes drawn. Let's consider, for contrast, another inquirer, one entirely different from Descartes's meditator. This inquirer is born native to our contemporary scientific world-view; she practices the modern descendants of the methods found wanting by Descartes. She begins from common sense, she trusts her perceptions, subject to correction, but her curiosity pushes her beyond these to careful and precise observation, to deliberate experimentation, to the formulation and stringent testing of hypotheses, to devising ever more comprehensive theories, all in the interest of learning more about what the world is like. She rejects authority and tradition as evidence, she works to minimize prejudices and subjective factors that might skew her investigations. Along the way, observing the forms of her most successful

[4] That is, the view that what I experience as the external world is really just the orderly flow of my subjective impressions ('ideas'). (I come back to Berkeley briefly in I.4.)

[5] On the 'standard reading' (Richardson [1998], pp. 10–13). See I.5 for more on Carnap.

[6] e.g., see Carnap [1963], p. 57: 'We assumed there was a certain rock bottom of knowledge, the knowledge of the immediately given, which was indubitable. Every other kind of knowledge was supposed to be firmly supported on this basis... Looking back at this view from our present position, I must admit that it was difficult to reconcile with certain other conceptions which we had at that time, especially in the methodology of science. Therefore the development and clarification of our methodological views led inevitably to an abandonment of the rigid frame in our theory of knowledge.' Baldwin [2003] traces the development of Russell's thinking.

theories, she develops higher-level principles—like the maxim that physical phenomena should be explained in terms of forces acting on a line between two bodies, depending only on the distance between them[7]—and she puts these higher-level principles to the test, modifying them as need be, in light of further experience.[8] Likewise, she is always on the alert to improve her methods of observation, of experimental design, of theory testing, and so on, undertaking to improve her methods as she goes.

We philosophers, speaking of her in the third person, will say that such an inquirer operates 'within science', that she uses 'the methods of science', but she herself has no need of such talk. When asked why she believes that water is H_2O, she cites information about its behavior under electrolysis and so on;[9] she doesn't say, 'because science says so and I believe what science says'. Likewise, when confronted with the claims of astrology and such like, she doesn't say, 'these studies are unscientific'; she reacts in the spirit of this passage from Richard Feynman on astrology:

> Maybe it's ... true, yes. On the other hand, there's an awful lot of information that indicates that it isn't true. Because we have a lot of knowledge about how things work, what people are, what the world is, what those stars are, what the planets are that you are looking at, what makes them go around more or less ... And furthermore, if you look very carefully at the different astrologers they don't agree with each other, so what are you going to do? Disbelieve it. There's no evidence at all for it. ... unless someone can demonstrate it to you with a real experiment, with a real test ... then there's no point in listening to them. (Feynman [1998], pp. 92–93)

My point is that our inquirer needn't employ any general analysis of what counts as 'scientific' to say this sort of thing, though we use the term 'science' in its rough and ready sense when we set out to describe her behavior.

Movies without much plot are sometimes called Character Studies; the conventions of the genre seem to dictate that it center on an otherwise inconspicuous person who undergoes some familiar life passage with terribly subtle, if any, reaction or results. If thesis is to philosophy as plot is to movie,

[7] This is the methodological principle Mechanism. See II.3.

[8] Mechanism was finally rejected with the rise of field theories. See II.3 for discussion and references.

[9] Wilson ([2006], pp. 427–429) analyzes the complexities of our usage of the terms 'water' and 'H_2O' in terms of his façade structures, described in II.6. The Second Philosopher's claim here should be understood as belonging loosely to analytic chemistry (or the analytical chemistry 'patch' of the façade), as opposed to, say, discussions of official standards for drinking water.

then perhaps what I'm up to here should be classified as a Character Study, with this inquirer as its Character, a mundane and unremarkable figure, as the genre requires. Following convention, I hope to tease out the hidden elements of her temperament by tracing her reactions to a familiar philosophical test: the confrontation with skepticism. How will she react to the challenge Descartes puts to his meditator? Does she know that she has hands?

In response to this question, our inquirer will tell a story about the workings of perception—about the structure of ordinary physical objects like hands, about the nature of light and reflection, about the reactions of retinas and neurons, the actions of human cognitive mechanisms, and so on. This story will include cautionary chapters, about how this normally reliable train of perceptual events can be undermined—by unusual lighting, by unusual substances in the bloodstream of the perceiver, and so on—and she will check as best she can to see that such distorting forces are not present in her current situation. By such careful steps she might well conclude that it is reasonable for her to believe, on the basis of her perception, that there is a hand before her. Given that it is reasonable for her to believe this, she does believe it, and so she concludes that she knows there is a hand before her, that she has hands.

But mightn't she be sleeping? Mightn't an Evil Demon be deceiving her in all this? Our inquirer is no more impressed by these empty possibilities than Descartes's meditator; with him, she continues to think it is far more reasonable than not for her to believe that she has hands, that she isn't dreaming, that there is no Evil Demon. The question is whether or not she will see the wisdom, as he does, in employing the Method of Doubt. Will she see the need 'once in [her] life, to demolish everything completely and start again' (Descartes [1641a], p. 12)?

This question immediately raises another, which we haven't so far considered, namely, what is it exactly that Descartes's meditator sees as forcing him to this drastic course of action? The only answer in the *Meditations* comes in the very first sentence:

> Some years ago I was struck by the large number of falsehoods that I had accepted as true in my childhood, and by the highly doubtful nature of the whole edifice that I had subsequently based on them. (Descartes [1641a], p. 12)

Our inquirer will agree that many of her childhood beliefs were false, and that the judgments of common sense often need tempering or adjustment in light of further investigation, but she will hardly see these as reasons to suspend her use of the very methods that allowed her to uncover those errors and make the required corrections! It's hard to see why the meditator feels differently.

The reason traces to Descartes's aim of replacing the reigning Scholastic Aristotelianism with his own Mechanistic Corpuscularism. As he was composing the *Replies* that were to be published with the first edition of the *Meditations*, he wrote to Mersenne:

> I may tell you, between ourselves, that these six Meditations contain all the foundations of my physics. But please do not tell people, for that might make it harder for supporters of Aristotle to approve them. I hope that readers will gradually get used to my principles, and recognize their truth, before they notice that they destroy the principles of Aristotle. (Descartes [1641b], p. 173)

To get a sense of the conflict here, notice that on the view Descartes comes to by the end of the *Meditations*, all properties of physical objects are to be explained in terms of the geometry and motions of the particles that make them up; the features we experience—like color, weight, warmth, and so on—exist, strictly speaking, only in us. For the Aristotelians, in contrast, physical objects themselves have a wide variety of qualities, which brings Aristotelianism into close alliance with common sense.

This background is beautifully laid out by Daniel Garber, who then takes the final step:

> Descartes thought [that] the common sense worldview and the Scholastic metaphysics it gives rise to is a consequence of one of the universal afflictions of humankind: childhood. (Garber [1986], p. 88)

On Descartes's understanding of cognitive development, children are 'so immersed in the body' (Descartes [1644], p. 208) that they fail to distinguish mind and reason from matter and sensation, and

> The domination of the mind by the corporeal faculties ... leads us to the unfounded prejudice that those faculties represent to us the way the world really is. (Garber [1986], p. 89)

So these are the 'childhood falsehoods' and Aristotelianism is the resulting 'highly doubtful edifice' that the meditator despairs of in the opening sentence of the *Meditations*.[10] As these errors of childhood are extremely difficult to

[10] As Broughton points out ([2002], p. 31), the meditator comes 'uncomfortably equipped with Cartesian theories' at the outset of the *Meditations*, though those theories aren't revealed to him until the end.

uproot in adulthood, only the Method of Doubt will deliver a slate clean enough to allow Descartes's alternative to emerge: the resulting principles of First Philosophy will be completely indubitable, and as such, strong enough to undermine the authority of common sense.[11]

Now our contemporary inquirer, unlike the meditator, has no such Cartesian reasons to believe that her most reasonable beliefs are problematic,[12] so she lacks his motivation for adopting the Method of Doubt. Still, if application of the Method does lead to First Philosophical principles that are absolutely certain, principles that may conflict with some of our inquirer's overwhelmingly reasonable, but ever-so-slightly dubitable beliefs, then she should, by her own lights, follow this course. Even if all her old beliefs re-emerge at the end, some of them might inherit the certainty of First Philosophy.[13] Though she quite reasonably regards such outcomes as highly unlikely, she might well think it proper procedure to read past the first Meditation, to see what comes next. The unconvincing arguments that follow will quickly confirm her expectation that there is no gain to be found in this direction.[14]

So our inquirer will continue her investigation of the world in her familiar ways, despite her encounter with Descartes and his meditator. She will ask traditionally philosophical questions about what there is and how we know it, just as they do, but she will take perception as a mostly reliable guide to the existence of medium-sized physical objects, she will consult her astronomical observations and theories to weigh the existence of black holes, and she will treat questions of knowledge as involving the relations between the world—as she understands it in her physics, chemistry, optics, geology, and so on—and human beings—as she understands them in her physiology, cognitive science, neuroscience, linguistics, and so on. While

[11] The need to undercut our most tenacious commonsense beliefs explains Descartes's interest in certainty: if *p* and *q* conflict, and there is some slight reason to doubt *p*, but *q* is certain, we take *q* to undermine *p*. See Broughton [2002], p. 51.

[12] She doesn't see the errors of childhood as based on a serious inability to distinguish mind from body, so she thinks her ordinary methods of inquiry can correct them.

[13] Not all of the new science will be indubitable, of course. See Garber [1986], pp. 115–116, and the references cited there. Even perceptual beliefs are only trustworthy when properly examined by Reason, so some room for error remains here as well (see the final two sentences of Descartes [1641a]).

[14] Recall that our Second Philosopher has no grounds on which to denounce First Philosophy as 'unscientific'. Open-minded at all times, she's willing to entertain Descartes's claim that the Method of Doubt will uncover useful knowledge. If, by her lights, it did generate reliable beliefs, she'd have no scruple about using it. But if it did, by her lights—that is, by lights we tend to describe as 'scientific'—then we'd also be inclined to describe the Method of Doubt as 'scientific'.

Descartes's meditator begins by rejecting science and common sense in the hope of founding them more firmly by philosophical means, our inquirer proceeds scientifically and attempts to answer even philosophical questions by appeal to its resources. For Descartes's meditator, philosophy comes first; for our inquirer, it comes second—hence 'Second Philosophy' as opposed to 'First'. Our Character now has a name: she is the Second Philosopher.[15] Let's continue our Study by turning her attention from Descartes's project to contemporary radical skepticism.

[15] The Second Philosopher is a development of the naturalist described in my [2001b] and [2003], building on [1997]; I adopt the new name here largely to avoid irrelevant debates about what 'naturalism' should be. Though some take naturalism to be a metaphysical doctrine—e.g., there are no abstracta or everything is physical—Second Philosophy is closer to various methodological readings—e.g., there are no extra-scientific means of finding out how the world is. Still, as we've seen, the Second Philosopher espouses no such doctrine: she is simply a certain type of inquirer (which it is the burden of Part I to delineate); the conclusions of her deliberations constitute Second Philosophy.

I.2

Neo-Cartesian skepticism

The Descartes we've been examining so far—let's call him Broughton's Descartes—regards the skeptical hypotheses as an invaluable tool in his search for a new foundation for science, [1] but contemporary epistemologists tend to entertain a more potent skepticism that takes center stage all on its own. To see how our Second Philosopher fares in this context, let's turn our attention to another Descartes, of whom Barry Stroud writes:

> By the end of his *First Meditation* Descartes finds that he has no good reason to believe anything about the world around him and therefore that he can know nothing of the external world. (Stroud [1984], p. 4)

The claim here is not merely that Descartes cannot be certain of the truth of his beliefs about the world; the claim is that he has no good reason to believe anything at all about the world, no good reason even to believe that it is more likely than not, on balance, that he has hands. This Descartes stands in clear conflict with common sense, with Broughton's Descartes, and with our Second Philosopher: though she may well admit that it's possible she has no hands—as a good fallibilist should[2]—she will insist that this is extremely unlikely.

Stroud's argument for this strong claim brings us back to the possibility of dreaming. The meditator realizes that the senses sometimes mislead him—when the light is bad, when he is tired, and so on—so he focuses on a best possible case: he sits comfortably by the fire with a piece of paper in his hand. At first, it seems to him impossible that he could be wrong about this—until he's hit by the thought that for all he knows he might be

[1] Both Broughton ([2002], pp. 13–15) and Garber ([1986], p. 82) would allow that Descartes has some interest in replying to the skeptical arguments current among his contemporaries, but they see this as a side benefit to carrying out his real project of revising the foundations of science.

[2] A fallibilist holds that we can't be absolutely certain that our reasonable beliefs about the world are true.

dreaming. 'With this thought,' Stroud writes, 'Descartes has lost the whole world' (Stroud [1984], p. 12). If this is correct, then Broughton's Descartes has misunderstood the force of his own skeptical scenario; he fails to realize that the dream possibility undercuts not only the certainty, but also the reasonableness of his belief that he's awake and has hands. If this is true, then the sensible-sounding approach of II.1—that it's far more reasonable than not to believe that I'm not dreaming, but I adopt the Method of Doubt for instrumental purposes—is in fact not fully coherent.

Challenged once again with the possibility that she might be dreaming, the Second Philosopher is tempted to answer in the spirit displayed by Descartes himself at the end of the *Meditations*:

> The exaggerated doubts ... should be dismissed as laughable ... especially ... my inability to distinguish between being asleep and being awake ... there is a vast difference between the two, in that dreams are never linked by memory with all the other actions of life as waking experiences are ... when I distinctly see where things come from and where and when they come to me, and when I can connect my perceptions of them with the whole of the rest of my life without a break, then I am quite certain that when I encounter these things I am not asleep but awake. (Descartes [1641a], pp. 61–62)[3]

Along these lines, the Second Philosopher points out that her experience is continuous and coherent: objects aren't popping in and out of existence (as they do in dreams); they have relatively stable identities (they don't morph one into another, as they do in dreams); ordinary expectations are fulfilled (animals don't speak, I don't fly, as happens in dreams), and so on. Furthermore, she continues, my thought process is deliberate—I can focus my attention—and sustained—I can follow a line of logical steps, or carry out a series of premeditated actions. All this clearly distinguishes my current experience from what I've experienced while dreaming.[4]

Descartes's First Meditation meditator also considers a reply along these lines:

> ... at the moment, my eyes are certainly wide awake when I look at this piece of paper; I shake my head and it is not asleep; as I stretch out and feel my hand I do

[3] If 'I am quite certain' is replaced with 'I have good reason to believe', it seems Descartes's meditator, on Broughton's reading, could have said this in the First Meditation.

[4] In my [2003], I forgo any sustained effort to rebut the dreaming challenge by ordinary means, on the grounds that the Evil Demon hypothesis is immune in principle to this style of response. It now seems to me important to explicitly distinguish the two challenges—ordinary dreaming and the Evil Demon—for reasons I hope will become clear in what follows.

so deliberately, and I know what I am doing. All this would not happen with such distinctness to someone asleep. (Descartes [1641a], p. 13)

His response, at this point, is brief:[5]

Indeed! As if I did not remember other occasions when I have been tricked by exactly similar thoughts while asleep! (Descartes [1641a], p. 13)

The 'similar thoughts' here must be those I have when I convince myself while dreaming that I'm not dreaming: I might, for example, shake my head in the dream, or dream that I'm carrying out a long and involved chain of reasoning. In Stroud's words, if

there is a test or circumstance or state of affairs that unfailingly indicates that he is not dreaming... In order to know that his test has been performed or that the state of affairs in question obtains Descartes would... have to establish that he is not merely dreaming that he performed the test successfully or that he established that the state of affairs obtains. (Stroud [1984], pp. 21–22)[6]

For that matter, I might even be dreaming my decisive test or state of affairs: for example, I might dream that I'm in a green room and that being in a green room is a reliable indicator of wakefulness, then conclude from these dream beliefs that I am awake.[7]

So the Second Philosopher is now challenged to show that her current impressions of continuous and coherent experience, of deliberate and sustained thought processes, aren't themselves dreamed, and that her inference from these features of her experience to the conclusion that, in all likelihood, she isn't dreaming isn't itself a dream delusion. Of course, she's already acknowledged that she often suffers from false convictions while dreaming; she now acknowledges, in particular, that she might, while dreaming, misapply her own criteria for wakefulness or apply incorrect criteria in their place. She knows, from past experience, what this would be like, what it would be like, for example, to apply the green room criterion: it would be a fleeting experience, lasting a few moments, in a general flux of confusion and disorder. In contrast, her current experience is part of

[5] Perhaps because, as on Broughton's reading, he actually thinks these ordinary considerations do in fact make it more reasonable than not to think that he's awake.

[6] For Stroud on 'ordinary methods', see his [1984], pp. 21–23, 46–48.

[7] Stroud [1984], p. 21, makes the point that Descartes must know, not dream, that his test or other criterion for wakefulness is reliable. (I'm grateful to Kyle Stanford for the 'green room' formulation.)

a much longer stream of experience that stretches into the past, with the memories of a lifetime, including episodes of dreaming and waking, talking with others about their dreams and comparing notes, performing or reading about experiments on sleeping subjects that correlate dream reports with REM movements, and so on. These explorations are of a piece with other observations, experiments, and theories that form a large body of beliefs about what the world is like, about what people are like, and about the place of these people in that world. Finally, this same elaborate stream of experience also projects into the future, with expectations and intentions, beliefs about what will happen, what may happen, about the actions she might take to influence these eventualities, and so on.

Here the Second Philosopher, in response to the second challenge—how do you know you aren't dreaming that you've applied your criteria for wakefulness, or dreaming up and applying some false criteria?—has merely fleshed out her response to the first challenge—are you dreaming that you have hands?—by elaborating on what it is about her current experience that makes it different from the dreaming she has experienced and studied. The skeptic will, of course, persist: 'yes, yes, I understand, but how do you know that *all this*, everything you describe, isn't itself a prolonged and intricate dream?' To which I think the Second Philosopher must reply: 'yes, I suppose, in some way, it might be. But if so, it's a dream unlike, say, the green room dream, from which I can awaken in the usual way, that I can come to recognize as deceptive in the usual way. If I were to awaken from the grand delusion you now ask me to imagine, I have no idea what kind of reality I would find. The delusion itself is so all-encompassing as to include everything I think I know about dreaming and waking, plus the overall picture of the world and people and myself in which that is embedded, in short, everything I've ever experienced or hope to or expect to or dread to experience in the future. Obviously, you're right—nothing I can point to would weigh for or against the possibility that the well-ordered experience and thought processes I'm now experiencing are parts of such a dream.'

What's happened here is that the hypothesis that I might be dreaming in the ordinary sense has been replaced by that of a dream delusion so powerful as to serve as the functional equivalent of the Evil Demon hypothesis:

> I will suppose ... some malicious demon of the utmost power and cunning has employed all his energies in order to deceive me. (Descartes [1641a], p. 15)

So, where does this leave the Second Philosopher? She continues to insist that she has grounds on which to be confident, though not certain, that she is not now dreaming in the ordinary sense, but this new challenge is different: by its very construction, it rules out appeal to any of her hard-won beliefs about the world; they are all brought into question at once by the Evil Demon-style hypothesis of extraordinary dreaming. With her hands so tied, she can't refute the hypothesis; indeed, she can't so much as show it to be unlikely, given that her judgments of likelihood also depend on ancillary beliefs about how the world is.[8] To answer the strong skeptical challenge, she must give up all her well-confirmed beliefs about the world and her place in it, surrender all her fine-tuned methods for finding out how things stand, and then justify ... well, it will hardly matter at this point what she is then asked to justify. She will freely affirm that she can't justify her beliefs about the world, can't explain the reliability of any belief-forming mechanism, without relying on her best methods of investigation. She realizes that those beliefs and methods are flawed in various ways, and she has and will continue to put every effort into uncovering and correcting those weaknesses, into strengthening safeguards and developing the most reliable tools, but she agrees with the skeptic that she can do nothing if she's required to set them all aside entirely.

Where the Second Philosopher and the skeptic disagree is on what follows from this. The Second Philosopher recognizes that the original challenge, from ordinary dreaming, was potentially serious: if I really have no good reason to believe that I'm not currently dreaming, she agrees that I also wouldn't have any good reason to believe that I have hands, or anything else. But this real challenge can be met, and the revised challenge seems much less troublesome; though she agrees that she can't rule out the possibility that she's being deceived by an Evil Demon or dreaming in the extraordinary sense, she denies that this fact undercuts the reasonableness of her belief that she has hands. That she can't justify all her beliefs ex nihilo doesn't surprise her, and seems much less unsettling.

Stroud's Descartes starkly disagrees, insisting that, in order to have reasonable beliefs about the world, we must be able to rule out the possibility

[8] Here the Second Philosopher makes no attempt to claim that the skeptical hypothesis is inherently less likely (in the jargon: that it has low a priori probability). See Putnam [1971], pp. 352–353, for this style of reply to the skeptic.

of Evil Demon-style extraordinary dreaming.[9] Of course, he admits that in ordinary life, we don't insist that the chemist's report include 'an account of how the experimenter determined that he was not simply dreaming that he was conducting the experiment' (Stroud [1984], p. 50).

And if a prosecutor were to ask, after

> I testify on the witness stand that I spent the day with the defendant, that I went to the museum and then had dinner with him, and left him about midnight... (Stroud [1984], p. 49)

whether I might not have dreamed the whole thing, everyone in the courtroom would consider the question 'outrageous'.[10] Does this show that Stroud's Descartes is operating with some extraordinary notion of what it takes to know something, that his ultra-refined worries have nothing to do with knowledge as we understand and use the term?

Stroud thinks not. He points out that it's being inappropriate to criticize the witness's or the chemist's knowledge claims in this way doesn't by itself show that ruling out the dream hypothesis isn't necessary for knowledge:

> The inappropriately-asserted objection to the knowledge-claim might not be an outrageous violation of the conditions of knowledge, but rather an outrageous violation of the conditions for the appropriate assessment and acceptance of *assertions* of knowledge. (Stroud [1984], p. 60)

The witness and the chemist make their claims to knowledge 'on just about the most favorable grounds one can have for claiming to know things' (Stroud [1984], p. 61), so it isn't appropriate to criticize them for failing to

[9] Stroud doesn't distinguish ordinary from extraordinary dreaming, which gives his presentation a rhetorical advantage: by phrasing his skeptical challenge in terms of a familiar phenomenon like dreaming, rather than explicitly invoking an Evil Demon-style hypothesis, he makes that challenge appear more commonsensical than it is. (Williams makes what may be a similar point in his [1988], p. 439.) This observation could help explain why we're so easily drawn in to the skeptical line of thought: it begins from a familiar and commonsensical possibility—I might be dreaming—that *would* undermine my purportedly reasonable belief in what I now seem to perceive if it couldn't be ruled out, but in the course of the argument, ordinary dreaming slides imperceptibly into extraordinary dreaming. This would explain the sensation that we're somehow, almost unconsciously, being led into a sort of philosophical game that leaves common sense behind.

[10] The Second Philosopher would say that such challenges are inappropriate or outrageous in everyday situations because they're silly: of course, the chemist and the witness weren't dreaming; it goes without saying! This analysis won't do if extraordinary dreaming is what's at issue, because that can't be ruled out, but in that case, the nature of the challenge being put to the chemist and the witness would have to be clarified, and we would no longer be describing 'ordinary life'.

rule out, or even to consider, the possibility that they're dreaming.[11] But this doesn't show that they do in fact know what they claim to know.

Having found this opening, Stroud's Descartes takes it: when there's no reason to suppose I might be dreaming, he thinks that it's appropriate for me to assert that I know, but that I still do not in fact know unless I can rule out that possibility. The reason for this discrepancy between conditions for knowledge assertions and conditions for knowledge lies in the contrast between the practical and the theoretical:

> It would be silly to stand for a long time in a quickly filling bus trying to decide on the absolutely best place to sit. Since sitting somewhere in the bus is better than standing, although admittedly not as good as sitting in the best of all possible seats, the best thing to do is to sit down quickly ... there is no general answer to the question of how certain we should be before we act, or what possibilities of failure we should be sure to eliminate before doing something. It will vary from case to case, and in each case it will depend on how serious it would be if the act failed, how important it is for it to succeed by a certain time, how it fares in competition on these and other grounds with alternative actions which might be performed instead, and so on. This holds just as much for the action of saying something, or saying that you know something, or ruling out certain possibilities before saying that you know something, as for other kinds of actions. (Stroud [1984], pp. 65–66)

The picture, then, is of a sliding scale of strictness on proper assertions of knowledge.

> From the detached point of view—when only the question of whether we know is at issue—our interests and assertions in everyday life are seen as restricted in certain ways. Certain possibilities are not even considered, let alone eliminated, certain assumptions are shared and taken for granted and so not examined. (Stroud [1984], pp. 71–72)

In ordinary life, then, we make knowledge claims loosely, for practical purposes, though in truth, we do not know. In contrast, when there are no mundane time pressures, when there is no limit on the amount of 'effort and ingenuity' (Stroud [1984], p. 66) we can bring to bear on the question of the truth of our claims—in such a context, we shouldn't claim to know

[11] Notice that the inappropriateness or outrageousness of the dream challenge is here traced to the idea that ruling it out is somehow too much to ask. Again this indicates that extraordinary dreaming is what's at issue. (See previous footnote.)

until we have ruled out every possibility that would preclude our knowing, and, in particular, we must rule out the possibility that we are dreaming (even extraordinary dreaming). So Stroud's Descartes hasn't changed the subject; he's simply working with the usual notion of knowledge in an unrestricted or theoretical context.[12]

Now there is considerable appeal in this notion of a sliding scale of stringency. The Second Philosopher imagines a shopkeeper concerned about the coins he takes in: are they pure metal or fakes?[13] He instructs his hired assistant to bite each coin to be sure, knowing that many counterfeits are laced with harder metals. He also knows that more sophisticated counterfeiters produce fake coins with hardness comparable to pure coins by a different, more difficult process, and that these finer fakes can be detected by an optical device he keeps in the back of his shop. But the fellows capable of this fine work are now in jail, so he doesn't bother to include this extra twist in his instructions to his assistant. Under these conditions, when the assistant says he knows a particular coin is pure metal, the shopkeeper realizes that the fellow doesn't really know, because he hasn't used the optical device in the back room and doesn't know that the coin isn't one of the finer fakes, but the knowledge claim is appropriate in the context, and the shopkeeper would be out of line to correct him.

Likewise, the chemist knows that there are impure metals that pass both the biting test and the optical test, so he can see that the shopkeeper's claim to know, after using his optical device, is also restricted, despite being appropriate in the given circumstances. Even the chemist's claim to know that the metal is pure will appear restricted to the physicist, who realizes that there are atomic variations undetectable by chemical means. And even the physicist may have to admit that there are possible variations he doesn't yet know how to test for, and he will always realize that there may be possibilities he's unaware of that will be uncovered by future scientists. So,

[12] Williams describes this nicely as a sort of 'vector addition': 'The concept of knowledge, left to itself so to speak, demands that we consider every logical possibility of error, no matter how far-fetched. However, the force of this demand is ordinarily weakened or redirected by a second vector embodying various practical or otherwise circumstantial limitations. The effect of philosophical detachment is to eliminate this second vector, leaving the concept of knowledge to operate unimpeded' (Williams [1988], p. 428).

[13] I use this example in place of Stroud's plane spotters (Stroud [1984], pp. 67–75) to bring out the role of scientific inquiry in the sliding scale. The plane spotters return later in Stroud's presentation (Stroud [1984], pp. 80–81) to what seems to me a different end; I take this up below.

even his claim to know that the metal is pure will be subject to the proviso, 'at least as far as current science can determine'.

All this gives the idea of a sliding scale of restrictiveness some initial plausibility. It does seem true that our standards of evidence are more stringent in the chemist's lab than in the shop, and so on, that when important legal judgments or fundamental scientific inquiry are at issue, we sift our evidence more carefully, require a higher degree of confidence. But, as the Second Philosopher will note, this doesn't show that there's anything lacking in a simple perceptual case like my seeing my hand before me under good conditions: here I'm not hampered by time pressure or ignorance or anything else; no further, more strenuous investigation or special expertise seems relevant. Likewise, the ideal of scientific research is to reach conclusions deliberately, with thorough and detached examination of all relevant data, etc.; it's hard to see, for example, what's lacking in our evidence that water is H_2O.[14] So the Second Philosopher fails to see how the fact that degrees of confidence fall along a sliding scale serves to undermine her claim to a reasonable belief that she has hands, that water is H_2O, and so on.

To see what's gone wrong here, let's return to that rapidly filling bus. The idea is supposed to be that when I claim to know I have hands under ordinary perceptual conditions, there's a risk I might be wrong—just as there's a risk I might not get the best seat on the bus if I sit down quickly—but it might still be best, in both cases, to take action despite that risk. But if Stroud's Descartes is right, it isn't that there's a small risk my knowledge claim might be wrong, as there's a small risk I won't get the best seat—if Stroud's Descartes is right, there's *no* chance that my knowledge claim is correct, because in fact I have *no* grounds on which to think my having hands is more likely than not!

These sliding scale considerations would make sense if Stroud's Descartes took the position that knowing requires certainty—our beliefs only amount to knowledge at the high endpoint of the sliding scale, when all possible care has been taken, when all competing possibilities, no matter how remote, have been ruled out—but this position wouldn't conflict with the Second Philosopher's claim to reasonable, though not certain, belief at various lower points. Furthermore, given that Stroud's Descartes insists that there isn't even reasonable belief, let alone knowledge, at lower points, it's

[14] See I.1, footnote 9.

hard to see what our everyday 'knowledge', improperly so called, has in common with that rarefied (indeed, non-existent) stuff at the top. On the one picture, where certainty is required, the chemist's reasonable beliefs are like knowledge, just not quite certain enough to qualify; on the second picture, Stroud's picture, the chemist has no reason to think his beliefs are more likely than not, until suddenly, at the very top, they turn to knowledge. This radical discontinuity counts against the claim that Stroud's Descartes is operating with our everyday notion of 'knowledge'.[15]

Of course, some contemporary discussions of skepticism do take certainty to be a requirement for knowledge,[16] but Stroud has been explicit in rejecting this approach. In reply to Michael Williams,[17] Stroud holds that the requirement of certainty isn't presupposed by the skeptical reasoning, but instead emerges from it:

> What some philosophers see as a poorly motivated demand for 'foundations' of knowledge looks to me to be the natural consequence of seeking a certain intellectual goal, a certain kind of understanding of human knowledge in general. (Stroud [1989], p. 104)

Stroud explains the key idea here, the special sort of insight into knowledge that's in question, with his analogy of the plane spotters: these fellows have a manual that tells them how to determine which type of plane they are seeing, but the reflective plane spotter realizes that he would be in a better epistemic position if he also checked the reliability of the manual. According to Stroud, Descartes's

[15] An unmentioned assumption here is that the 'ordinary notion of knowledge' is unified and determinate enough to provide a fact of the matter on questions like these. In addition to the reservations expressed in the text, the Second Philosopher might well doubt that a real world analysis of the semantics of the word 'know' would turn up any such thing. (See II.6 for discussion of Wilson [2006], a powerful antidote to the conviction that our words typically work by being affixed to concepts with stable and determinate extensions.) Williams ([1988], p. 428) seems to make a similar suggestion, though in a different argumentative setting. The contextualist Lewis ([1996]) finds the concept complex, but still more strictly codifiable than seems likely for a rough and ready notion like knowledge.

[16] e.g., David Lewis takes the idea of fallible knowledge to be 'madness' (Lewis [1996], p. 221). Williams, on the other hand, holds that 'there is no obvious route from fallibilism ... to skepticism' (Williams [1988], p. 430). I tend to agree with Williams that the skeptical challenge isn't of much interest (unless as a method, as for Broughton's Descartes) if it rests on a requirement of certainty.

[17] Stroud [1996]. The central topic of debate between Stroud and Williams is the status of 'foundationalism' (i.e., the view that all knowledge rests on some indubitable basic beliefs)—whether it's a presupposition or a consequence of the skeptical reasoning—rather than an explicit requirement of certainty, but clearly these are closely related. Stroud mentions certainty directly in [1989], p. 104.

conception of our own position and of his quest for an understanding of it is parallel to this reflective airplane-spotter's conception. (Stroud [1984], p. 81)

We aspire in philosophy to see ourselves as knowing all or most of the things we think we know and to understand how all that knowledge is possible. (Stroud [1994], p. 296)

In other words, I must investigate the reliability of my entire 'manual', that is, my entire store of beliefs and belief-forming methods.

The road from here to what Stroud calls 'the philosophical problem of the external world' (Stroud [1984], p. 82) is short: I believe there is a hand before me—On what grounds?—Because I perceive it—But mightn't I be dreaming? At this point in the argument, we imagined the Second Philosopher appealing to ordinary evidence that she wasn't dreaming, but with this new understanding of the problem, that road is closed:

... how, given that we do perceive what we do, do we know [we have hands]? This ... is ... a straightforward question which simply awaits an answer. ... I think we believe we could give good answers to those questions. We would appeal to many other things we know to explain the connection in [this particular case] between what we see and what we claim to know.

But in philosophy we want to understand how *any* knowledge of an independent world is gained on *any* of the occasions on which knowledge of the world is gained through sense-perception. So, unlike those everyday cases, when we understand the particular case in the way we must understand it for philosophical purposes, we cannot appeal to some piece of knowledge we think we have already got about an independent world. (Stroud [1996], p. 132)

And, as we've seen, the Second Philosopher agrees that she can't justify anything without appeal to her familiar beliefs and methods.

Here it's hard to avoid the impression that the skeptic has shifted his ground:[18] before, it was the special skeptical hypotheses, like extraordinary dreaming or the Evil Demon, that undercut the Second Philosopher's appeal to her ordinary justifications; now it's the distinctive features of the philosophical, as opposed to everyday, question of knowledge that do that job. On this new version of the problem, it's clear how the certainty requirement falls out rather than being presupposed: as soon as I realize

[18] I suspect my insensitivity to some of Stroud's argumentation is more to blame for this impression than any inconsistency in his thinking. I follow up these two trains of skeptical thought in the text, for the sake of argument, because they seem distinct to me.

there's room for any sort of doubt, I'm lost, because I'm denied access to the sort of collateral information I'd need in order to show that the grounds for doubt are unlikely. The doubt in question could be the possibility that I'm dreaming, in the ordinary sense, or it could be even more familiar concerns: am I too far away to judge properly?, is the lighting deceptive?, am I under the influence of some strong medication?[19] But for our purposes here, perhaps the most striking difference is this: before, the sliding scale argument was mounted to show that the skeptic's worries are, in fact, of a piece with our ordinary worries about particular knowledge claims; now, it's freely admitted that the skeptic is engaged in a peculiarly philosophical project, distinct from mundane concerns.[20]

From the Second Philosopher's point of view, the situation looks like this. She has various methods of finding out what the world is like, beginning with observation, and as she builds and tests her theories, she also tests and refines those methods themselves. She has seen, in her day, implementations of various bad procedures for finding out about the world, like astrology and creationism, and she can explain in detail where and how these methods go wrong. She constantly works to conduct her inquiries in a detached and unhurried way, as unimpeded as possible by practical limitations and lingering prejudices. When she claims to know that she has hands, she can't conclusively rule out the possibility that she's dreaming, but she can confidently argue that it's quite unlikely, that her belief in her hands is reasonable.

Then Stroud's Descartes presents her with an alternative hypothesis: perhaps she is dreaming in an extraordinary sense, perhaps her whole life has been a long and elaborate delusion, perhaps there is an Evil Demon who has made it seem to her that what she thinks she knows is true when it is not. The very structure of these skeptical hypotheses guarantees that none of her tried and true beliefs and methods can be brought to bear on them

[19] Indeed, I might simply note that I have room to doubt that perception is a reliable means of forming beliefs: it errs sometimes, after all—just how good *is* its track record? This highlights the connection to the perfectly general demand that the Second Philosopher defend her belief-forming methods without appeal to any of those methods.

[20] Cf. Stroud [1996], p. 133: 'I think the special generality we seek in philosophy, combined with the "truism" about the perceptual source of all human knowledge, and with the introduction of certain possibilities of error which are not normally raised in everyday life is what together makes the [skeptic's] question impossible to answer satisfactorily.' Here the extraordinary character of the philosophical question of knowledge is explicitly acknowledged as a presupposition of the skeptic's argument. What I don't understand is why the extraordinary possibilities of error are also needed when the problem is posed in this 'philosophical' sense; ordinary possibilities of error would seem to be enough.

one way or the other. In the face of this challenge, the Second Philosopher must acknowledge that she has no case against such scenarios, indeed, that she can't rule anything out or in, likely or unlikely, reasonable or unreasonable when all her methods of investigation are rendered irrelevant. She acknowledges, in other words, that she can't justify anything without using the methods of inquiry she's developed for that purpose; the very suggestion that she attempt this seems wrong-headed to her.[21] Where she and Stroud's Descartes disagree is on whether this admission renders her current beliefs unreasonable, based as they are on methods she can't justify independently. It's hard to avoid the suspicion that there may be no fact of the matter to ground a judgment either way, that what we're faced with here is a decision on how best to employ the honorifics 'knowledge', 'justification', 'reasonable', and so on.

Now the skeptic reappears in a different guise, insisting that

> *All* of my knowledge of the external world is supposed to have been brought into question in one fell swoop ... I am to focus on my relation to the whole body of beliefs which I take to be knowledge of the external world and to ask, from 'outside' as it were ... whether and how I know it ... (Stroud [1984], p. 118)

In this terminology, the Second Philosopher's account of how and when perception is a reliable guide, her study of various methods of reasoning, of theory formation and testing, and so on, are all relentlessly 'from the inside'. To illuminate what he takes to be the shortcoming of the Second Philosopher's internal efforts, Stroud invites us to imagine a pseudo-Cartesian inquirer who gives the following account of his knowledge of the world:[22] 'I know because I have a clear and distinct idea, and God makes sure that I only have clear and distinct ideas about things that are true; furthermore, I came to believe this about God by means of clear and distinct ideas, so I have good reason to believe I am right.' This account is to run parallel to the Second Philosopher's: 'I know because my belief is generated by such-and-such methods, and such-and-such methods are reliable; furthermore, I

[21] When Broughton's Descartes proposed that she adopt the Method of Doubt, that she reject her familiar beliefs and methods in order to uncover a deeper methodology that would place science on a firmer foundation, the Second Philosopher was willing to give this a try, despite her strong conviction that the project wouldn't work. Now Stroud's Descartes is asking that she undertake a project similar to Broughton's Descartes, that she attempt to justify her own methods ex nihilo—an undertaking she thought unreasonable all along—without giving her any fresh motivation.

[22] This is adapted from Stroud [1994], a reply to externalism. See also Stroud [1989].

came to believe that they are reliable by means of such-and-such methods, so I have good reason to believe that I'm right.' We may be inclined to think that the Second Philosopher *is* right—that perception and her other methods of belief formation *are* reliable—and that the pseudo-Cartesian is wrong—that there is no such accommodating God—but, as Stroud points out, the best either of these inquirers can say is: 'If the theory I hold is true, I do know ... that I know ... it, and I do understand how I know the things I do' (Stroud [1994], p. 301). Given that all our knowledge is being called into question at once, neither the Second Philosopher nor the pseudo-Cartesian can detach the antecedent, so neither can give a philosophically satisfying account of their knowledge.

So this time around, the challenge isn't merely to justify my claim to know that I have hands; in addition, that justification must be 'philosophically satisfying', which is to say, it must come 'from the outside'. This the Second Philosopher admits she cannot do, much as she cannot rule out extraordinary dreaming or the Evil Demon hypothesis—that is, she can't explain her knowledge without using her methods of explanation—but Stroud has enhanced the rhetorical force of this admission with the suggestion that she's in no better position than this woeful pseudo-Cartesian. Of course, she doesn't see it that way; to her, the pseudo-Cartesian is just another in a long line of the benighted—like the astrologer and the creationist—all of whom she can dispatch on straightforward grounds. What Stroud's comparison invites her to attempt is an explanation of the pseudo-Cartesian's errors that uses none of her methods, a task that seems to her no more reasonable than the original challenge to explain her knowledge using none of her methods.

In sum, then, by whichever route, we end up with the same insurmountable skeptical challenge. This challenge has not been shown to be implicit in our ordinary notions of knowledge and justification,[23] and adopting the philosophical perspective on the problem of knowledge is an undertaking quite remote from our ordinary dealings. The Second Philosopher's position comes down to this: she has some well-honed ways of trying to find out what the world is like; they have delivered a picture of the world that is stable, predictively useful, admirably coherent, and explanatory; these methods are fallible, always subject to improvement;

[23] That is, the sliding scale considerations are not persuasive.

indeed, they can't be defended at all against various, carefully constructed skeptical scenarios, nor can they be justified 'from the outside'; and finally, no one, including the skeptic, has proposed a more promising way of going about her investigations. We observers are left to determine which of our two characters—the skeptic or the Second Philosopher—has a stronger claim to the notion of 'reasonable belief'.

The Second Philosopher's reactions in all this are reminiscent of G. E. Moore, who famously answered the skeptic with this 'proof of an external world':

> I can prove now ... that two human hands exist. How? By holding up my two hands, and saying, as I make a certain gesture with the right hand, 'Here is one hand', and adding, as I make a certain gesture with the left, 'and here is another'. (Moore [1939], pp. 145–146)[24]

In such reasonings, Moore, like the Second Philosopher, sticks to the 'internal' or 'everyday' versions of the skeptic's questions. Stroud writes:

> It is precisely Moore's refusal or inability to take his own or anyone else's words in [the] 'external' or 'philosophical' way that seems to me to constitute the philosophical importance of his remarks. He steadfastly remains within the familiar, unproblematic understanding of those general questions and assertions with which the philosopher would attempt to bring all our knowledge of the world into question. He resists, or more probably does not even feel, the pressure towards the philosophical project as it is understood by the philosophers he discusses[25] ... But how could Moore show no signs of acknowledging that [those questions] are even intended to be taken in a special 'external' way derived from the Cartesian project of assessing all our knowledge of the external world at once? That is the question about the mind of G. E. Moore that I cannot answer. (Stroud [1984], p. 119, 125–126)

Here the Second Philosopher must sympathize with Stroud. Though she, like Moore, is disinclined to succumb to the 'lure' of the philosophical project, she surely realizes that those who do so are intending that the question of the external world be understood in a sense that explicitly marks off everything she has to offer as beside the point. She may even

[24] See also Moore [1925]. (I should note that Moore, unlike the Second Philosopher, thinks he is certain he has hands.)

[25] Stroud notes ([1984], p. 120) that 'even Homer nods': there are places where Moore leans farther than perhaps he should toward the 'external' understanding.

sympathize with their quixotic wish for a more satisfying answer than hers. For this reason, she, unlike Moore, makes no claim to have answered the skeptic's challenge.

Speaking for himself, rather than the skeptic, Stroud writes:

> I think reflection on this kind of reflection [that is, on the reasoning that leads to skepticism] can be expected to reveal something interesting and deep about human beings, or human aspiration. (Stroud [1996], p. 124)

This seems right. Our Second Philosopher freely acknowledges one poignant aspect of the human condition: we can't step outside our system of beliefs and methods and justify them from an external perspective; the only perspective we can occupy is our own. Her equanimity in the face of this admission may frustrate us, insofar as we are subject to a familiar aspiration:

> ... a desire to understand ourselves in a certain way, to get into a certain position with respect to human knowledge and perhaps the human condition generally. It takes the form of a desire to get outside that knowledge and that condition, as it were, while somehow retaining all the resources needed to see them as they are. (Stroud [1996], p. 138)

One instance of this aspiration is our desire to say what's wrong with the thinking of Stroud's pseudo-Cartestian and what's right about our own, and to do so without opening ourselves to the charge: but that's just how it looks according to *your* methods! The Second Philosopher sets aside this objection with the now-familiar observation that her methods are the best she knows, justified by considerations so-and-so, but many of us aspire to a more conclusive refutation, a case whose force the pseudo-Cartesian must feel.[26]

Perhaps, after all, the proper upshot of reflection on the skeptic's reasoning should be a particular brand of humility, once recommended by David Hume:

> ... could such dogmatical reasoners become sensible of the strange infirmities of human understanding ... such a reflection would naturally inspire them with more modesty and reserve, and diminish their fond opinion of themselves, and their prejudice against antagonists. ... And if any of the learned be inclined, from their natural temper, to haughtiness and obstinacy, a small tincture of [skepticism] might abate their pride, by showing them, that the few advantages, which they may have

[26] It seems to me that many professed naturalists can't resist the temptation to rise to this challenge (see I.7, IV.1).

obtained over their fellows, are but inconsiderable, if compared with the universal perplexity and confusion, which is inherent in human nature. In general, there is a degree of doubt, and caution, and modesty, which, in all kinds of scrutiny and decision, ought forever to accompany a just reasoner. (Hume [1748], p. 208)

Of course, Hume himself is often described as a naturalist,[27] so perhaps we can find other points of contact for the Second Philosopher.

[27] 'Naturalism' is a blanket term for views in the same generally science-friendly family as Second Philosophy (see I.1, footnote 15). Hume clearly rejects the supernatural, but our focus, again, will be more methodological.

I.3
Hume's naturalism

The inspiration for the characterization of Hume as a naturalist isn't hard to find; his *Treatise of Human Nature* bears the subtitle: 'Being an Attempt to Introduce the Experimental Method of Reasoning into Moral Subjects' (Hume [1739], p. 1).

The proposed 'science of man' is to describe

> the extent and force of human understanding, and ... explain the nature of the ideas we employ, and of the operations we perform in our reasonings. (Hume [1739], introd., para. 4)

This study includes morals, criticism, and politics, but begins with 'logic' whose sole end ... is to explain the principles and operation of our reasoning faculty, and the nature of our ideas (Hume [1739], introd., para. 5).

Because even mathematics and natural science 'lie under the cognizance of men, and are judged by their powers and faculties ...' our pursuit of the Science of Man can be expected to lead to 'changes and improvements ... in these sciences' (Hume [1739], introd., para. 4). As befits a naturalist, Hume's method for this important study is thoroughly empirical:

> As the science of man is the only solid foundation for the other sciences, so the only solid foundation we can give to this science itself must be laid on experience and observation. (Hume [1739], introd., para. 7)

The introduction closes with some musings on the difficulties of devising experiments in this new science, recommending instead 'cautious observation[s] of human life ... judiciously collected and compared' (Hume [1739], introd., para. 10).

Now the Second Philosopher wouldn't follow Hume in placing the science of man (what she thinks of as psychology, sociology, linguistics, etc.) at the foundation of Natural Philosophy (what she thinks of as physics, chemistry, etc.)—she tends to regard these various natural sciences as complementary parts of one big puzzle—but leaving that aside, she would be surprised and disturbed to learn that these 'cautious observation[s] of human life ... judiciously collected and compared' in fact lead to the skepticism Hume reaches at the end of Book 1. If Hume is right that 'philosophical decisions are nothing but the reflections of common life, methodized and corrected',[1] then his case for skepticism might show how it arises directly from common sense, contrary to the conclusions of I.2.[2] So we need to examine how he got from here—an empirical inquiry into human cognition—to there—skepticism.

Some commentators see the skeptical Hume as distinct from the naturalistic, commonsensical Hume; for example, P. F. Strawson suggests

> One might speak of two Humes: Hume the skeptic and Hume the naturalist; where Hume's naturalism ... appears as something like a refuge from his skepticism. ... Hume ... is ready to accept and to tolerate a distinction between two levels of thought: the level of philosophical critical thinking which can offer us no assurances against skepticism; and the level of everyday empirical thinking, at which the pretensions of critical thinking are completely overridden ... (Strawson [1985], pp. 12–13)

How, then, does the philosophical Hume reach his skeptical conclusions? Michael Williams, another advocate of the two Humes, thinks he

> echoes Descartes's thought that we come face to face with the possibility of skepticism, not in the course of everyday practical affairs, but in the context of an extraordinary form of inquiry in which we have no aim beyond knowing the truth: skepticism arises when we set aside all particular interests and attachments and attempt to command an objective view of the world and our place in it. ... The first fatal step on the road to skepticism is taken *as soon as we ask the basic epistemological questions.* (Williams [1996], pp. 6–7)[3]

Leaving Descartes aside, what's clearly echoed here is Stroud's peculiarly philosophical question of knowledge, asked from a completely general, 'external' point of view. As we've seen, this does lead to skepticism, as do

[1] See Hume [1748], p. 208. This passage is cited on p. 10 of Broughton [2003], discussed below.

[2] Recall that Stroud's journey from common sense to skepticism derailed at the flawed 'sliding scale' argument. The journey reaches its destination only if the subject is changed from reasonable belief to certain knowledge, or common sense is left behind for the peculiarly philosophical perspective.

[3] I should note that Strawson thinks Hume, the naturalist, can overcome Hume, the skeptic, while Williams does not. For our purposes, what matters is that they both see Hume's skepticism as arising non-naturalistically.

extraordinary dreaming or the Evil Demon hypothesis, but none of these routes begins in common sense.

Others, by way of contrast, see a single naturalistic Hume; for example, Stroud writes:

> Of all the ingredients of lasting significance in Hume's philosophy I think this naturalistic attitude is of the greatest importance and interest. ... He was interested in human nature, and his interest took the form of seeking extremely general truths about how and why human beings think, feel and act in the ways they do. ... These questions were to be answered in the only way possible—by observation and inference from what is observed. Hume saw them as empirical questions about natural objects within the sphere of human experience, so they could be answered only by an admittedly general, but none the less naturalistic, investigation. He thought we could understand what human beings do, and why and how, only by studying them as part of nature, by trying to determine the origins of various thoughts, feelings, reactions and other human 'products' within the familiar world. (Stroud [1977], p. 222)

What interests us here is the possibility of reaching skepticism from this starting place in scientific common sense.

The argument is complex, but it should be enough for our purposes to sketch the outline in three steps. The first is Hume's analysis of percepts, that is, the items of which we're directly aware in perception. He argues that these cannot be external objects:

> When we press one eye with a finger, we immediately perceive all objects to become double, and one half of them to be removed from their common and natural positions. But as we do not attribute a continued existence to both these perceptions, and as they are both of the same nature, we clearly perceive, that all our perceptions are dependent on our organs, and the disposition of our nerves and animal spirits. This opinion is confirmed by the seeming encrease and diminution of objects, according to their distance; by the apparent alterations in their figure; by the changes in their colour and other qualities from our sickness and distempers; and by an infinite number of other experiments of the same kind; from all of which we learn, that our sensible perceptions are not possessed of any [external] existence. (Hume [1739], 1.4.2.45)

We might, like Berkeley, attempt to 'reject the opinion, that there is such a thing in nature as [external] existence' and to hold that only our mind-dependent percepts exist, but our human nature won't allow us 'sincerely to believe it' (Hume [1739], 1.4.2.50).

> The second step marks our reaction to this conflict: we contrive a new hypothesis ... of ... double existence ..., and distinguish betwixt perceptions and objects, of which the former are supposed to be interrupted, and perishing, and different at every return; the later to be uninterrupted, and to preserve a continued existence and identity. (Hume [1739], 1.4.2.52, 46)

The question, then, is what we can know about those external objects on the basis of our fleeting perceptions. In his third step, Hume reminds us

> That there are three different kinds of impressions conveyed by the senses. The first are those of figure, bulk, motion and solidity of bodies. The second those of colours, tastes, smells, sounds, heat and cold. The third are the pains and pleasures, that arise from the application of objects to our bodies, as by the cutting of our flesh with steel, and such like. (Hume [1739], 1.4.2.12)

No one regards perceptions of the third variety as representing qualities of the object itself, and Hume endorses the well-known argument that so-called secondary qualities—colors, sounds, etc.—are also dependent on the perceiver: noting first that perceptions of color, for example, differ from person to person, from one viewing angle to another, and so on, he concludes that, at least in these instances, the color doesn't resemble anything in the object itself (as the object can't have two different colors at the same time); furthermore, 'from like effects we presume like causes', so

> Many of the impressions of colour, sound, and co., are confest to be nothing but internal existences, and to arise from causes, which no way resemble them. These impressions are in appearance nothing different from other impressions of colour, sound, and co. We conclude, therefore, that they are, all of them, derived from like origin. (Hume [1739], 1.4.4.4)

Thus, secondary qualities, as well as pains and pleasures, arise in us without representing anything actually in the object. Hume then argues that

> If colours, sounds, tastes, and smells be merely perceptions, nothing we can conceive is possessed of a real, continued, and independent existence; not even motion, extension and solidity, which are the primary qualities chiefly insisted upon. (Hume [1739], 1.4.4.6)

The claim is that the ideas of these so-called primary qualities are so intertwined with those secondary qualities that 'when we exclude [secondary qualities] there remains nothing in the universe with [external] existence' (Hume [1739], 1.4.2.15).

Of course there is no one way of reading this general line of thought as issuing from Humean naturalism; for illustrative purposes, I consider here the contrasting analyses of two by-now familiar commentators: Broughton and Stroud. Among the naturalistic features[4] of Broughton's Hume is that he begins, not from a detached philosophical perspective, but unproblematically equipped with our ordinary cognitive standards. We've already seen a hint of this in his admonitions to proceed by 'careful and exact experiments' and 'render our principles as universal as possible' (Hume [1739], introd., para. 8), and Broughton identifies more, which she calls the norm of clarity (trace the origins of ideas back to their corresponding sense impressions); norms of causal reasoning (a list of eight appears in Hume [1739], 1.3.15); and the norm of consistency. According to Broughton, Hume 'never doubts whether our beliefs are justified if they meet our norms ... our norms do not call for grounding', and prior to his brush with skepticism, 'he regards his adherence to these norms as leaving our commonsense assumptions unscathed' (Broughton [2003], p. 13).[5]

From this perspective, the skeptical reasoning sketched above reveals an incompatibility between our most fundamental cognitive norms. We have, on the one hand, our norms of causal reasoning; one of these—'from like effects we presume like causes'—figures prominently in the argument sketched above. But, on the other hand, if our ordinary cognitive norms lead us to believe anything, it's that there is an external world around us:

> 'Tis this principle [our ordinary cognitive norms], which makes us reason from causes and effects; and 'tis the same principle, which convinces us of the continued existence of external objects, when absent from the senses. But though these two operations be equally natural and necessary in the human mind, yet in some circumstances they are directly contrary, nor is it possible for us to reason justly and regularly from causes and effects, and at the same time believe the continued existence of matter. (Hume [1739], 1.4.7.4)

[4] Broughton, like the Second Philosopher above, notes various senses in which Hume is not typically naturalistic, e.g., because natural science (what Hume calls 'natural philosophy') is secondary to his science of man, the science of man doesn't appeal to 'concepts or results from well-established empirically based disciplines' (Broughton [2003], p. 7).

[5] As skepticism (i.e., for our purposes, skepticism about the external world) doesn't arise until 1.4 of Hume [1739], this means that Hume's famous discussion in 1.3.6 is not intended to undercut the reasonableness of causal beliefs based on appropriate norms. This explains why he goes on, after 1.3.6, to distinguish reasonable from unreasonable causal inferences, e.g., in 1.3.15.1, 'to fix some general rules, by which we may know when [apparent causal relations] really are so'. See Broughton [1983].

This is the shipwreck that leads Hume to his forlorn cries in the final section of Book 1:

> The wretched condition, weakness, and disorder of the faculties ... the impossibility of amending or correcting these faculties, reduces me almost to despair, and makes me resolve to perish on the barren rock, on which I am at present ... This sudden view of my danger strikes me with melancholy; and as 'tis usual with that passion, above all others, to indulge itself; I cannot forebear feeding my despair, with all those desponding reflections, which the present subject furnishes me with in such abundance ... I am confounded with all these questions, and begin to fancy myself in the most deplorable condition imaginable, invironed with the deepest darkness, and utterly deprived of the use of every member and faculty. (Hume [1739], 1.4.7.1, 1.4.7.8)

Hume ends with the hope that his encounter with skepticism will cure any 'dogmatical spirit' or 'conceited idea of my own judgment' (Hume [1739], 1.4.7.15)—in other words, he embraces the modesty or humility touched on at the end of I.2.

For all the captivating drama of this intellectual journey, the Second Philosopher is apt to find Hume's reaction misdirected. Her spirit of open inquiry suggests a different course: if our cognitive norms lead to a contradiction, shouldn't we re-examine our cognitive norms? Why should we assume it is impossible to amend or correct them? Granted, some of the beliefs these norms generate seem difficult to reject—''tis vain to ask, *Whether there be body or not?*' (Hume [1739], 1.4.2.1)—but might it not be possible to question others? Perhaps there is some distortion lurking in the assumption that we are 'directly aware' of something or other when we perceive;[6] perhaps the dependence of our perceptions on our 'animal spirits' doesn't guarantee that they are all deceptive; perhaps the norm 'like effects, like causes' is too simple or has been misapplied; perhaps we're assuming, incorrectly, that the factors relevant to the reasonableness of a given perceptual belief are all available to the perceiver; and so on. Regarded second-philosophically, the despairing Hume hasn't fully lived up to his guiding naturalistic impulse.[7]

[6] See Broughton [1992], p. 156.

[7] This is our first, but not last, illustration of how easy it is to depart, all unawares, from the naturalistic path (see, e.g., I.4, I.6).

Let me leave this thought for the moment and turn to Stroud's quite different characterization of the role of skeptical reasoning for the naturalistic Hume. On Stroud's reading, the argument doesn't begin from ordinary cognitive norms, isn't given in Hume's full voice; rather, it's an attempt to reduce to absurdity a particular theory of the human subject that reached its highest articulation in Descartes:

> According to the ancient definition, man is a rational animal. ... Descartes, for example, believed that non-human animals have no souls—they are physical automata all of whose behavior can be given a purely naturalistic, even mechanistic, explanation. A human being, on the other hand, is partly a 'spiritual substance', and therefore has a free and completely unlimited will. (Stroud [1977], p. 11)

This human free will is what accounts for the error and evil in the world. To avoid these, man must strive to be as rational as possible, to act and believe only on the best of reasons. Furthermore,

> This view was thought to have the consequence that human thought and behavior ... cannot be explained as part of the natural causal order ... distinctively human thought and behavior is forever beyond the scope of ... the science of man as Hume envisages it. (Stroud [1977], p. 13)

According to Stroud, it is this picture of 'the detached rational subject' that Hume aims to reduce to skeptical absurdity:

> He argues that none of our beliefs or actions can ever be shown to be rational or reasonable in the way the traditional theory requires. ... But of course we all do believe all kinds of things all the time. ... It follows that either humans are not such detached rational agents at all or, if they are, they never in fact manage to be rational in any of their beliefs or actions, and hence never perform in a way that is distinctively human. ... Hume thinks we can find out what rationality and true humanity really are only by examining the creatures that actually exemplify them. Everything about man must be subject to naturalistic, scientific investigation. (Stroud [1977], p. 14)

The skeptical reasoning shows that reason cannot found belief; this makes room for Hume's naturalistic analysis of how we actually come to believe the things we do.

If this is right, then why is Hume driven to despair at the end of Book 1; why doesn't he simply celebrate the triumph of his naturalistic vision over that of the detached rational agent? One answer[8] to be found in Stroud's text begins like this:

> If we remain within the traditional philosophical theory we will inevitably regard ourselves as worse off in ordinary life than we would have originally supposed. But ... if we see that we simply do not, and cannot, operate according to the traditional philosophical conception of reasonable belief and action, it is just possible that our dissatisfactions will then be directed onto that conception itself, and not onto our ordinary life which is seen not to live up to it. (Stroud [1977], p. 117)

This is the thought process of our imagined, non-despairing Hume: simply pleased to have uncovered the shortcomings of the Cartesian rational subject. So, what goes wrong?

> The Cartesian picture is certainly more than a mere *a priori* prejudice; there are powerful considerations in its favor. ... What is needed, then, and would be completely in the spirit of Hume's 'experimental' examinations of human nature, is an alternative description of how we actually proceed in everyday life, and what we regard as essential to the most reasonable beliefs and actions we find there. (Stroud [1977], p. 117)

The trouble is that Hume hasn't actually given us this 'more naturalistic and hence more palatable conception of how and why we think and act as we do':

> Hume does not suggest even the beginning of such a quest, probably because the theory of ideas makes it unthinkable to him, but once we escape the theory of ideas there is nothing in Hume's general picture of the proper study of man that would rule out an alternative to the traditional philosophical picture. (Stroud [1977], p. 117)

On this reading, Hume falls into despair because something—perhaps the assumption that we are directly aware only of mind-dependent percepts (the theory of ideas)—keeps him from finding a fully satisfying replacement for the traditional picture.

[8] I'm not sure this is Stroud's answer, because the final section of his book returns to something like the philosophical perspective of I.2: 'to philosophize is perhaps inevitably to try to see the world and oneself in it "from outside" ' and 'refusing to theorize with complete generality would leave us with no understanding of the kind we seek about ourselves' (Stroud [1977], p. 249). If this is Hume's problem, it seems his skeptical argumentation isn't just a reductio, and his despair is easier to understand. But, as we've seen (in I.2), the fact that skepticism can be reached from this philosophical standpoint is not a threat to common sense.

Despite the stark differences between Broughton's and Stroud's accounts, the Second Philosopher's reaction to Broughton's Hume and Stroud's reaction to his own Hume end up in the same vicinity: something or other is blocking Hume from the full flower of his naturalism. I'm in no position to speculate on what that something or other is—Broughton and Stroud disagree, for example, on the extent to which Hume accepts the theory of ideas[9]—but however it is best understood, it is this element that opens the road to skepticism,[10] that plays the role erstwhile assumed by the certainty requirement, or the extravagant doubts of hyperbolic dreaming or the Evil Demon, or Stroud's peculiarly philosophical question of knowledge. Once again, common sense and its scientific refinements have not been convicted of undercutting the reasonableness of their own methods.[11]

Let me close this section by returning to some of Strawson's remarks on the naturalistic Hume. He notes with approval Hume's assurance that

> [even] the skeptic ... must assent to the principle concerning the existence of body.... Nature has not left this to his choice, and has doubtless esteemed it an affair of too great importance to be trusted to our uncertain reasonings and speculations.... 'Tis vain to ask, *Whether there be body or not?* That is a point, which we must take for granted in all our reasonings. (Hume [1739], 1.4.2.1)

then observes

> ... having said that the existence of body is a point which we must take for granted in *all* our reasonings, he then conspicuously does *not* take it for granted in the reasonings he addresses to the [question: what causes us to believe in external objects?]. (Strawson [1985], p. 12)

[9] Compare Broughton [1992] and Stroud [1977].

[10] For Broughton's Hume. For Stroud's Hume, it opens the road to despair after the skeptical reductio of the Cartesian rational agent.

[11] Philosophers of science most often associate Hume not with the external world skepticism discussed in 1.4 of Hume [1739] but with the problem of induction from 1.3.6: what grounds our belief 'that instances, of which we have had no experience, must resemble those, of which we have had experience' (Hume [1739], I.3.6.4), or as it's often phrased, what grounds our belief that the future will be like the past? The Second Philosopher would first remark that in many ways the future *isn't* like the past—plants get bigger, people get older, skyscrapers rise up, old buildings crumble—so the proper question is what grounds our belief that the future is like the past in various respects. So, e.g., why do we believe that the sun will rise tomorrow? Because the solar system is like this, the earth rotates like that, unless some unforeseen catastrophe prevents it, the sun will (appear to) rise tomorrow. Why do we believe that bismuth will melt at 271° C. while we aren't so sure that this sample of wax will melt at 91° as that one did? (The example comes from Norton [2003].) Because bismuth is an element and elements tend to behave consistently for such-and-such reasons, while samples of wax vary widely in composition. In this way, from the Second Philosopher's perspective, the problem of induction dissolves into a series of ordinary scientific questions.

Of course, Hume himself is aware of this:

I begun this subject with premising, that we ought to have an implicit faith in our senses, and that this would be the conclusion, I should draw from the whole of my reasoning. But to be ingenuous, I feel myself *at present* of a quite contrary sentiment, and am more inclined to repose no faith at all in my senses, or rather imagination [which provides our belief in external objects] than to place in it such an implicit confidence. (Hume [1739], 1.4.2.56)

Conducted by some oversight down his skeptical course, Hume loses his naturalistic faith.

But, to be fair to Hume, recall the attribution to Nature in the first quotation above:

Nature has not left this to his choice, and has doubtless esteemed it an affair of too great importance to be trusted to our uncertain reasonings and speculations. (Hume [1739], 1.4.2.1)

We today can hardly read this passage without thinking of the evolutionary pressures on our ancestors that most likely contributed to our ability to perceive external objects, but for Hume, of course, Darwin was still in the future. With the benefit of foresight, Hume might well have appreciated the possibility that an explanation of the origins of a belief might also explain its reasonableness. Strawson suggests:

An exponent of a more thoroughgoing naturalism could accept the question, *What causes induce us to believe in the existence of body?* as one we may well ask, as one that can be referred to empirical psychology, to the study of infantile development; but would do so in the justified expectation that answers to it would in fact take for granted the existence of body. (Strawson [1985], p. 12)

Again, in fairness to Hume, he was certainly among the principal forebears of the very empirical psychology that Strawson here recommends to his attention! In any case, the Second Philosopher undertakes just this sort of inquiry, complete with infant studies (in III.5).

I.4

Kant's transcendentalism

As is well known, it was Hume's philosophy that awakened Immanuel Kant from his 'dogmatic slumber' (Kant [1783], p. 260).[1] Kant's reply to Hume's skepticism involved a version of idealism (in the tradition of Berkeley), but for our project here—illuminating the contours of Second Philosophy—the most salient feature of Kant's system is that it epitomizes, at least on one reading, a striking structural feature: it involves two distinct levels of inquiry. Theories of this sort have turned up frequently in Kant's wake—as we'll see below[2]—and the Second Philosopher can be characterized in part by her puzzled reaction to them. Not on principle, but relentlessly in practice, her investigations are pursued on one level, as part and parcel of the single mosaic of natural science.

To see how this contrast goes, we need to sketch the outlines of Kant's critical philosophy.[3] Kant's is not a philosophy about which it's easy to be brief, and distinguished commentators disagree even on fundamental points. To give a general sense of what's at issue, my plan is to contrast two styles of interpretation that lie at opposing extremes, one of which will then become our focus.

For background, we reach back before Hume, to John Locke and the aforementioned Berkeley. Kant praises 'the famous Locke', as he repeatedly calls him,[4] for 'having first opened the way' (B119) with his '*physiology* of the human understanding' (Aix), the attempt to trace all human knowledge to its source:

[1] By 'dogmatism', Kant means the 'worm-eaten' (Ax) metaphysical theories that preceded him, those pursued by 'pure reason *without an antecedent critique of its own capacity*' (Bxxxv)—which, obviously, his *Critique of Pure Reason* was to provide. (I follow the standard practice of referencing passages to the 1781 A version and the 1787 B versions of Kant [1781/7].)

[2] e.g., in I.5 (Carnap), I.7 (Putnam), IV.1 (van Fraassen).

[3] Having this synopsis on the table will also be useful in III.2.

[4] See Aix, A86/B119, A94/B127.

How comes [the mind] to be furnished? Whence comes it by that vast store which the busy and boundless fancy of man has painted on it with an almost endless variety? Whence has it all the *materials* of reason and knowledge? To this I answer, in one word, from EXPERIENCE. (Locke [1690], II.i.2)

Pursuit of this project led Locke to Hume's doctrine of double existence (see I.3), now called a representative theory of perception: we are directly aware of our sensory experiences; these represent for us various objects in the external world. This theory leads immediately to skeptical worries, as Locke himself realized:

It is evident the mind knows not things immediately, but only by the intervention of the ideas it has of them. Our knowledge, therefore, is real only so far as there is a *conformity* between our ideas and the reality of things.... How shall the mind, when it perceives nothing but its own ideas, know that they agree with things themselves? (Locke [1690], IV.iv.3)

Locke held that our ideas of primary qualities (extension, shape, etc.), as opposed to secondary qualities (color, texture, etc.), do represent real aspects of things, but we've seen how Hume disputed this point, as did Berkeley before him (Berkeley [1710], §§9–15).

So, how can we infer the properties of external things from those of our sensory experiences? Berkeley, as we've seen, solved the problem by 'the simple, but extraordinarily dramatic, expedient' (Pitcher [1977], p. 92) of eliminating the extra-mental world altogether and identifying objects with collections of sensations:

By sight I have the ideas of light and colours with their several degrees and variations. By touch I perceive for example hard and soft, heat and cold, motion and resistance... Smelling furnishes me with odours; the palate with tastes, and hearing conveys sounds to the mind in all their variety of tone and composition. And as several of these are observed to accompany each other, they come to be marked by one name, and so to be reputed as one thing. Thus, for example, a certain colour, taste, smell, figure and consistence having been observed to go together, are accounted one distinct thing, signified by the name *apple*... (Berkeley [1710], §1)

There is no difficulty as to how we know about apples, so characterized, as our sensations are precisely the things of which we are directly aware.

So much for backdrop. Turning now to Kant, let's begin with what's undisputed between our two opposing interpretations. Kant takes human beings to enjoy two distinct cognitive faculties:

If we call the **receptivity** of our mind to receive representations insofar as it is affected in some way **sensibility**, then on the contrary the faculty for bringing forth representations itself, or the **spontaneity** of cognition, is the **understanding**. (A51/B75)

In a simple cognition, the passively receptive sensibility provides an intuition, which 'is immediately related to the object and is singular', while the active understanding supplies a concept, whose relation to the object 'is mediate, by means of a mark [a feature or property], which can be common to several things' (A320/B377). Both faculties are required:

Neither concepts without intuition corresponding to them in some way nor intuition without concepts can yield a cognition.... Without sensibility no object would be given to us, and without understanding none would be thought. Thoughts without content are empty, intuitions without concepts are blind. (A50–51/B74–75)

The sensibility combines what it passively receives—the 'matter' of experience, the rush and flux of unprocessed sensation—under its 'form', which orders the matter enough for the understanding to actively apply its concepts.

Now 'that within which the sensations can alone be ordered' cannot itself be found in the a posteriori matter of sensation, so the form of sensibility 'must all lie ready ... in the mind *a priori*' (A20/B34).[5] Space and time, Kant argues, are the two a priori elements of our sensibility, the forms of our intuitions. Likewise, as concepts are needed for any cognition, some concepts must be available before experience, again a priori; these are the pure concepts of the understanding, the categories. No human experience is possible without these forms and categories, so we know in advance that the world as we experience it will conform to them. Thus we know a priori various mathematical and scientific facts about the world of experience:

[5] 'A posteriori' is a flexible term applied to sensations, intuitions, concepts, beliefs given in or arising from experience. The contrast is with 'a priori' elements, present independently of experience.

that it will be spatiotemporal (in accord with the forms of intuition), that it will involve individual objects with various properties (in accord with the category object-with-properties), that there will be causal relations (in accord with the category cause-and-effect),[6] and so on.

The first of our contrasting interpretations, the 'harsh reading',[7] begins with an examination of the relation between objects we experience—appearances, phenomena, or empirical objects—and objects independent of us—things in themselves, noumena, or transcendental objects.[8] That these are distinct items is suggested by Kant's insistence that the former are spatiotemporal and the latter are not:

> Space represents no property at all of any things in themselves nor any relation of them to each other ... Space is nothing other than merely the form of all appearances of outer sense. (A26/B42) ... what we are talking about is merely an appearance in space and time, neither of which is a determination of things in themselves, but only of our sensibility. (A493/B522)

Furthermore, this spatiotemporality, enjoyed by appearances but not by things in themselves, is purely ideal:

> Space itself, however, together with time, and, with both, all appearances, are **not things**, but rather nothing but representations, and they cannot exist at all outside our mind. (A492/B520)

These mental appearances are all we know, but the extra-mental things in themselves are somehow responsible for them:

> ... we have to do only with our representations; how things in themselves may be (without regard to representations *through which they affect us*) is entirely beyond our cognitive sphere. (A190/B235) The sensible faculty of intuition is really only a receptivity *for being affected in a certain way* with representations ... The non-sensible *cause* of these representations is entirely unknown to us. (A494/B522, emphasis mine)[9]

[6] Cause-and-effect is actually a schematized category; the corresponding pure category is ground-consequent. See III.2 for discussion.

[7] For versions of the harsh interpretation, see Prichard [1909], Strawson [1966], Guyer [1987]. Allison [1983] calls it 'the standard picture' (p. 3); in Allison [2004] (p. 4 and footnote 3), he switches to 'the anti-idealist reading'.

[8] There are subtle differences in Kant's use of the three terms in each group (see Bird [1962], pp. 76–80), or Allison [1983], pp. 242–246, [2004], pp. 51–64), but I think they can safely be ignored for our purposes.

[9] See also A44/B61, B72, A288/B344, A358, A393.

Things in themselves are unknowable, but they affect our sensibility to produce appearances, which we can and do know.

The result of this analysis is a mixed and unsavory stew. Empirical objects are strictly mental, essentially Berkeleian congeries of ideas. Things in themselves are Locke's external objects, except that Kant takes primary as well as secondary properties to be features of our experience that correspond to no actual features of the object experienced:[10]

> That one could, without detracting from the actual existence of outer things, say of a great many of their predicates: they belong not to these things in themselves, but only to their appearances and have no existence of their own outside our representation, is something that was generally accepted and acknowledged long before *Locke's* time, though more commonly thereafter. To these predicates belong warmth, color, taste, etc.... I, however, even beyond these, include (for weighty reasons) also among mere appearances the remaining qualities of bodies, which are called [primary]: extension, place, and more generally space along with everything that depends on it (impenetrability or materiality, shape, etc.). (Kant [1783], §13, note II)

We're left with an uneasy combination of Berkeleian idealism with a Lockean representative theory, a combination that preserves the bitter of Berkeley—his idealism—without the sweet—his reply to the skeptic. A supporter of this interpretation remarks that 'Kant ... is closer to Berkeley than he acknowledges' (Strawson [1966], p. 22), and a detractor concludes that 'Kant is seen as a ... skeptic *malgré lui*' (Allison [1983], pp. 5–6, [2004], p. 6).

Even staunch opponents of the harsh reading[11] admit that there is much in Kant that might seem to support it; for example, one writes, 'it would be foolish to deny that Kant can be interpreted in this way' (Matthews [1969], p. 205).[12] In response, they cite passages where Kant rejects Berkeleian idealism,[13] rejects the representative theory of perception,[14] and defends himself against the charge of skepticism.[15] The trick is to use these and

[10] As did Hume and Berkeley, though for different reasons (see I.3).

[11] e.g., see Bird [1962] in reply to Prichard, Matthews [1969] and Allison [1983] in reply to Strawson [1966], and Allison [2004] in reply to Guyer [1987].

[12] See also Bird ([1962], p. 3): 'it is also true that some things which Kant says appear quite strongly to support Prichard's interpretation.'

[13] e.g., the 'dogmatic idealism' of Berkeley, 'who declares ... things in space to be merely imaginary' (B274). See also B70–71.

[14] '... appearances ... do not represent things in themselves' (A276/B332).

[15] 'the complaints "**That we have no insight into the inner in things**" ... are entirely improper and irrational' (A277/B333).

similar passages to suggest reinterpretations of the passages most supportive of the harsh reading. To see how this goes, let me sketch in the second promised analysis of transcendental idealism, the 'benign' reading.[16]

The reorientation begins with a return to Kant's notion of appearance. These are not, the benign theorists point out, mere sensations; rather, as we've seen, they involve both the a posteriori matter of sensation and the a priori forms of sensibility. These forms are space and time, so appearances are necessarily spatiotemporal, external to us.[17]

> If I say: in space and time intuition represents ... outer objects ... as each effects our senses, i.e., as it **appears**, that is not to say that these objects would be mere **illusion**. ... I do not say that bodies merely **seem** to exist outside me ... if I assert that the quality of space and time ... lies in my kind of intuition and not in these objects in themselves. (B69)

To the harsh theorist, it sounds as if Kant is insisting that a batch of my mental contents is located outside me, in space![18] The benign theorist dissolves this apparent absurdity with a fundamental distinction between two levels of inquiry: empirical and transcendental.

The key idea here is that questions about appearance and reality are ambiguous: they can be posed, considered, and answered either empirically or transcendentally. At the empirical level, we draw the familiar distinction between the real and the illusory:

> ... we would certainly call a rainbow a mere appearance in a sun-shower, but would call this rain the thing in itself, and this is correct, as long as we understand the latter concept in a merely physical [empirical] sense. (A45/B63)

The rainbow is a mere optical phenomenon and the rain itself is real. But if we inquire into the reality of the rain at the transcendental level:

> ... not only these drops are mere appearances, but even their round form, indeed even the space through which they fall are nothing in themselves, but only mere modifications ... of our sensible intuition; the transcendental object, however, remains unknown to us. (A46/B63)

[16] Leading benign theorists are listed in footnote 11.

[17] The objects of outer sense, that is. Inner sense is another matter, which I set aside here.

[18] See Allison [1983], p. 5, [2004], p. 6, where this reading is attributed primarily to Prichard, with 'echoes in Strawson'. Guyer also finds Kant in danger of '*identif[ying]* objects possessing spatial and temporal properties with mere mental entities' (Guyer [1987], p. 335). Cf. Allison [2004], p. 8.

In short, the rain is empirically real, but transcendentally ideal.[19]

Once this vital distinction is in place, we see that the appearance and the thing in itself aren't two separate items—one mental, one extra-mental—but a single object regarded in two different ways: 'we can have cognition of no object as a thing in itself, but only insofar as it is an object of sensible intuition, i.e., as an appearance' (Bxxvi). The object as appearance is subject to our human forms and categories; it is non-mental, spatiotemporal, subject to causal laws. Thus Kant is an empirical realist, as opposed to Berkeley's empirical idealist, who holds that empirical objects only appear to be extra-mental. On the other hand, the object as it is in itself is not subject to our forms and categories; these are impositions of our minds, not features of the transcendental object. Thus Kant is also a transcendental idealist, as opposed to a Lockean transcendental realist, who holds that things in themselves are spatiotemporal.

This reading goes a long way toward reinterpreting the passages favored by the harsh theorists, passages where Kant refers to 'mere appearances' or objects 'in the mind': he is to be understood as speaking transcendentally rather than empirically. And there is considerable textual support for this distinction:

> Our expositions accordingly teach the **reality** (i.e., the objective validity) of space in regard to everything that can come before us externally as an object, but at the same time the **ideality** of space in regard to things when they are considered in themselves through reason, i.e., without taking account of the constitution of our sensibility. We therefore assert the **empirical reality** of space (with respect to all possible outer experience), though to be sure at the same time its **transcendental ideality**, i.e., that it is nothing as soon as we leave out the condition of the possibility of all experience, and take it as something that grounds the things in themselves. (A27–28/B44)

This benign reading clearly undermines the harsh interpretation of Kant's appearances as akin to Berkeley's congeries of ideas.

What then of the second component of the harsh reading: the representative theory of perception with things in themselves somehow producing appearances by their action on our senses? Though the benign reading has abandoned the harsh two-object line of thought, it might still be that

[19] Or largely so, as the matter of appearance isn't ideal.

the way things are in themselves affects us so as to produce the way things appear. Indeed, if we confine ourselves to the empirical level, something like a Lockean analysis is possible: objects in the empirical realm of appearances cause certain responses in the sensory organs of human beings, again regarded empirically; some of these responses produce veridical beliefs, others produce illusions; these facts can be described and explained scientifically, as part of empirical psychology.[20] It's important to realize that Kant's transcendental inquiry doesn't compromise or even affect the ordinary practice of science at the empirical level.[21]

But what should we say about the representative theory when speaking transcendentally? Do things as they are in themselves affect us to produce appearances? The various Kantian passages that suggest an affirmative answer to this question[22] raise two of the oldest problems for his sympathetic interpreters: how can we know that things in themselves do this when we can know nothing about them?, and how can things in themselves 'affect' or 'cause' appearances when causation is a category of the understanding, and as such, only applicable to the world of phenomena? Harsh theorists cite these passages as evidence of deep inconsistencies in Kant; benign theorists attempt other readings.

To see how one of these other readings might go, recall this recalcitrant passage:

> The sensible faculty of intuition is really only a receptivity for being affected in a certain way with representations ... The non-sensible cause of these representations is entirely unknown to us. (A494/B522)

Kant elaborates:

> ... therefore we cannot intuit it as an object; for such an object would have to be represented neither in space nor in time (as mere conditions of our sensible representation), without which conditions we cannot think any intuition.

[20] See, e.g., A28 for an empirical contrast between the way colors and tastes depend on 'the particular constitution of sense in the subject', as opposed to space, which is empirically real, or A213/B260, where Kant writes of 'the light that plays between our eyes and the heavenly bodies'. Cf. Allison [1983], p. 249, [2004], p. 67. This distinction considerably weakens the analogy between primary and secondary qualities in the above quotation from Kant [1783], as Kant himself notes (see A29–30/B45). Cf. Allison [2004], p. 8.

[21] e.g., 'observation and analysis of the appearances penetrate into what is inner in nature, and one cannot know how far this will go in time' (A278/B334). See also A39/B56. I come back to this point below.

[22] e.g., see A190/B235 and A494/B522, quoted above, and the other passages listed in footnote 9.

Meanwhile we can call the merely intelligible cause of appearances in general the transcendental object, merely so that we may have something corresponding to sensibility as a receptivity. (A494/B522)

The key here is the final phrase. The sensibility is a capacity for being affected in a certain manner; the affecting agent provides the matter for the resulting appearance; this matter becomes an appearance only after being organized by the forms of intuition. Given this scenario, the affecting agent cannot be the appearance, already imbued with spatiotemporal form; it must be something as yet untouched by this intuitive processing, in other words, the thing as it is in itself. There is only one object, considered either as experienced or as it is in itself; what affects us must be as yet unaffected by us, and so, must be considered as it is itself. To say this is not to gain any contentful information about things in themselves, but merely to follow out the concepts involved.[23] In this sense, to speak of a transcendental object is a harmless way of alluding to the receptive character of sensibility.[24]

The third and final component of the harsh reading is the accusation of skepticism: Kant admits we know only things as they appear, not things as they are in themselves. To this objection, Kant gives a direct reply:

If the complaints '**that we have no insight into the inner in things**' are to mean that we do not understand through pure reason what the things that appear to us might be in themselves, then they are entirely improper and irrational; for they would have us be able to cognize things, thus intuit them, even without senses, consequently they would have it that we have a faculty of cognition entirely distinct from the human not merely in degree but even in intuition and kind, and thus that we ought to be not humans but beings that we cannot even say are possible, let alone how they are constituted. Observation and analysis of the appearances penetrate into what is inner in nature, and we cannot know how far this will go. (A277–278/B333–334)

The final sentence reminds us that our ordinary empirical inquiry does give us knowledge of external, spatiotemporal objects and their causal

[23] In other words, the claim is analytic (cf. footnote 35 below). See Allison [1983], pp. 247–254, [2004], pp. 64–73, especially pp. 67–68, 72. Allison gives a similar response to another puzzle: how do we know the thing in itself isn't spatiotemporal (as opposed to the agnostic position that we can't know whether it is or isn't)? See Allison [1976], [1983], pp. 27, 104–114, [1996], pp. 8–11, [2004], pp. 128–132. For a spirited dissent, see Guyer [1987], pp. 336–342.

[24] Another benign reading would be to understand the idea (or Idea) of the transcendental object as operating in a methodological, rather than a factual sense: to say that the transcendental object causes appearances is to commit ourselves to an unending pursuit of deeper and deeper causal factors. See Bird [1962], pp. 68–69, 78–80.

interconnections; there is no opening for skepticism here. What the objection demands, instead, is transcendental knowledge of the thing in itself, which is, on Kant's analysis, to ask to know what an object is like when it is not known. To reject this demand is good judgment, not skepticism.

In sum, then, the benign reading goes like this. Two levels of inquiry must be distinguished: empirical and transcendental. From the empirical perspective, external objects are perfectly real, in contrast with sensory illusions; from the transcendental perspective, objects as they appear are (partly)[25] ideal, in contrast with objects as they are in themselves. Neither appearances nor things in themselves are Berkeleian congeries of ideas and neither is a Lockean object inscrutable behind its veil of perception; rather they are the same thing, considered in two different ways. Kant's notions of thing in itself, noumenon, and transcendental object have various uses, but never to give us contentful knowledge of matters beyond the world of experience. The empirical world is the world, and we have ever-improving knowledge of it.

Leaving aside the question of how Kant is most accurately interpreted, I suspect most observers would find the benign view a more attractive piece of philosophy than the harsh. In any case, it will be my focus here, as a foil for a recurring naturalistic theme central to Second Philosophy. To get at this, we need to look more closely at the two levels of inquiry. First the empirical.

I suggested earlier that empirical inquiry, for Kant, should be understood as the ordinary pursuit of scientific knowledge. As the world of appearance is closely tied to our experience, it might seem that investigation of the properties of empirical objects cannot go beyond what we actually observe, as in Berkeley's subjective idealist credo 'to be is to be perceived'.[26] Harsh theorists, for example, might be heartened by this passage:

> We cognize the existence of a magnetic matter penetrating all bodies from the perception of attracted iron filings, although an immediate perception of this matter is impossible for us given the constitution of our organs. For in accordance with the laws of sensibility and the context of our perceptions we could also happen upon the immediate empirical intuition of it in an experience if our senses, the crudeness of which does not affect the form of possible experience in general,[27] were finer. (A226/B273)

[25] See footnote 19. [26] 'Their *esse* is *percipi*', Berkeley [1710], §3.

[27] That is, the spatiotemporality of our experience is independent of the relative 'fineness' of our sense organs.

In fact, this calls to mind not so much Berkeley as his phenomenalistic descendants, who hold that an object is a collection of actual and possible sensations[28] or that a statement about an object can be translated into a statement about what sensations would occur under which circumstances.[29] But our benign theorist[30] takes the reference here to what would happen if my senses were more powerful as inessential, as Kant also writes, just before the above passage:

> Cognizing the **actuality** of things requires **perception**, thus sensation of which one is conscious—not immediate perception of the object itself the existence of which is to be cognized, but still its connection with some actual perception in accordance with the analogies of experience, which exhibit all real connection in an experience in general. (A225/B272)

To know an object exists, we must perceive something, not necessarily the object itself, but something connected to it 'in accordance with the analogies of experience'. As these last include the law of cause and effect, this means we can know something exists by perceiving things causally connected with it. Thus, Newton's extensive findings result from 'the empirical use of the understanding' (A257/B313), and presumably an early twentieth-century Kant could have granted empirical reality to, say, atoms, despite our inability to perceive them, because they are suitably connected to what we can perceive.[31]

Assuming, then, that Kant's empirical inquiry is ordinary scientific investigation, how should we understand transcendental inquiry? At this level, we speak of appearances and things in themselves, or better, of things considered as they appear to us and things considered as they are in themselves. As we've seen, the transcendental counterpart to appearance is an increasingly empty notion;[32] the thing in itself is not a separate entity, but a sort of featureless placeholder[33] that calls attention to the receptivity of our senses, to the role of the forms of intuition and the categories in our knowledge, perhaps to

[28] e.g., the 'permanent possibilities of sensation' of Mill [1865], chapter II.

[29] e.g., Ayer [1936], pp. 63–68.

[30] See Allison [1983], pp. 30–34, [2004], pp. 38–42.

[31] I have in mind observations of Brownian motion and related phenomena (see my [1997], pp. 135–143). This case comes up again in I.5, I.6, IV.1, and IV.5.

[32] See A255/B310: 'The domain outside of the sphere of appearances is empty (for us).'

[33] See A104: 'What does one mean, then, if one speaks of an object corresponding to and therefore also distinct from the cognition? It is easy to see that this object must be thought of only as something in general = X, since outside our cognition we have nothing...'

certain methodological principles.[34] Such claims as we make about them in the course of transcendental inquiry are analytic[35]—mere logical consequences of the concepts involved—not contentful information about the world. In fact, Kant sometimes suggests that none of our investigations at the transcendental level give us knowledge, properly speaking:

Such a thing would not be a **doctrine**, but must be called only a **critique** of pure reason, and its utility would really be only negative, serving not for the amplification but only for the purification of our reason, and for keeping it free of errors ... (A11/B25)

This investigation, which ... does not aim at the amplification of the cognitions themselves but only at their correction ... is that with which we are now concerned. (A12/B26)

Perhaps there is no such thing as transcendental knowledge, no higher-level counterpart to empirical knowledge!

Alas, it's hard to know how to take this claim seriously, given that the *Critique* is obviously filled with apparent knowledge claims: human knowledge is discursive; there are two forms of human sensibility; or one of my personal favorites:

The **transcendental unity** of apperception is that unity through which all of the manifold given in an intuition is united in a concept of the object. (B139)

If these are not knowledge claims, we need some general information on how they should be read; if they are knowledge claims, we need to know how they can be fitted into Kant's own account: are concepts being applied to intuitions, as is required of all human knowledge? Are the claims analytic or synthetic? A priori or a posteriori? Meditation on such questions has led even sympathetic commentators like Lewis White Beck to a less than satisfying 'meta-critique of pure reason':

Not only is it [the claim that 'the only intuition available to us is sensible'] not proved, it is not even a well-formed judgment under the rubrics allowed in the *Critique*, for it is neither analytic nor a posteriori, and if it is synthetic yet known

[34] See footnote 24.

[35] Broadly speaking, analytic truths are conceptual, or logical, or follow from the meanings of the words involved; the canonical examples nowadays are trivial definitional claims like 'all bachelors are unmarried'; everything else is synthetic. The distinction originated in Kant, then went through a series of metamorphoses at the hands of Frege, the logical positivists, Quine, and so on. Kant's version is discussed in III.2; Carnap's in I.5.

a priori, none of the arguments so painfully mounted in the *Critique* to show that such knowledge is possible[36] has anything to do with how we know this (if indeed we do know it). (Beck [1976], p. 24)

The benign theorist H. E. Matthews writes of another such transcendental claim:

The statement ... can certainly not be given any factual content, since the conditions for the empirical application of the concept ... cannot be met. But the statement is not self-contradictory, and may well have a function, that of expressing the limitations of our experience, which gives it some kind of meaning. The difficulty which is met here is one which arises whenever one tries to talk about the limits of human knowledge ... The only way in which one can really present the limits of human thought is by showing the confusions and contradictions which arise when one tries to overstep the limits (as Kant does in the Antinomies). But if one does try to *state* the limits (rather than just *showing* them[37]), then the statement, despite its factual appearance, should be interpreted as having a different function. (Matthews [1969], pp. 218–219)

It's hard to see how any transcendental investigation at all is consistent with the very conclusions about human thought that Kant himself draws (at the transcendental level) in the *Critique*.[38]

In addition to this internal tension, difficulties also arose external to the *Critique*, in the developments of nineteenth-century mathematics and twentieth-century physics. In mathematics, the pursuit of Kant's prized Euclidean geometry expanded into complexified, projective, and Riemannian realms that challenged the reach of a priori intuition. By the late 1800s, a range of schools of neo-Kantianism had developed, from the psychologistic Helmoltz to the methodological Marburgians. In physics, relativity theory

[36] The idea is that analytic claims are a priori, because it takes no experience to know what's contained in our concepts or meanings; common sense and scientific claims about the world are synthetic (not merely conceptual, going beyond meanings) a posteriori (known by experience). Kant holds that mathematical and some physical claims are synthetic a priori, and the whole of the *Critique* is aimed at showing how this is possible (e.g., geometric claims about the world are a priori, because they can be gleaned from our forms of intuition without any appeal to actual experience, and those forms contain structure that goes beyond the merely analytic).

[37] Readers of Wittgenstein [1921] will recognize an allusion to the distinction there between saying and showing, which Matthews makes explicit in his discussion.

[38] Allison might argue that such claims are all analytic, but even if 'discursive intellects employ both concepts and intuitions' merely follows out the concepts involved, it's hard to see how 'human beings are discursive intellects' or 'there are two forms of human intuition' could be other than a substantive claim about us. See footnote 49.

denied the objectivity of Kant's time sequence (its observer independence), replaced his a priori forms of intuition, space and time, with a new conception of spacetime, and most famously, substituted a posteriori Riemannian manifolds for his a priori Euclidean space. These concerns were confined to the forms of intuition, but with the rise of quantum mechanics, difficulties spread to the categories: the ubiquity of cause and effect was undermined by the probabilistic nature of subatomic events, and the simple conception of objects with properties is hard-pressed to deal with the behavior of sub-atomic particles.[39] Those most involved in the effort to square Kant with relativity theory were the very philosophers who eventually founded logical empiricism, logical positivism, and the Vienna circle.[40]

The path of Hans Reichenbach is illustrative. He began by separating the Kantian a priori into two notions—that which is certain to be true, and that which is prior to experience—where the second of these is understood in a generally Kantian manner as that which is (partly) constitutive of experience.[41] In these terms, he hoped to retain the idea of constitutive principles while allowing that these can be revised on empirical grounds (as Euclidean geometry was replaced by Riemannian). In reply, Moritz Schlick argued that any properly Kantian position must combine these two:

> Now I see the essence of the critical viewpoint in the claim that these constitutive principles are *synthetic a priori judgements*, in which the concept of the *a priori* has the property of apodeicticity (of universal, necessary and inevitable validity) inseparably attached to it. (Schlick [1921], p. 323)

Reichenbach came to agree that claims subject to empirical confirmation or discomfirmation can hardly be considered a priori:

> The evolution of science in the last century may be regarded as a continuous process of disintegration of the Kantian synthetic *a priori*. (Reichenbach [1936], p. 145)

> the synthetic principles of knowledge which Kant had regarded as *a priori* were recognized as *a posteriori*, as verifiable through experience only and as valid in the restricted sense of empirical hypotheses. (Reichenbach [1949], p. 307).

In place of the constitutive quasi-a priori, Reichenbach now sets out to isolate the definitional or conventional elements of our scientific theorizing.[42]

[39] See III.4 for discussion and references. [40] See Coffa [1991], chapter 10.

[41] See Reichenbach [1920]. Support for the notion of neo-Kantian constitutive principles survives to this day; see, e.g., Friedman [2001].

[42] See Reichenbach [1928]. For a summary, see Reichenbach [1936], p. 146.

Reichenbach's new perspective includes some strikingly naturalistic themes. Where Kant placed a separate level of transcendental analysis over and above ordinary scientific inquiry, Reichenbach came to oppose those who believe 'That philosophical views are constructed by means other than the methods of the scientist' (Reichenbach [1949], p. 289).

Instead, he holds that

> Modern science ... has refused to recognize the authority of the philosopher who claims to know the truth from intuition, from insight into a world of ideas or into the nature of reason or the principles of being, or from whatever super-empirical source. There is no separate entrance to truth for philosophers. (Reichenbach [1949], p. 310)
>
> It should be clear that this philosophy treats genuine philosophical problems by modern scientific methods. (Reichenbach [1931], p. 383)[43]

Alas, even a cursory examination of Reichenbach's own philosophical work reveals departures from these high ideals: his treatment of human knowledge was divorced from empirical psychology;[44] his theory of meaning was uninformed by empirical linguistics;[45] indeed, his analysis of science method was entirely a priori.[46] As we saw in Hume's case, consistent adherence to naturalist principle is harder than it might seem!

Which returns us at last to the point all this is designed to highlight: the Second Philosopher's reaction to Kant's two-level approach to human knowledge. From Kant's point of view, the Second Philosopher is stuck at the empirical level, investigating the external world of spatiotemporal objects, their causal relations, etc. There's nothing wrong with this, according to Kant: unlike Descartes,[47] who thinks ordinary scientific theorizing needs justification and revision, Kant takes scientific methods to be entirely

[43] I'm indebted to Kevin Zollman for this final quotation.

[44] e.g.: 'Epistemology ... considers a logical substitute rather than the real processes. For this logical substitute the term *rational reconstruction* has been introduced; it seems an appropriate phrase to indicate the task of epistemology in its specific difference from the task of psychology' (Reichenbach [1938], pp. 5–6). Let me acknowledge that this brief sketch of Reichenbach's views is oversimplified. In fact, it seems a good bet that our contemporary historians of analytic philosophy will soon present us with a more complete, more nuanced, and almost certainly more interesting portrayal of his contributions than is now readily available.

[45] e.g., the verification theory of meaning (Reichenbach [1938], part I). To be fair to Reichenbach, scientific linguistics as we now understand it dates to the 1950s.

[46] e.g., his defense of induction in Reichenbach [1938], part V. For discussion, see Friedman [1979].

[47] I mean here the Descartes of I.1, out to improve science, not the Descartes of I.2, who takes skepticism more seriously.

in order, for the purposes of empirical inquiry. Kant's message to the Second Philosopher is not that she needs to reform her empirical investigations, but that she should add to them a level of transcendental inquiry. Where Descartes is only selling a new method, Kant is promoting a new purpose, beyond the purposes of empirical inquiry, and a new method to go with it. This is why I call Kant's, but not Descartes's, a two-level approach.

The difficulty of this sales pitch is easily illustrated if we imagine Kant presenting his case. He tells the Second Philosopher that he has performed a transcendental analysis of her knowledge claims and determined that they are partly constituted by a priori forms of sensibility and categories of the pure understanding; he tells her, in particular, that the objects of her world are spatiotemporal, considered from her empirical perspective, but aren't spatiotemporal, considered from a transcendental perspective. He doesn't mean that they aren't spatiotemporal in some absolute sense—he doesn't say that her claims are false, that objects only appear to be spatiotemporal—but he does say there is an important sense in which they aren't spatiotemporal.

How will this strike the Second Philosopher? If the spatiotemporality she attributes to the world is actually a projection of her cognitive processing, she certainly wants to know this, but as far as her empirical analysis of human sensory and neurological processing goes, she sees no reason to think it's true. Kant of course agrees, as what she's doing is merely empirical psychology, like 'the famous Locke', but he insists that there is another level of analysis. Reichenbach, as we've described him, might object that the proposed transcendental psychology is unscientific, and therefore unacceptable, but we've seen that the Second Philosopher doesn't appeal to a distinction of that sort: the contrast between empirical and transcendental inquiry is Kant's not hers; as she sees it, she is simply engaged in inquiry. In contrast to this version of Reichenbach, she responds by respectfully asking Kant to explain what it is that his transcendental psychology studies, and how this study is to be conducted.

Well, Kant answers, transcendental psychology is the study of the nature of the discursive intellect, that is, any intellect whose cognition requires both concepts and intuitions, both pure categories of the understanding and some forms of sensible intuition.[48] This study bears on human knowledge because humans are such intellects, with a sensibility formed, in particular,

[48] See III.2 for more on the structure of the discursive intellect.

by space and time. All very interesting, replies the Second Philosopher, but how can Kant know that human cognition is discursive[49] or that space and time are forms of our sensibility without empirical inquiry, without doing empirical psychology? The debate will continue in this unsatisfactory way, with the Second Philosopher consistently asking for a type of evidence that Kant takes to be inappropriate to the task.

If Kant is to remove the empirical blinders from the Second Philosopher's eyes, he must explain to her why an extra-empirical investigation is needed, what purposes it will serve. The Kantian opus provides various answers to this question,[50] but the one most likely to impress the Second Philosopher is the one fundamental to the *Critique*: 'This transcendental reflection is a duty from which no one can escape if he would judge anything about things *a priori*' (A263/B319). To found a priori knowledge, we must identify transcendentally necessary conditions of experience. Ordinary empirical investigation can at best uncover empirically necessary conditions, but this would only tell us how we are bound to regard objects, not how they must really be; in fact, part of the pay-off of such a theory would be to help us better understand how they really are, independent of our cognitive limitations. Transcendental investigation, on the other hand, tells us how things we experience must be, a priori. So the extra-empirical perspective is essential if we are 'to supply the touchstone of the worth or worthlessness of all cognitions *a priori*' (A12/B26).

This plea will no doubt strike the Second Philosopher as misguided. Any judgment about whether or not we have a priori knowledge of the world can only be supported by serious investigation of the world and our cognitive access to it. For example, from a thorough study of infant cognition and of evolutionary pressures, we might conclude that human brains come equipped to perceive medium-sized physical objects, and we might think it reasonable to describe ourselves as knowing something about the world—for example, that objects continue to exist when they disappear behind barriers (see III.5)—before experience, a priori. Of course, this is not at all what Kant has in mind!

[49] Though it often seems the discursivity of human cognition is an unargued premise for Kant, Allison holds that 'he gives us the requisite materials' for a defense (Allison [2004], p. 452, note 37). But the defense itself rests on premises that the Second Philosopher would regard as requiring empirical support. See Allison [2004], pp. 13–14.

[50] e.g., in his solutions to the cosmological antinomies or in his moral philosophy.

I think it's clear that the Second Philosopher finds in Kant's project no compelling motivation for his extra-empirical inquiry, no puzzle about a priori knowledge that demands a new approach. Furthermore, were she to set this aside for the sake of open-mindedness, Kant would be hard pressed to defend the reliability of the methods of that extra-empirical inquiry: even without the external problems arising from post-Kantian mathematics and physics, we've seen that there are the serious internal difficulties in explaining how transcendental inquiry is carried out and the status of its conclusions. Worries of these two sorts—on the motivation for the new inquiry and on the nature of its methods—are typical of second-philosophical reaction to two-level views in general, as we'll soon see.

I.5

Carnap's rational reconstruction

With his critique of the pretensions of 'dogmatic philosophy' and his respect for the Newtonian mathematics and science of his day, Kant stands as a landmark for succeeding generations of anti-metaphysical, scientific philosophers. With Reichenbach (see I.4), we caught a glimpse of how the school of logical positivism grew out of an effort to adapt Kantian thought to the science of the early twentieth century, especially the physics of relativity theory, but the full flowering of this tendency appears in the work of Rudolf Carnap, a leading member of the Vienna Circle.[1] In his doctoral dissertation, Carnap argued that space is, as Kant proposed, a form of intuition, that Kant was only wrong about the details of its structure.[2] This purely Kantian view was soon abandoned, as Carnap's thinking evolved in stages toward the mature position that will be the focus here.[3]

[1] Coffa [1991] pioneered the analysis of positivism as arising out of the Kantian tradition; until then, Carnap had been read as an empiricist, following after Russell. In Quine's words, 'Russell had talked of deriving the world from experience by logical construction. Carnap, in his *Aufbau* [Carnap [1928]], undertook the task in earnest' (Quine [1970a], p. 40). Richardson ([2004], p. 69) remarks: 'Quine's Carnap is a very traditional philosopher—he is simply a technically competent empiricist. If Hume had had logic, he would have been Carnap.' In fairness to Quine, Carnap himself acknowledges the influence of 'radical empiricism' and of Russell in particular on one strand of thought in the *Aufbau* (Carnap [1963], p. 18). What Quine's interpretation misses is Carnap's neutrality between this strand and other possibilities (Carnap [1963], pp. 16–19), an attitude foreshadowing the Principle of Tolerance (see below).

[2] Carnap writes: 'Kant's contention concerning the significance of space for experience is not shaken by the theory of non-Euclidean spaces, but must be transferred from the three dimensional Euclidean structure, which was alone known to him, to a more general structure. ... the spatial structure possessing experience-constituting significance (in place of that supposed by Kant) can be precisely specified as topological intuitive space with indefinitely many dimensions.' See Friedman [1995] for discussion (the passage just cited appears on p. 46 in Friedman's translation).

[3] After the dissertation came the *Aufbau* (Carnap [1928], see footnote 1); I concentrate here on the post-*Aufbau* Carnap, a period dominated by the Principle of Tolerance, especially *Logical Syntax of Language* (Carnap [1934]) and the well-known and influential 'Empiricism, semantics and ontology' (Carnap [1950a]). Carnap interpretation is currently in a state of lively controversy; here I trace a reading that serves my goal of illuminating Second Philosophy, borrowing liberally (in ways

Both Reichenbach and Carnap set out to update Kant's idea that some components of our knowledge are constitutive, that these elements must be present for any knowledge to be possible at all, but Carnap also shares a larger, overarching motivation with Kant, namely, the hope of saving philosophy from pseudo-questions. Kant bemoans metaphysics as

A wholly isolated speculative cognition of reason that elevates itself entirely above all instruction from experience ... where reason thus is supposed to be its own pupil ... it is so far from reaching unanimity in the assertions of its adherents that it is rather a battlefield, and indeed one that appears to be especially determined for testing one's powers in mock combat; on this battlefield no combatant has ever gained the least bit of ground. (Bxiv–xv)

He traces the problem to ill-posed questions:

It is already a great and necessary proof of cleverness or insight to know what one should reasonably ask. For if the question is absurd in itself and demands unnecessary answers, then, besides the embarrassment of the one who proposes it, it also has the disadvantage of misleading the incautious listener into absurd answers, and presenting the ridiculous sight (as the ancients said)[4] of one person milking a billy-goat while the other holds a sieve underneath. (A58/B82–83)

As we've seen, for Kant, the understanding achieves knowledge only in cooperation with sensibility, but unfortunately

It is very enticing and seductive to make use of these pure cognitions of the understanding and principles by themselves, and even beyond all bounds of experience, which however itself alone can give us the matter (objects) to which those pure concepts of the understanding can be applied ... the understanding falls into the danger of ... empty sophistries. (A63/B87–88)

After setting out the proper uses of sensibility and understanding, Kant devotes hundreds of pages of the *Critique* to unmasking such 'hyperphysical' misuses of the understanding—'an unfortunately highly prevalent art among the manifold works of metaphysical jugglery'—'to uncover the

they shouldn't be assumed to condone!) from Friedman [1988], [1997], [1999a], Richardson [1996], [1997], [1998], [2004], Goldfarb and Ricketts [1992], Ricketts [1994], [2003].

[4] In their notes to Kant [1781/7] (p. 724), Guyer and Wood trace this image to Lucian, who writes, of the sage: 'Once when he came upon two uncouth philosophers inquiring and wrangling with one another—one of them putting absurd questions, the other answering perfectly irrelevantly—he said "Don't you think, my friends, that one of these guys is milking a he-goat and the other putting a sieve underneath it?" '

false illusion of their groundless pretensions' and guard the understanding 'against sophistical tricks' (A63–64/B88).

In a similar tone, we find Carnap confessing that

> Most of the controversies in traditional metaphysics appeared to me sterile and useless. When I compared this kind of argumentation with investigations and discussions in empirical science or [logic], I was often struck by the vagueness of the concepts used and by the inconclusive nature of the arguments. I was depressed by disputations in which the opponents talked at cross purposes; there seemed hardly any chance of mutual understanding, let alone of agreement, because there was not even a common criterion for deciding the controversy. (Carnap [1963], pp. 44–45)

Eventually, he came to see that

> Many theses of traditional metaphysics are not only useless, but even devoid of cognitive content. They are pseudo-sentences, that is to say, they seem to make assertions because they have the grammatical form of declarative sentences, and the words occurring in them have many strong and emotionally loaded associations, while in fact they do not make any assertions ... and are therefore neither true nor false. (Carnap [1963], p. 45)

Many of the questions asked by philosophers, Carnap classifies as 'pseudo-questions', answers to which could very well be described, in Kant's phrase, as 'the same old worm-eaten dogmatism' (Ax).

In the foreword to *Logical Syntax of Language*, Carnap announces: 'To eliminate ... the pseudo-problems and wearisome controversies ... is one of the chief tasks of this book' (Carnap [1934], pp. xiv–xv). The position Carnap embraces in his effort to accomplish this inspired a famous, sustained critique from his admiring student W. V. O. Quine,[5] and that critique in turn led Quine to his naturalism, surely the most widely discussed and influential contemporary version of the position. I trace this line of development in this section and the next, noting relations with and reactions from the Second Philosopher along the way.[6]

[5] Quine writes: 'Carnap is a towering figure. I see him as the dominant figure in philosophy from the 1930s onward ... Carnap was my greatest teacher ... I was very much his disciple for six years. In later years his views went on evolving and so did mine, in divergent ways. But even where we disagreed he was still setting the theme; the line of my thought was largely determined by problems that I felt his position presented' (Quine [1970a], pp. 40, 41).

[6] If it weren't so pretentious, I might say, with similar admiration and gratitude, that Second Philosophy was largely determined by problems that I felt Quine's naturalism presented.

The story begins with Carnap's central tool, the linguistic framework:

If someone wishes to speak in his language about a new kind of entities, he has to introduce a system of new ways of speaking, subject to new rules; we shall call this procedure the construction of a linguistic *framework* for the new entities in question. (Carnap [1950a], p. 242)

So, for example, there is a linguistic framework for talking about observable things and events in space and time, which Carnap calls 'the thing language'. This language includes the usual syntactic apparatus—names, predicates, connectives ('and', 'or', 'not', 'if/then',...), variables, quantifiers ('all', 'some',...), functions, etc.—some logical axioms and rules, and evidential rules specifying what counts as evidence for what, in particular, which sensory experiences count as evidence for which claims about physical objects:

Results of observations are evaluated according to certain rules as confirming or disconfirming evidence for possible answers [to questions like] 'Is there a piece of white paper on my desk?', 'Did King Arthur actually live?', 'Are unicorns and centaurs real or merely imaginary?' (Carnap [1950a], pp. 242–243)[7]

There are also linguistic frameworks for numbers, including variations with weaker or stronger logics; linguistic frameworks for unobservable entities like atoms, with evidential rules spelling out what observations would count as evidence for and against; mathematics-heavy linguistic frameworks for relativity theory, for quantum mechanics, and so on.

Suppose, then, that we're deciding whether or not to speak the thing language,[8] or whether or not to add to it the number language or the atom language. Carnap warns us that

Many philosophers regard a question of this kind as an ontological question that must be raised and answered *before* the introduction of the new language forms. The latter introduction, they believe, is legitimate only if it can be justified by

[7] Cf. Carnap [1950a], pp. 243–244: 'To accept the thing world means nothing more than to accept a certain form of language, in other words, to accept rules for forming statements and for testing, accepting, or rejecting them.'

[8] Carnap recognizes that 'In the case of this particular example, there is usually no deliberate choice because we all have accepted the thing language early in our lives as a matter of course. Nevertheless, we may regard it as a matter of decision in this sense: we are free to choose to continue using the thing language or not' (Carnap [1950a], p. 243).

an ontological insight supplying an affirmative answer to the question of reality. (Carnap [1950a], p. 250)

These thinkers hold that a decision on whether or not to adopt a given linguistic framework should be based on a prior assessment of the accuracy of that framework—on whether or not physical objects, numbers, or atoms exist; whether or not the framework's logic is correct; whether or not what's cited in the evidential rules is good evidence—but Carnap thinks they are wrong in this. Questions of existence, of logic, of evidence and truth, he insists, can only be asked within a framework, as it is the framework alone that gives them cognitive significance.

Within the thing language, then, I correctly assert that there is a tree outside my window—or that I have hands, for that matter!—because the evidential rules of that framework guarantee that my sensory experiences are good evidence. Of course, this answer, internal to the framework, isn't what Stroud's Descartes (in I.2) has in mind when he asks if I'm justified in believing these things! But Carnap holds that the skeptic's question, asked external to the thing framework, 'cannot be solved because it is framed in a wrong way': 'To be real [is] to be an element of the system; hence this concept cannot be meaningfully applied to the system itself' (Carnap [1950a], p. 243). The skeptic's external question lacks cognitive significance because the framework is what supplies the necessities for any inquiry to be meaningful in the first place.

Unfortunately, these philosophers have so far not given a formulation of their question in terms of the common scientific language. Therefore our judgment must be that they have not succeeded in giving to the external question and to the possible answers any cognitive content. Unless and until they suppose a clear cognitive interpretation, we are justified in our suspicion that their question is a pseudo-question. (Carnap [1950a], p. 245)

Here the linguistic framework plays a constitutive role analogous to that of Kant's forms and categories: these elements must be in place before questions of existence, evidence, truth, and falsity can be posed, let alone answered. Asking such a question outside a linguistic framework is comparable to asking about things as they are in themselves, independent of our modes of cognition—it cannot be done.

Still, there is a legitimate question that can be asked outside the framework, a legitimate question with which the philosopher's illegitimate external question is sometimes confused, namely,

a practical question, a matter of a practical decision concerning the structure of our language. We have to make a choice whether or not to accept and use the forms of expression in the framework in question. (Carnap [1950a], p. 243)

We decide on pragmatic grounds whether or not to adopt the framework in question; we ask, is it effective, simple, fruitful, efficient, 'conducive to the aim for which the language is intended' (Carnap [1950a], p. 250).

However, it would be wrong to describe this situation by saying: 'the fact of the efficiency of the thing language is confirming evidence for the reality of the thing world'; we should rather say instead: 'This fact makes it advisable to accept the thing language'. (Carnap [1950a], p. 244)

The very notion of confirmation here is available only internal to the framework.

It must be emphasized that nothing objective, no facts, constrain our choice here; we are free to adopt any linguistic conventions we like. This freedom is encapsulated in Carnap's

Principle of Tolerance: It is not our business to set up prohibitions, but to arrive at conventions ... In logic, there are no morals. Everyone is at liberty to build up his own logic, i.e. his own form of language, as he wishes. All that is required of him is that, if he wishes to discuss it, he must state his methods clearly, and give syntactical rules instead of philosophical arguments. (Carnap [1934], pp. 51–52, emphasis in the original)

Any linguistic form is admissible (here Carnap departs mightily from Kant's single set of necessary forms and categories[9]), but some linguistic framework or other is required before any inquiry can begin (there must be *some* constitutive elements). So, if someone wants to debate with us, he must first stipulate the framework in which he sees the debate as taking place, which includes specifying the grounds on which it can properly be resolved. Without this, he is posing a pointless pseudo-question, analogous to those of Kant's dogmatic philosophers.[10]

[9] From Carnap's point of view, Kant's transcendental inquiry is in the business of answering external pseudo-questions.

[10] Allison ([2004], chapter 2) argues that Kant takes all other philosophers to be transcendental realists, that is, he takes them to confuse the object as experienced (the spatiotemporal object) with

Perhaps the Kantian analogy can be pressed further, into an account of a priori knowledge. Recall that on Kant's two-level view, a question like 'is the raindrop real?' is ambiguous: empirically, yes; transcendentally, no. For Carnap, a question like 'do physical objects exist?' is similarly ambiguous. When it's posed within the framework of the thing language, my sensory experiences are enough to establish the existence of a tree outside my window, the existence of my hands, and so on, and hence the existence of physical objects; outside the framework, asking about existence boils down to asking about the advantages and disadvantages of adopting the thing language. In the case of an internal question about numbers, the rules of the language alone are enough to establish their existence: I have a name '0' and a function '*S*' and a predicate 'is a number'; I have a rule that says, '0 is a number and if x is a number, so is Sx'; from these I conclude, without any experience at all, that 0 and $S0$ are numbers and thus that there are numbers. So the existence of numbers is a priori in the number language, while the existence of physical objects is a posteriori in the thing language. For Kant, certain claims are objective a priori truths at the empirical level, but dependent on our cognitive structures when viewed transcendentally; for Carnap, certain claims are objective a priori truths within the appropriate framework, but dependent on pragmatically motivated, conventional choices when viewed externally.

To press this reading a bit further, let's suppose we've adopted a linguistic framework for simple scientific observation and generalization—perhaps an elaboration of the thing language—and we're wondering whether or not to embrace a new range of entities, say atoms. As our current language has no terms for such things, no predicate 'is an atom', no evidential rules with which to settle questions of their existence or nature, Carnap holds that this is not a question that can be asked or answered internally, that we must step outside our linguistic framework and address it pragmatically, as a conventional decision about whether or not to adopt a new linguistic framework. This new framework would include new evidential rules linking various indicators to the presence of atoms, just as the thing language includes evidence rules linking various experiences to the existence of ordinary objects.

the thing as it is in itself (they think things in themselves are spatiotemporal). Carnap's accusation can be seen as running parallel: philosophers confuse objects as understood inside the framework with objects as they are independently of all frameworks, i.e., they think questions that make sense for the former—questions of existence, etc.—can be asked of the latter.

Now I think it must be admitted that the Second Philosopher will find this analysis odd.[11] Starting from her elaboration of the thing language, she sees, of course, that she needs new methods for settling the existence or non-existence of atoms.[12] Indeed, reputable scientists once rejected atomic theory on the grounds that it could not be confirmed, in other words, because there were no evidential rules that could settle the question.[13] The young Albert Einstein set himself the problem of devising a theoretical test: 'My major aim ... was to find facts which would guarantee as much as possible the existence of atoms of definite finite size' (Einstein [1949], p. 47). Even when he had succeeded, Einstein doubted that actual experiments of sufficient accuracy could be designed and carried out.[14] Thus the meticulous and decisive work of Jean Perrin on Brownian motion came as a welcome surprise. In circumstances like these, where the new evidential rules are such elusive and hard-won scientific achievements, the Second Philosopher is unlikely to agree with Carnap that their adoption is a purely pragmatic matter, a conventional choice of one language over another. Instead, she insists that the development of the Einstein/Perrin evidence was of a piece with her standard methods of inquiry, that it required careful examination and justification of the usual sorts.[15]

This reaction is reminiscent of one of Quine's famous criticisms: that Carnap's central distinction between analytic (true by virtue of language) and synthetic (true by virtue of the way the world is) cannot be drawn.[16] Perhaps somewhat less skeptically, the Second Philosopher allows that the

[11] I talk mostly about the transition to the atom language in the text, but readers with worries about this example should substitute the parallel concern about the status of the evidential rules of the thing language.

[12] For discussion of the historical debate over atoms, including references, see my [1997], pp. 135–143. This case comes up again in I.6, IV.1, and IV.5.

[13] e.g., the famous chemist Jean-Baptiste Dumas condemned atomic theory because 'it goes beyond experience; and never in chemistry ought we to go beyond experience' (quoted by Glymour [1980], p. 257).

[14] See Nye [1972], p. 147.

[15] I've been assuming that a typical evidential rule of the thing language is something like 'if I have experiences so-and-so, then there's a tree outside my window' and that a typical counterpart in the atom language would be something like 'if such-and-such results from this Perrin experiment, then there are atoms in the solution'—neither of which should count as linguistic, according to the Second Philosopher—but it might be argued that the atom language rule would be more like 'all things are made of atoms', so that the existence of a thing implies the existence of atoms. On the latter construal, the work of Einstein/Perrin motivates the pragmatic decision to adopt the atom language but is not enshrined in its evidential rules, and the Second Philosopher's objection would be even closer to Quine's, that is, she would challenge the distinction between change of language and change of theory.

[16] The classic statement is Quine [1951].

empirical study of human language use might justify some notion of purely linguistic truth, but she doubts that a distinction so grounded would put the relevance of Einstein/Perrin's work to the existence of atoms on the linguistic side of the ledger.[17] Other parts of the Quinean critique focus on the methodological point that changes of framework are conventional or pragmatic, not empirical or theoretical, another distinction subjected to concerted Quinean attack.[18] Again, the Second Philosopher, unlike Quine, recognizes that some, not all, parts of her theories are present for pragmatic reasons[19]—for example, she sees the use of continuous manifolds in our representation of spacetime as unjustified apart from the fact that we have no other equally effective mathematics to the job[20]—but again, she doesn't see the Einstein/Perrin evidential rules in this way.

The parallel with Kant might be furthered here, comparing these worries about Carnap with the difficulties raised for Kant by subsequent developments in mathematics and science: in both cases, we might say with Quine that the desired distinction between the a priori and a posteriori parts of science, between changes of language and changes of theory, 'begins to waver and dissolve' (Quine [1954], p. 122).[21] This view of Carnap is

[17] Cf. Quine [1954], p. 129: 'One quickly identifies certain seemingly transparent cases of synonymy... Conceivably the mechanism of such recognition, when better understood, might be made the basis of a definition of synonymy and analyticity in terms of linguistic behavior... I see no reason to expect that the full-width analyticity which Carnap and others make such heavy demands upon can be fitted to such a foundation even in an approximate way.' Also Quine [1960], p. 66: 'Let us face it: our socialized stimulus synonymy and stimulus analyticity are still not behaviorist reconstructions of intuitive semantics, but only a behaviorist ersatz.' The Second Philosopher is no behaviorist—she turns to linguistics and cognitive science—but she and Quine agree that there's little hope of anything from scientific quarters that would do the job Carnap wants done.

[18] See, e.g., Quine [1936], [1948], [1954].

[19] e.g., various sorts of idealizations and mathematizations (see IV.2.i). Notice that for the Second Philosopher, an aspect of a theory added for pragmatic reasons will not be regarded as confirmed: unlike Quine, she isn't a holist (see I.6).

[20] e.g., Einstein ([1949], p. 686) writes: 'Adhering to the continuum originates with me not in a prejudice, but arises out the fact that I have been unable to think up anything organic to take its place.' Cf. Dedekind, just after introducing our modern notion of continuity: 'If space has at all a real existence it is *not* necessary for it to be continuous; many of its properties would remain the same even were it discontinuous' (Dedekind [1872], p. 12). For discussion, my [1997], pp. 143–157, or IV.2.i below.

[21] Let me note here two niceties I intend to neglect. First, it might be that rules aren't statements, and thus aren't candidates for analytic status despite being part of the language (so that changing them is a change of language). Second, in his [1934], Carnap separates the rules of his two sample languages into L-rules (logical rules) and P-rules (physical rules), the latter usually being 'general laws' (p. 316) like Maxwell's equations (p. 319); though changing such a P-rule would count as a change of language, P-rules are nonetheless synthetic, because predictions following from them can be tested; thus 'can't be changed without a change of language' and 'analytic' don't match up perfectly. (For discussion, see Ricketts [1994], pp. 189, 192, Friedman [1999a], pp. 218–220.) Fortunately, it seems clear that the

tempting[22]—but ultimately hard to sustain. The mathematical and scientific troubles for Kantianism developed after Kant's own day: he can surely be forgiven for failing to foresee the downfall of Euclidean geometry, much less the rise of quantum mechanics; presumably if he'd known the shape of science-to-come, his philosophical position would have been different. But the Second Philosopher's complaints to Carnap are mere commonplaces, perfectly obvious to Carnap himself. For that matter, Carnap happily accedes to the substance of Quine's attacks:

> The main point of his criticism seems... to be that the doctrine [that logic is analytic] is 'empty' and 'without experimental meaning'. With this remark I would certainly agree, and I am surprised that Quine deems it necessary to support this view by detailed arguments. (Carnap [1963], p. 917)

This is hardly the response that a strong analogy with Kant on a priori knowledge would lead us to expect! To see what's gone wrong I think it helps to turn the situation on its head: instead of considering the Second Philosopher's reactions to Carnap, let's imagine Carnap's response to the Second Philosopher.

Suppose, then, that an early twentieth-century Second Philosopher is convinced of the existence of atoms by the Einstein–Perrin evidence. Carnap holds that she has adopted a new language, with new evidential rules, and that this decision, coupled with the results of Perrin's experiments, leads to her justified belief in atoms. Recognizing that the Second Philosopher doesn't see it this way, that she doesn't take the adoption of the new evidential rules to be a linguistic decision, what is Carnap's reaction? It seems the Carnap we've been describing so far would insist that she is wrong, that she doesn't understand what she's doing as well as he does. This Carnap would presumably feel the force of Quine's demand for 'mental or behavioral or cultural factors relevant to analyticity' (Quine [1951], p. 36). He would explain to Quine and the Second Philosopher what pre-analytic notion of 'language' and 'meaning' he's trying to capture

evidential rules of Carnap [1950a] aren't P-rules and aren't synthetic, because they must be in place before any testing can be done (perhaps more like the C-rules of Carnap [1966]). In sum, I pretend here that there's no mismatch between being analytic and being 'part of the language' (changeable only by a 'change of language').

[22] I've succumbed to this temptation many times, e.g., in my [2000].

in his formal theory of linguistic frameworks; he would argue that the Second Philosopher's change of mind qualifies as linguistic in these terms.[23]

But in fact Carnap does none of these things. Quine's arguments and demands seem to puzzle him, but perhaps the Second Philosopher's reaction is easier to digest: if, as Carnap claims, she is wrong to think that her new evidential rules aren't true by virtue of meaning, she wants to understand her error, so she asks on what grounds his criticism is based; what she wants, of course, is a defense based on empirical study of human language use. In contrast, Carnap explicitly distinguishes his pursuit of 'the logic of science' from the empirical study of language:

> Apart from the questions of the individual sciences, only the questions of the logical analysis of science, of its sentences, terms, concepts, theories, etc., are left as genuine scientific questions. We shall call this complex of questions the *logic of science*. (We shall not employ here the expression 'theory of science'; if it is to be used at all, it is more appropriate to the wider domain of questions which, in addition to the logic of science, includes also the empirical investigation of scientific activity, such as historical, sociological, and, above all, psychological inquiries.) (Carnap [1934], p. 279)
>
> That part of the work of philosophers which may be held to be scientific in its nature—excluding the empirical questions which can be referred to empirical science—consist of logical analysis. (Carnap [1934], p. xiii)

From this perspective, the Second Philosopher's worry is misguided; Carnap isn't classifying her new evidential rule as linguistic in the empirical sense she has in mind. Like Kant, Carnap thinks there is nothing amiss in her proceedings: when she decides, on what she takes to be sound theoretical and experimental grounds, to add atoms to her theorizing, he agrees that this is a good idea; when she denies it is a linguistic matter, in the sense of empirical linguistics, he agrees with this, too.[24] Just as empirical psychology was irrelevant to the Kantian claim, her empirical study of human language is irrelevant to the Carnapian.

[23] Or to pick up the methodological strand of the debate, he would offer an account of 'convention' that differentiates some scientific beliefs or decisions from others, and argue that the Second Philosopher's change of mind meets these criteria. I think the morals of this methodological variation run parallel to the line of thought in the text.

[24] This may be a tactful expression of Carnap's attitude. See next footnote.

All this means, so far, is that Carnap should be understood as presenting a truly two-level position. Recall the contrast (from I.4) between Descartes and Kant: both engage in what we might call 'extra-scientific' inquiry—first philosophical and transcendental, respectively—but for Descartes, this inquiry is capable of correcting ordinary science, while for Kant, ordinary science is perfectly in order for its own empirical purposes. The Second Philosopher, who employs no general litmus test for science vs. non-science, largely misses the analogy between Descartes and Kant; instead, what moves her is the stark methodological disanalogy. We see this in her reactions to the two: to Descartes, she says, 'I'm always eager to improve my theories and methods. Show me how your Method of Doubt works... '; to Kant, she says, 'you say my theories and methods are all in order, for my purposes, but that you have other purposes. Tell me what those purposes are, and how you intend to pursue them.' Kant's is a genuinely two-level view, because the practice of ordinary science—his empirical level—is methodologically independent of transcendental analysis.

If Carnap stands on Kant's side of this divide rather than Descartes's, he doesn't disagree with anything the Second Philosopher has to say about the empirical status of her evidential rules. When Quine insists that he explain what empirical phenomenon his analytic/synthetic distinction is designed to capture, he will demur, because it isn't an empirical phenomenon that he has in mind. The actual discourse of a scientific community doesn't embody any linguistic framework; 'The logical syntax of [the framework] is imposed like a grid on an investigator's used language, on her speech habits' (Ricketts [2003], p. 263). But if Carnap's apparatus of linguistic frameworks, of internal and external questions, of analytic and synthetic is not supposed to describe pre-existing features of actual scientific language,[25] what is the point of imposing

[25] The interpreter of Carnap would seem to have a choice between two broad options: on the one hand, take a linguistic framework as something actually spoken, or as a skeletal structure implicit in natural language, or as a neatening up of a fuzzy structure implicit in a natural language, or something along these lines. This approach invites Quine's challenge to explain what aspect of scientific discourse the analytic/synthetic distinction (or the conventional/pragmatic vs. empirical/theoretical distinction) is designed to capture. The other option is the one now under consideration: take a linguistic framework as a structure imposed on actual discourse. On this second approach, the ordinary scientist has only speech dispositions; her use of words like 'evidence', 'true', 'logic' is hopelessly vague and indeterminate; evidential relations, truth and falsity, specification of what counts as logic and mathematics, come only with the imposed framework; as Ricketts [2003], pp. 263, 274, puts it: 'with this grid in place, we represent an investigator's acceptance and rejection of sentences as the epistemic evaluation of hypotheses... only this selection imposes definite standards of right or wrong

this 'grid'? To repeat the Second Philosopher's question to Kant: if my theories and methods are proper for my purposes, but you have other purposes, what are those other purposes and how are they to be pursued?

On our previous (mis-)understanding of Carnap, his goal ran parallel to (one of) Kant's: he was out to explain how a priori knowledge is possible. As with Kant, the higher-level analysis—the choice of linguistic framework is pragmatic, conventional—explains how we have a priori knowledge at the lower level—once the framework is chosen, a priori truths follow from the language alone, without any experiential input. Indeed, on this (mis-)reading, Carnap does Kant one better: Carnap's lower-level inquirer can explain how she knows her a priori truths—'they are true by virtue of the structure of my language'—while Kant's lower-level inquirer can only say—'they are true because the world really is that way (spatiotemporal, etc.)'—with no account of how we come to know this a priori.[26]

But, as we've seen, the distinction between 'part of language' and the rest, between analytic and synthetic, is only defined for formal linguistic frameworks—actual scientific language admits various such 'grids', with the line between analytic and synthetic drawn in various places—so Carnap's system does not underwrite an ordinary claim that, say, ' "2+2 = 4" is true by virtue of meaning'. Carnap's policy is to replace such ordinary sentences—which he called 'pseudo-object sentences' (Carnap [1934], §74)—with properly syntactic versions; when this is done, we end up with something like ' "2+2 = 4" follows from the rules of the number framework'. This, of course, no one doubted; the problem was to explain the ground of a priori truths like those embodied in the axioms and rules of the number framework. When we imagined Carnap to be capturing a natural language notion of analyticity and using it to account for the source of a priori knowledge, we characterized him as attempting to answer what for him is an external pseudo-question.[27]

on amorphous linguistic behavior.' This seems hard to square with the inspiration Carnap drew from, e.g., Einstein; in Richardson's words, 'Carnap both took for granted and sought philosophically to comprehend and advance the conceptual techniques of the exact sciences, for they are the locus of best knowledge for him' (Richardson [2004], p. 74). Promising work in progress by Sam Hillier proposes to read Carnap as embracing both projects, perhaps not clearly distinguishing them at all times, but for present purposes I pursue the second approach.

[26] This disanalogy was pointed out by Hillier.

[27] To put the point another way, for Carnap it makes no sense to inquire into the ground of logical truth, because logic must be in place before any inquiry can begin. For the empiricist Quine, of course, the question is perfectly in order. See Richardson [1998], chapter 9.

To return, then, to the Second Philosopher's question: if the goal isn't to give an account of a priori knowledge, what *is* the point of the elaborate syntactic apparatus provided by *Logical Syntax of Language*? Carnap answers that this machinery is 'a first working-tool' (Carnap [1934], p. 333) for a different project, the project of rational reconstruction. When the practitioner of the logic of science takes a scientific community to be speaking the thing language or the number language or the atom language, this isn't a straightforward description, but a way of structuring, of organizing, of rationalizing their actual practice.[28] But what is the point of this exercise? To echo the Quinean cry, 'why ... all this make believe?' (Quine [1969a], p. 75).

The answer to this fundamental question brings us back to our opening remarks on Carnap's Kantianism. We now see that Carnap's project falls in the Kantian tradition, not because he's trying to ground a priori knowledge, not because he's out to isolate underlying constitutive elements implicit in ordinary scientific practice, but because he aims to rid philosophy of its pseudo-questions.[29] When Carnap recalls his philosophical discussions, since his student days, he reports a formative insight:

> Only much later ... did I become aware that in talks with my various friends I had used different philosophical languages, adapting myself to their ways of thinking and speaking. With one friend I might talk in a language that could be characterized as realistic or even materialistic ... Not that my friend maintained or even considered the thesis of materialism; we just used a way of speaking which might be called materialistic. In a talk with another friend, I might adapt myself to his idealistic kind of language ... With some I talked a language which might be labelled nominalistic, with others again Frege's language of abstract entities of various types ... a language which some contemporary authors call Platonic.
>
> I was surprised to find that this variety in my way of speaking appeared to some as objectionable and even inconsistent. ... When asked which philosophical positions I myself held, I was unable to answer. ... Only gradually, in the course of years, did I recognize clearly that my way of thinking was neutral with respect to the traditional controversies: e.g., realism vs. idealism, nominalism vs. Platonism ..., materialism vs. spiritualism, and so on. (Carnap [1963], pp. 17–18)

[28] Cf. Ricketts [2003], p. 264: 'The projection of calculi onto actual ... used languages is a stipulation solely for the purposes of [the logic of science]. ... The coordination of calculi and languages yields an understanding of the linguistic activity of scientists as the formulation and empirical testing of theories.'

[29] Cf. Friedman [1999a], p. 213: 'It cannot be stressed too much, I think, that this diagnosis and transformation of characteristically philosophical problems constitutes the main point of both the principle of tolerance and the method of logical syntax more generally.'

Here Carnap doesn't see himself as professing various conflicting philosophical positions, one after another or all at once, but as adopting now one, now another 'kind of language', one of which 'might be labeled' realistic, another idealist, and so on.

The tools of logical syntax allow him to make these ideas precise: each of the various forms of natural language that he spoke with his friends can be rationally reconstructed in various ways; the choice between these frameworks and variations is purely conventional, entirely pragmatic; we should be flexible and tolerant toward all.[30] Faced with our ordinary talk about medium-sized physical objects, the 'realist' reconstructs it more or less as Carnap's thing language, so that its evidential rules guarantee that our sensory experience is good evidence for the existence of the external world. The 'skeptic', by way of contrast, projects a different structure, with weaker evidential rules, with the result that we cannot know whether or not we have hands. Similarly, the 'scientific realist', who believes that atoms exist, reconstructs scientific theorizing along the lines of the atom language, once again with evidential rules strong enough to allow us to justify our belief in what we cannot directly observe. For his counterpart, the 'instrumentalist', the projected evidential rules are weaker, and we can never know such things. Carnap's point is that what seems a serious philosophical question, the locus of heated debate—can we know that we have hands? that atoms exist?—actually hinges on no more than a conventional choice of rational reconstruction.[31]

On this view, many philosophical claims are pseudo-object sentences ('Five is a number' and ' "2+2 = 4" is true by virtue of meaning'); they are to be replaced by syntactic claims (' "Five" is a number-word', ' "2+2 = 4" follows by the rules of the number language'). Prior to replacement, these cause confusion because their implicit language relativity is suppressed: is the underlying claim that 'five' is a number word in all linguistic frameworks?, in

[30] Cf. 'I had acquired insights valuable for my own thinking from philosophers and scientists of a great variety of philosophical creeds' (Carnap [1963], p. 17). Richardson puts it nicely: 'People tend to be tolerant of different tools. No one refutes a screwdriver by using a hammer, and no one feels compelled to use a screwdriver to drive in a nail by the truth of screwdrivers and the falsity of hammers' (Richardson [2004], p. 71).

[31] To put another way: Carnap holds that the actual language the 'realist' and the 'skeptic' are speaking is hopelessly imprecise, amorphous; that it only makes sense to speak of evidence in the context of a rational reconstruction; that their shared language can be reconstructed in various ways, some conducive to realism and some to skepticism; that there's no fact of the matter at stake in the choice between these possible reconstructions, only a pragmatic choice.

some specific framework?, in whatever framework (as yet underdetermined) we chose for scientific language? .... [32] So, for sensible debate '*Translatability into the formal [syntactic] mode of speech constitutes the touchstone for all philosophical sentences*' (Carnap [1934], p. 313). The parties to a philosophical controversy are called upon to heed the Principle of Tolerance:

> If he wishes to discuss it, he must state his methods clearly, and give syntactical rules instead of philosophical arguments. (Carnap [1934], p. 52)

In auspicious cases, the combatants will agree on a shared linguistic framework, and thus on the standards by which their disagreement can be properly adjudicated. In cases like that of the realist and the skeptic, an apparently serious disagreement will dissolve into a matter of conventional choice. Finally some philosophical theses, like empiricism or logicism, turn out to be pragmatic recommendations, that a certain sort of framework be preferred.[33] In these and various related ways,[34] pseudo-questions are eliminated.

Thus Carnap's reply to the Second Philosopher's query: her methods are proper for her purposes, describing the world scientifically (as he not she would put it), but he has another goal, namely, to move philosophy beyond its endless and pointless debates, to put its questions on a proper syntactic footing. To return to the case of atomic theory, adopting Carnap's logic of science promises to relieve us of such philosophical debates as that between the scientific realist and the instrumentalist. Now if this were an ordinary scientific debate over the efficacy of the Einstein/Perrin evidence, then the Second Philosopher could engage it in her usual ways, but in fact this is not the question that is endlessly contested in philosophical settings: there what's at issue is whether or not the conclusions drawn by ordinary science methods should be regarded as definitive. So, for example, the scientific realist might argue that the best explanation for the success of the scientific enterprise is the assumption that the entities it describes really do exist, while the instrumentalist denies that this is so. These issues come up for closer scrutiny in IV.1, but for present purposes, it's enough to note that this sort of debate flies over the head of the Second Philosopher. From

[32] See Carnap [1934], §78.

[33] Carnap in fact recommends these two: logicism is the proposal that our rational reconstruction classify what's normally thought of as math and logic as analytic; empiricism the proposal that all synthetic claims be testable by protocol sentences.

[34] Some improper philosophical questions become straightforward questions of logical syntax, e.g., questions of intertheoretic reduction. See Carnap [1934], §82.

her point of view, one either accepts her evidence or explains in her terms why it is inadequate, something neither the realist nor the instrumentalist purports to do. What they are squabbling about escapes her in the first place—and for that reason, she doesn't feel the draw of Carnap's cure.[35] As with Kant's transcendental analysis, the advertised pay-offs of pursuing the new enterprise hold no attractions for her; once again, the Second Philosopher sees no point in the second level of a two-level view.

No doubt Carnap, following Kant, deserves to be considered a 'scientific' philosopher:

> He who wishes to investigate the questions of the logic of science must ... renounce the proud claims of a philosophy that sits enthroned above the special sciences, and must realize that he is working in exactly the same field as the scientific specialist, only with a somewhat different emphasis: his attention is directed more to the logical, formal, syntactic connections. Our thesis that the logic of science is syntax must therefore not be understood to mean that the task of the logic of science could be carried out independently of empirical science or without regard to its empirical results. (Carnap [1934], p. 332)

Recalling (from I.4) the difficulty of understanding what methods are in play at Kant's transcendental level of analysis—he seems to make claims that are ill formed on his own account of human knowledge—we note that, for Carnap, there is a corresponding obscurity about the methodology of the logic of science. Our evaluation of alternative rational reconstructions takes place against the backdrop of the Principle of Tolerance, which assures us that there is no fact of the matter at stake, that only pragmatic considerations are relevant. But why should we believe this? Presumably Carnap would answer that Tolerance itself is a meta-principle chosen on pragmatic grounds[36]—so as to eliminate apparently unproductive philosophical debate—but if so, it must also be possible—though perhaps not, Carnap

[35] Cf. Friedman [1999a], pp. 214–215: 'Carnap thus adopts a deflationary stance towards traditional philosophy, but it is nonetheless a characteristically philosophical form of deflationism. Carnap does not simply leave philosophy behind in favor of the standpoint of the "working scientist" ... This transformation and reformulation of traditional philosophy involves Carnap himself in a philosophical task.'

[36] Coffa argues, to the contrary, that for Carnap, 'behind the first-level semantic conventionalism there is a second-level semantic factualism' (Coffa [1991], p. 322). On our initial (mis)reading, the one that has Carnap attempting to provide an account of a priori knowledge as true-by-virtue-of-meaning, this might be so, but such a position seems hard to square with the rest of Carnap (see above). Goldfarb [1997] gives a persuasive rebuttal to Coffa on this point.

would argue, advantageous—to choose various intolerant meta-principles in its place. But this suggestion seems to border on incoherence: how could the conventional choice of a dogmatic meta-principle concerning rational reconstructions make it the case that, bring it about that there are truths (of existence, of logic, of confirmation relations) outside of rational reconstructions?![37]

Setting this puzzle aside, Carnap's logic of science would seem to rest on more firmly scientific ground than Kant's transcendental analysis, because the methods Carnap employs at his higher level are explicitly drawn from among those of ordinary science: the tools of modern mathematical logic. The catch—what loses the Second Philosopher—is that he applies those ordinary scientific methods in the service of a peculiarly philosophical project that from her perspective has no discernible point.[38]

[37] Indeed, it's hard to avoid the suspicion that for all its purported openness and neutrality, the Principle of Tolerance embodies a substantive and controversial claim about the nature of logic not unlike the one Quine attacks: we are free to choose any logic we like without risk of offending against any facts (as opposed to other claims that are up for empirical confirmation or disconfirmation within a framework). This is true, we're told, because questions of confirmation and disconfirmation can't even be raised until our language, and hence our logic, is in place; in other words, because logic is constitutive in some descendant of Kant's sense. But *this* distinction, between the constitutive and the non-constitutive part of our theory, is just the sort of division (like analytic/synthetic and conventional/empirical) that the spirit of Quine's critique calls into question. The only way to avoid these conclusions, as far as I can see, is to posit the stark methodological disconnect sketched in the text: what the Second Philosopher, the ordinary scientist, is doing is perfectly appropriate for her purposes; the project of rational reconstruction has different purposes; from the perspective of rational reconstruction, the discourse of ordinary science is amorphous, mere speech dispositions upon which various 'grids' can be imposed without conflicting with any pre-existing fact. This reading has the strange consequence that the Second Philosopher's apparently 'dogmatic' position on the nature of (rudimentary) logical truth (in Part III) doesn't in fact conflict with Carnap's 'tolerance'.

[38] See IV.1, footnote 13, for a similar case.

I.6
Quine's naturalism

There can be no doubting Quine's deep admiration for Carnap:

Carnap is a towering figure. I see him as the dominant figure in philosophy from the 1930s onward, as Russell had been in the decades before. (Quine [1970a], p. 40)

Still, as we've seen (in I.5), even this homage miscasts Carnap as a descendant of Russell, as carrying on the empiricist tradition.[1] Carnap's central task was not, as Quine imagined it, to found science or to account for a priori knowledge on the empiricist premise that 'whatever evidence there *is* for science *is* sensory evidence' (Quine [1969a], p. 75); in fact, no claim that logic and mathematics are 'true by virtue of' analyticity or convention or pragmatic decision is to be found in Carnap. Though he believed that one benefit of studying 'the logical syntax of science' would be to clarify what empiricism comes to—namely, the adoption of a certain sort of language[2]—Carnap's principle of tolerance precludes the outright embrace of empiricism just as it would any other form of dogmatism.

Quine's attack, then, doesn't touch the Carnap of I.5. But Quine himself *was* an empiricist, *was* concerned to explain how science arises out of sense experience, *was* concerned to account for logical and mathematical knowledge, and his critique did, at the very least, convince him and many others that the maneuvers of Carnap-as-Quine-understood-him don't do this job. (The Carnap of I.5 escapes Quine's critique, not because he successfully solves Quine's problems, but because he never addresses them

[1] This excerpt doesn't explicitly place Carnap in this role, but Quine goes on to say, 'Russell had talked of deriving the world from experience by logical construction. Carnap, in his *Aufbau*, undertook the task in earnest' (Quine [1970a], p. 40). We've seen (in I.5) how Quine's criticisms of the post-*Aufbau* Carnap mistakenly assumed that Carnap's goal was to give an empiricist account of logical and mathematical knowledge.

[2] See Ricketts [1994], pp. 194–195.

in the first place.) In the face of this failure, Quine begins to doubt the efficacy of rational reconstruction:

> But why all this creative reconstruction, all this make-believe? The stimulation of his sensory receptors is all the evidence anybody has had to go on, ultimately, in arriving at his picture of the world. Why not just see how this construction really proceeds? Why not settle for psychology? (Quine [1969a], p. 75)

So begins Quine's naturalistic turn: why look to some extra-scientific level of logical analysis; why not settle for science itself?[3]

Of course, the answer to this question, for the empiricist, is that he had hoped to found science on the indubitable cornerstone of pure sensory experience—and appealing to science itself would obviously undercut any effort toward this goal. The goal itself is just one manifestation of the widespread aspiration to build science on some secure foundation, the same aspiration that motivated Descartes (see I.1). Bemoaning the sad history of failures in this project, from Hume to Russell to (his version of) Carnap, Quine reluctantly gives it up:

> I am of that large minority or small majority who repudiate the Cartesian dream of a foundation for scientific certainty firmer than scientific method itself. (Quine [1990], p. 19)

He realizes that 'repudiation of the Cartesian dream is no minor deviation' (Quine [1990], p. 19), but once it's accomplished and

> we seek no firmer basis for science than science itself ... we are free to use the very fruits of science in investigating its roots. (Quine [1995], p. 16)[4]

All our investigations then take place 'from the point of view of our own science ... the only point of view I can offer' (Quine [1981b], p. 181).

Thus Quinean naturalism is the 'abandonment of the goal of a first philosophy':

[3] My goal here is to draw out the naturalistic thread in Quine's thought as a foil for the Second Philosopher, so I ignore many of the less naturalist-friendly strands of Quinean thought (e.g., indeterminacy of translation, inscrutability of reference, ontological relativity, which do come up briefly in II.2). See Fogelin [1997] for discussion of these distinct Quinean trends.

[4] Cf.: 'If the epistemologist's goal is validation of the grounds of empirical science, he defeats his purpose by using psychology or other empirical science in the validation. However, such scruples against circularity have little point once we have stopped dreaming of deducing science from observations' (Quine [1969a], pp. 75–76).

It sees natural science as an inquiry into reality, fallible and corrigible but not answerable to any supra-scientific tribunal, and not in need of any justification beyond observation and the hypothetico-deductive method. (Quine [1975], p. 72)

Quine often appeals to a beloved image:

Neurath has likened science to a boat which, if we are to rebuild it, we must rebuild plank by plank while staying afloat in it. The philosopher and the scientist are in the same boat. (Quine [1960], p. 3)

Thus his 'naturalistic philosopher'

Begins his reasoning within the inherited world theory as a going concern. He tentatively believes all of it, but believes also that some unidentified portions are wrong. He tries to improve, clarify, and understand the system from within. He is the busy sailor adrift on Neurath's boat. (Quine [1975], p. 72)

These bold Quinean sentiments have inspired a generation of naturalistic philosophers.[5]

Already, though, there are hints that the Quinean naturalist may not be a true Second Philosopher. As we saw in I.1, the Second Philosopher doesn't object, in principle, to Descartes's first philosophical goals: if he were able to produce the certainty he aims for by using the Method of Doubt, the Second Philosopher would be delighted. She doesn't classify his methods as 'extra-scientific', nor does she declare that science is not answerable to any such tribunal; she simply follows his reasonings and finds them wanting. Similarly, she doesn't denounce as 'unscientific' the desire of Stroud's skeptic (in I.2) to find an independent justification for her methods; she simply sees no way to do this. Nowhere does she repudiate, on principle, any inquiry or method. On the flip side, she is not driven to her position by 'despair' at the failure of any or all attempts to 'ground' science, as Quine's naturalist seems to be: 'One [source of naturalism] is despair of being able to define theoretical terms generally in terms of phenomena' (Quine [1975], p. 72). She is more aptly described, from birth, as the 'busy sailor', not as someone who later elects to enlist, perhaps in reaction to some deep disappointment. This may seem a fine point, but it's important to maintain the distinction between 'I believe in atoms because I

[5] Cf. I.5, footnote 6.

believe in science and it supports their existence' (as the enlistee might say) and 'I believe in atoms because Einstein argued so-and-so, and Perrin did experiments such-and-such, with these results' (as the Second Philosopher says).

To return to the Quinean story, Carnap's attempt to classify ontological questions—like the existence of atoms or numbers—as matters of language also falls to the generally scientific approach:

Our acceptance of an ontology is, I think, similar in principle to our acceptance of a scientific theory, say a system of physics ... Our ontology is determined once we have fixed upon the over-all conceptual scheme which is to accommodate science in its broadest sense; and the considerations which determine a reasonable construction of any part of that conceptual scheme, for example, the biological or the physical part, are not different in kind from the considerations which determine a reasonable construction of the whole. (Quine [1948], pp. 16–17)

Ontological questions ... are on a par with questions of natural science. (Quine [1951], p. 45)

At this early stage of his career, Quine retains some Carnapian impulses:

The question what ontology actually to adopt still stands open, and the obvious counsel is tolerance and an experimental spirit. Let us by all means see how much of the physicalistic conceptual scheme can be reduced to a phenomenalistic one; still, physics also naturally demands pursuing, irreducible *in toto* though it may be. (Quine [1948], p. 19)

But eventually he takes the second source of his naturalism, after despair, to be

Unregenerate realism, the robust state of mind of the natural scientist who has never felt any qualms beyond the negotiable uncertainties internal to science. (Quine [1975], p. 72)

The proceedings of the naturalistic philosopher do 'not differ in method from the special sciences' (Quine [1975], p. 72). Thus ontological questions, and metaphysical questions generally, are naturalized, that is, they are posed and answered within science.

Alongside metaphysics, the second great division of philosophy is epistemology, and here, too, Quine naturalizes:

Naturalism does not repudiate epistemology, but assimilates it to empirical psychology. Science itself tells us that our information about the world is limited to irritations of our surfaces, and then the epistemological question is in turn a question within science: the question how we human animals can have managed to arrive at science from such limited information. Our scientific epistemologist pursues this inquiry ... Evolution and natural selection will doubtless figure in this account, and he will feel free to apply physics if he sees a way. (Quine [1975], p. 72)[6]

The Second Philosopher understands the project on similar lines: how do human beings—as described by physiology, psychology, evolutionary biology, linguistics, and the rest—come to reliable knowledge of the world—as described by physics, chemistry, astronomy, geology, botany, and so on? Physics, in the form of optics, will no doubt be part of the story! For that matter, appeals to evolutionary factors might well undercut Quine's pure empiricism: some of our information about the world could be encoded in our innate cognitive machinery, and thus not arise directly from 'irritations of our surfaces'.[7] Surely the Quinean naturalist, as described so far, would take this as a friendly amendment.

Critics often claim that the naturalistic approach to epistemology cannot account for the normative, that it can only describe how we in fact come to believe, not how we should come to believe. Quine disagrees:

Our speculations about the world remain subject to norms and caveats, but these issue from science itself as we acquire it. Thus one of our scientific findings is the very fact ... that information about the world reaches us only by forces impinging on our nerve endings; and this finding has normative force, cautioning us as it does against claims of telepathy and clairvoyance. The norms can change somewhat as science progresses. For example, we were once more chary of action at a distance than we have been since Sir Isaac Newton. (Quine [1981b], p. 181)[8]

The Second Philosopher also recognizes the role of methodological norms: Mechanism, for example, arose out of scientific practice and was eventually undercut by scientific progress.[9] Closer to home, her analysis

[6] Cf. Quine [1969a], p. 82: 'Epistemology still goes on, though in a new setting and a clarified status.'

[7] See III.5.

[8] Cf.: 'They are wrong in protesting that the normative element, so characteristic of epistemology, goes by the board' (Quine [1990], p. 19); 'A normative domain within epistemology survives the conversion to naturalism, contrary to widespread belief' (Quine [1995], p. 49). I won't try to sort out whether or not Quine's notion of normativity diverges from the straightforward second-philosophical notions rehearsed here (as, e.g., Putnam [1981c], pp. 244–245, and Johnsen [2005] suggest).

[9] Mechanism is the admonition to seek explanations of physical phenomena solely in terms of particularly simple forces acting between objects. For discussion of its rise and fall, see my [1997], pp. 111–116. I come back to this methodological maxim in II.3.

of perception—calling on physiology, psychology, neuroscience, optics, etc.—indicates why perceptual beliefs are largely reliable, and therefore reasonable, under certain conditions, and largely unreliable, and therefore unreasonable, under others.

Alas the apparent harmony between Quinean epistemology naturalized and the Second Philosopher's normative inquiry into the highs and lows of human belief formation masks stark differences whose sources are sometimes difficult to trace. In fact, the Quinean opus embraces many themes, some in apparent conflict with his naturalism,[10] and many statements and restatements, so that a fair and complete assessment of agreement and disagreement would be an arduous undertaking. To illustrate the discrepancies, let me note that Quine, unlike the Second Philosopher, thinks that he can reply to the philosophical skeptic in terms drawn from science and common sense; when asked if he knows he isn't 'a brain-in-a-vat with all his experiences fed into him by a clever neurophysiologist'—a contemporary version of Descartes's Evil Demon or Stroud's extraordinary dreaming—he replies:

> I would think in terms of naturalistic plausibility. What we know, or what we firmly believe ... is that it would really be an implausible achievement, at this stage anyway, to rig up such a brain. And so I don't think I am one.[11]

We've seen that the Second Philosopher doesn't pursue this style of response, because any consideration to which she might appeal in defending such a judgment of likelihood or probability has itself been called into question by hypotheses like extraordinary dreaming, the Evil Demon, or the brain-in-a-vat; the force of such hypotheses is to replace the ordinary question of knowledge with Stroud's peculiarly philosophical question. In contrast, Quine, like Moore, persists in addressing the 'internal' or 'everyday' version of the question.

More seriously, Quine often characterizes the very project of epistemology naturalized quite differently from the Second Philosopher, a point dramatized by Stroud's analysis. From Stroud's perspective, any appeal to science is akin to Moore's reliance on common sense:

[10] Again, see Fogelin [1997] for an introduction to these complexities. See footnote 12 for an example.

[11] Fogelin ([1997], p. 549) quotes this from a panel discussion. The quoted description of the brain-in-a-vat hypothesis comes from Fogelin himself, on the same page.

What Moore says is perfectly legitimate and unassailable ... The results of an independently-pursued scientific explanation of knowledge would be in the same boat. (Stroud [1984], p. 230)

As we've seen (in I.2), Stroud thinks 'there is wisdom in this strategy' (Stroud [1984], p. 248), though it doesn't answer the skeptic's challenge as he understands it. The trouble comes in Quine's distinctive conception of the scientific undertaking: 'We are studying how the human subject ... posits bodies ... from his data' (Quine [1969a], p. 83), where '[what] can be said ... in common-sense terms about ordinary things are ... far in excess of any available data' (Quine [1960], p. 22). For Quine, the naturalized epistemologist studies

The relation between the meager input and the torrential output ... in order to see how evidence relates to theory, and in what ways one's theory of nature transcends any available evidence. (Quine [1969a], p. 83)

Though Quine replaces the traditional empiricist's indubitable sensory given with 'the limited impingements' of our sensory surfaces (Quine [1974], p. 3), he persists in the language of 'evidence', 'information', and 'data'.

Stroud's concern is that this way of describing the project provides a new foothold for the skeptic. If I regard my beliefs about the external world as the result of my own positing, a positing that could have gone any number of different ways without coming into conflict with my sensory evidence, it's hard to see that I can properly use those very beliefs to explain how I come to know what the world is like:

Countless 'hypotheses' or 'theories' could be 'projected' from those same slender 'data', so if we happen to accept one such 'theory' over others it cannot be because of any objective superiority it enjoys over possible or actual competitors ... our continued adherence to our present 'theory' could be explained only by appeal to some feature or other of the knowing subjects rather than of the world they claim to know. And that is precisely what the traditional epistemologist has always seen as undermining our knowledge of the external world. (Stroud [1984], p. 248)

Though Quine hopes to use ordinary science in his epistemological project, the project itself is formed by 'the old epistemological problem of bridging a gap between sense data and bodies'; it is 'an enlightened persistence ... in the original epistemological problem' (Quine [1974], pp. 2–3). Stroud's point is that Quine's enlightenment will not save him. As soon as he allows 'a completely general distinction between everything we get through the senses, on the one hand, and what is or is not true of the external world, on the other' he is 'cut ... off forever from knowledge of the world around us' (Stroud [1984], p. 248).[12] Because Quine's line on positing and underdetermination is supposed to have resulted from scientific inquiry, Quine's science has undermined itself from within.[13]

But a commitment to science and common sense doesn't force us to conceive the problem of naturalized epistemology in Quine's way. Faced with the same question—how do we come to know about the world?—the Second Philosopher turns to contemporary cognitive science; she is less inclined to speak of 'data' and 'positing' and more inclined to cite studies of how prelinguistic infants come to perceive and represent physical objects (see III.5). The best psychology of Quine's day was behavioristic, but even that difference can't account for his posture in the passages that trouble Stroud. In this mood, Quine is not appealing to empirical psychology or evolutionary biology; rather, he is pursuing the empiricist's philosophical project, after Russell and Carnap:

> Despite this radical shift in orientation and goal [i.e., the switch to naturalism], we can imitate the phenomenalistic groundwork of Carnap's *Aufbau* in our new setting. (Quine [1995], p. 16)

In Robert Fogelin's delightfully apt phrase, 'Quine's inspiration comes from the library, not the laboratory' (Fogelin [1997], p. 561).

[12] Oddly enough, Quine, in his less naturalistic moments, might agree with this diagnosis: his views on proxy functions suggest that the world could be made of numbers instead of physical objects, for all our evidence tells us. Stroud needn't take an external perspective and declare that all these ontologies are equally good, as Quine suggests (Quine [1981b], p. 21); he need only point out that science itself has told us that its evidence doesn't support its ontology over many rivals. Thus it's hard to see how Quine has 'defend[ed] science from within, against its own self-doubts' (Quine [1974], p. 3). Quine replies that his 'only criticism of the skeptic is that he is over-reacting' when he 'repudiates science' (Quine [1981c], p. 475), which I take to mean that the skeptic denies science is knowledge. But Quine himself denies that the world can be 'said to deviate from ... a theory that is conformable to every possible observation' (Quine [1981c], p. 474), which seems to amount to denying there is any fact of the matter about ontology that we can be said to know or fail to know. In the end, it's hard to resist Fogelin's conclusion (Fogelin [1997]) that Quine's naturalism sits ill with his ontological relativity. Surely ordinary science holds that there is a fact of the matter about whether the world is composed of physical objects, as opposed to numbers.

[13] See Stroud [1984], pp. 225–234.

Of course, the Second Philosopher is born native to the laboratory. She is interested in explaining how a causal process beginning with light falling on and reflecting off an object, continuing through the stimulations of our sense organs, proceeding through various levels of cognitive processing often results in reliable belief about the external world, and in such a story, we find nothing about 'data' or 'theory', no grounds for identifying one episode in the causal chain—the 'irritation' of our 'physical receptors'—as 'data' or 'information' or 'evidence' that radically underdetermines the rest. Ironically, Quine himself, at other times, counsels us to drop such talk of 'epistemic priority' (Quine [1969a], p. 85), but if we do so

> We are left with questions about a series of physical events, and perhaps with questions about how those events bring it about that we believe what we do about the world around us. But in trying to answer these questions we will not be pursuing in an 'enlightened' scientific way a study of the relation between 'observation' and 'scientific theory' or of the 'ways one's theory of nature transcends any available evidence'... (Stroud [1984], p. 252)

Any suggestion that we are addressing the traditional epistemological problem or the skeptic's original challenge now evaporates.

From our ordinary knowledge of the external world, Quine's naturalist and the Second Philosopher turn their attention to a scientific study of the methods of science itself. Here Quine is struck by a simple but important observation of Pierre Duhem:

> The physicist can never subject an isolated hypothesis to experimental test, but only a whole group of hypotheses; when the experiment is in disagreement with his predictions, what he learns is that at least one of the hypotheses constituting this group is unacceptable and ought to be modified; but the experiment does not designate which one should be changed. (Duhem [1906], p. 187)

This phenomenon undermines the picture of a single scientific claim enjoying 'empirical content' by itself, and leads Quine to holism and his famous 'web of belief':

> our statements about the external world face the tribunal of sense experience not individually but only as a corporate body... The totality of our so-called knowledge or beliefs, from the most casual matters of geography and history to the profoundest laws of atomic physics... is a man-made fabric which impinges on experience only along the edges. (Quine [1951], pp. 41–42)

Somewhat later, Quine tempers this holism to something more 'moderate':

> It is an uninteresting legalism ... to think of our scientific system of the world as involved *en bloc* in every prediction. More modest chunks suffice ... (Quine [1975], p. 71)

But the moral—that particular scientific theories are tested and confirmed as wholes—remains intact.

Faced with a failed prediction, then, Quine notes that strictly speaking,

> Any statement can be held true come what may, if we make drastic enough adjustments elsewhere in the system ... Conversely, by the same token, no statement is immune to revision. (Quine [1951], p. 43)

Practically speaking, we are guided by the 'maxim of minimum mutilation' (Quine [1990], p. 14), 'our natural tendency to disturb the total system as little as possible' (Quine [1951], p. 44), so we quite properly prefer to alter simple statements about observable physical objects—deciding that the swami only seems to levitate—rather than highly general laws—here the law of universal gravitation—if this is at all possible. In the image of the web, altering a statement closer to the experiential edges causes less widespread disturbance than revising a centrally located generality.

Granting that confirmation accrues holistically to scientific theories, on what sort of evidence is this confirmation based? On what grounds, for example, do we adopt atomic theory? Quine addresses this question as he continues his pursuit of a 'scientific understanding of the scientific enterprise' (Quine [1955], p. 253):

> The benefits ... credited to the molecular doctrine may be divided into five. One is simplicity ... Another is familiarity of principle ... A third is scope ... A fourth is fecundity ... The fifth goes without saying: such testable consequences of the theory as have been tested have turned out well, aside from such sparse exceptions as may in good conscience be chalked up to unexplained interferences. (Quine [1955], p. 247)

Writing with J. S. Ullian,[14] Quine gives a slightly different list of theoretical virtues—conservatism, generality, simplicity, refutability, modesty, plus conformity with observation—and elsewhere (Quine [1990], p. 95), he lists economy and naturalness as examples, but the general flavor is

[14] Quine and Ullian [1970], chapter V.

the same throughout. Finally, as Quine notes, the various virtues can conflict; they must be balanced off against one another in particular cases.

Quine acknowledges that such a defense of atomic theory is indirect, and he considers the possibility that

> the benefits conferred by the molecular doctrine give the physicist good reason to prize it, but afford no evidence of its truth ... Might the molecular doctrine not be ever so useful in organizing and extending our knowledge of the behavior of observable things, and yet be factually false? (Quine [1955], p. 248)

Quine begins his response by pushing this skeptical line of thought even further, calling into question the tendency to 'belittle molecules ... leaving common-sense bodies supreme':

> What are given in sensation are variformed and varicolored visual patches, varitextured and varitemperatured tactual feels, and an assortment of tones, tastes, smells and other odds and ends; desks [and other common-sense bodies] are no more to be found among these data than molecules. (Quine [1955], p. 250)

This line of thought tempts us to conclude that

> In whatever sense the molecules in my desk are unreal and a figment of the imagination of the scientist, in that sense the desk itself is unreal and a figment of the imagination of the race. (Quine [1955], p. 250)

The upshot would be that only sense data are real, but this conclusion

> is a perverse one, for it ascribes full reality only to a domain of objects for which there is no autonomous system of discourse at all ... Not only is the conclusion bizarre; it vitiates the very considerations that lead to it. (Quine [1955], pp. 254, 251)

We can hardly see ourselves as positing objects to explain our pure sense data when that sense data can't even be described without reference to objects.

All this, Quine counts as a reductio: 'Something went wrong with our standard of reality' (Quine [1955], p. 251). To correct the situation, he urges that we turn this thinking on its head:

> We became doubtful of the reality of molecules because the physicist's statement that there are molecules took on the aspect of a mere technical convenience in smoothing the laws of physics. Next we noted that common-sense bodies are

epistemically much on a par with the molecules, and inferred the unreality of the common-sense bodies themselves. (Quine [1955], p. 251)

But surely 'the familiar objects around us' are real if anything is; 'it smacks of a contradiction in terms to conclude otherwise'. So,

Having noted that man has no evidence for the existence of bodies beyond the fact that their assumption helps him organize experience, we should have done well, instead of disclaiming the evidence for the existence of bodies to conclude: such, then, at bottom, is what evidence is, both for ordinary bodies and for molecules. (Quine [1955], p. 251)

This, then, is Quine's conclusion: the enjoyment of the theoretical virtues is, at bottom, what supports all our knowledge of the world.

We've seen that the Second Philosopher resists the characterization of her commonsense beliefs about ordinary physical objects as inferred from some sensory 'data'; it now emerges that she also departs from Quine's naturalistic analysis of higher scientific theorizing. Taking up his example, it happens that historically the existence of atoms was finally confirmed to the satisfaction of most observers in the opening years of the twentieth century. In telegraphic summary, the story goes like this: in the first half of the nineteenth century, the atomic hypothesis swept through chemistry, explaining the laws of proportion, combining volume and substitution, and elaborating such notions as isomer and valence, but conflicts between various methods of determining precise atomic weights produced serious doubts. In 1858, this problem was solved, leading to the widespread acceptance of atomic theory at a crucial international conference in 1860. Thereafter, with the rise of kinetic theory, the success of the atomic hypothesis spread into physics. If being confirmed is enjoying the theoretic virtues, atomic theory was amply well confirmed by 1860 and even more so by 1900.

The sad surprise for the orthodox Quinean is that leading scientists at that time did not so regard it. Poincaré, Duhem, Mach, and the prominent chemist Ostwald were among the skeptics, and even supporters like Perrin admitted their doubts were not baseless:

It appeared to them more dangerous than useful to employ a hypothesis deemed incapable of verification ... the skeptical position ... was for a long time legitimate and no doubt useful. (Perrin [1913], pp. 15, 216)

Indeed, we saw (in I.5) that Einstein himself felt the need to 'find facts which would guarantee ... the existence of atoms' (Einstein [1949], p. 47). The sort of 'experimental verification' sought by all these observers was finally achieved with Perrin's experiments on Brownian motion (in 1908–11), confirming Einstein's theoretical predictions (from 1905). This was regarded as the direct detection of atoms; Poincaré, Ostwald, and many others were immediately convinced. In sum, these actual scientists did not regard the existence of atoms as having been established holistically, as Quine would have it, simply by their indispensable occurrence in a theory enjoying all theoretical virtues.[15]

There is room here for the Quinean to argue that Poincaré and the rest were simply wrong to have remained skeptical in 1900, that Perrin was wrong to have taken their doubts to be reasonable, and that Einstein was wrong to have imagined a decisive confirmation was still needed. This could be done within the bounds of naturalism if these charges are made on scientific grounds; one might, for example, argue that the skeptics were inappropriately influenced by some version of purely philosophical verificationism.[16] Perhaps this was true, for example, of Mach, who refused to accept the existence of atoms even after 1910,[17] but it seems an unlikely account of Poincaré and Ostwald's reservations, given that they took the Einstein/Perrin evidence to be conclusive,[18] and even less plausible as an explanation of Einstein and Perrin's acceptance of the skeptical challenge in the first place.[19]

What's gone wrong with the Quinean picture is confirmational holism:[20] this case suggests that we cannot regard a scientific theory as a homogeneous

[15] See my [1997], pp. 135–143, for further discussion and references. I return to this case briefly in IV.1 and in slightly more depth in IV.5.

[16] Colyvan [2001], pp. 99–101, takes this line.

[17] The other holdout, Duhem, took religious revelation, not science, to be the source of insight into the structure of the world (in the appendix to Duhem [1906], see my [1996], pp. 320–323, for discussion). Wilson ([2000b], [2006], pp. 154–157, 356–369, 654–659) points out that both Mach and Duhem also had legitimate scientific worries about mechanical hypotheses in general and the atomic hypothesis in particular. (Cf. II.6, footnote 16, IV.4, footnote 48, IV.5, footnote 35.)

[18] Presumably their skepticism would have survived, like van Fraassen's (see IV.1), if it had been based on extra-scientific philosophy.

[19] Notice that Einstein and Perrin didn't accuse their opponents of bad methodology; they took their skepticism to be a reasonable view and set out to disprove it.

[20] Azzouni holds that Quine's criterion of ontological commitment not his holism is at fault, that it makes sense to claim, e.g., that 'there are so-and-so's' is true while denying 'ontological commitment' to so-and-so's. I think much of the working structure of Azzouni's position can be translated into my terms: where Azzouni asks if a given true existential claim carries ontological import, I ask if it is confirmed. See the discussion of Azzouni in IV.5 for references.

whole, confirmed as a unit, that a consistent naturalist must recognize various different types of evidence and various different roles hypotheses can play in our theorizing. The Second Philosopher sees the evidential relations of modern science in its many branches as complex and varied, to be studied and assessed in their particular contexts of inquiry, not obviously subject to general characterization like 'observation and the hypothetico-deductive method' (Quine [1975], p. 72). This divergence from Quinean orthodoxy, combined with the others touched on above, ramifies into disagreements in the philosophy of logic (see Part III), the philosophy of mathematics (see IV.2, IV.4), and scientific metaphysics in general (see IV.5). Still, for all that, Second Philosophy should be regarded as a variety of post-Quinean naturalism.[21]

[21] Another version is due to John Burgess (see my [2005a]).

I.7
Putnam's anti-naturalism

I conclude this extended exercise in compare and contrast with the views of Hilary Putnam, a philosopher whose work casts a troubled shadow over contemporary discussions of naturalism.[1] It isn't that he's alone in opposing naturalistic approaches: Quine may be right that 'a large minority or small majority' renounce the Cartesian dream of an extra-scientific foundation for science, but I suspect a large majority retains its faith in alluring philosophical enterprises that stand uncomfortably close to Descartes's.[2] What sets Putnam apart is his early, eloquent trumpeting of his own naturalist attitudes. Closely allied with Quine in the early 1970s, he writes:[3]

> It is silly to agree that a reason for believing that *p* warrants accepting *p* in all scientific circumstances, and then to add 'but even so it is not *good enough*'. Such a judgment could only be made if one accepted a trans-scientific method as superior to the scientific method; but this philosopher, at least, has no interest in doing *that*. (Putnam [1971], p. 356)

Ten years later,[4] he strongly opposes both 'contemporary attempts to "naturalize" metaphysics' and 'attempts to naturalize the fundamental notions of the theory of knowledge' (Putnam [1981c], p. 229). Thus Putnam's harsh critique carries the added sting of coming from a former sympathizer.

Much of Putnam's early, more naturalist-friendly work centers on the topic of scientific realism, which he describes this way:

[1] As with Carnap and Quine, I won't try to capture the entire range of Putnam's still-evolving thought. I focus here on the anti-naturalism of Putnam [1981b] and [1981c] and the change of views that led up to it (especially Putnam [1971], [1976b], [1976c], [1981a]).

[2] As noted in I.2, I suspect many professed naturalists seek a reply to Stroud's pseudo-Cartesian that isn't open to the response: but that depends on your methods! See IV.1.

[3] This comes in the course of a defense of Quine's indispensability argument for mathematical realism (see IV.2.i).

[4] In Putnam [1981b] and [1981c].

The statements of science are ... either true or false (although it is often the case that we don't know which) and their truth and falsity does not consist in their being highly derived ways of describing regularities in human experience. Reality is not part of the human mind; rather the human mind is part—and a small part at that—of reality. (Putnam [1979], p. vii)

Others, myself included, might prefer to understand scientific realism as an ontological position—holding that the theoretical entities of science exist and are as they are independently of us—with an epistemological component—science is a partly successful effort to find out about these things. But Putnam, under the influence of Michael Dummett,[5] prefers a formulation in terms of truth:

A realist (with respect to a given theory or discourse) holds that (1) the sentences of that theory or discourse are true or false; and (2) that what makes them true or false is something *external*—that is to say, it is not (in general) our sense data, actual or potential, or the structure of our minds, or our language, etc. (Putnam [1975b], pp. 69–70)

Putnam's 'scientific realism' is just this sort of realism about scientific theorizing. The central contrast with anti-realism then hinges on their respective theories of truth:

Views of truth can be divided into two kinds: 'realist' views, which interpret truth as some kind of correspondence to what is the case, and 'verificationist' views, which interpret truth as, for example, what would be *verified* under ideal conditions of inquiry. (I choose Peirce's form as my example of a verificationist view because it seems to me the most tenable; but, of course, there are many versions of both the realist view and the verificationist view in the literature.) (Putnam [1978b], p. 1)

Verificationist views might be called 'epistemic', as they understand truth, somehow or other, in terms of our ways of coming to know.[6]

The shift in Putnam's views developed out of his growing worries about the viability of the correspondence theory of truth. The turning point came sometime in the second half of 1976. In May of that year, in a talk delivered in Jerusalem, Putnam envisions a notion of correspondence arising in the course of an ordinary empirical investigation of the efficacy of human language use:

[5] e.g., see Dummett [1963].

[6] I prefer 'epistemic' because 'verificationist' carries more specific connotations. See Blackburn [1994], pp. 392–393.

... the *success* of the 'language-using program' [i.e., human language use] may well depend on the existence of a suitable correspondence between the words of the language and things, between sentences of the language and states of affairs ... we say certain things, conduct certain reasonings with each other, manipulate materials in a certain way, and finally we have a bridge that enables us to cross a river that we couldn't cross before. And our reasoning and discussion is as much part of the total organized behavior-complex as is our lifting of steel girders with a crane ... [what needs explaining is] the *contribution* of our linguistic behavior to the success of our total behavior. (Putnam [1976b], pp. 100–101)

A 'correspondence' between words and sets of things ... can be viewed as part of an *explanatory model* of the speakers' collective behavior ... let me refer to ... this sort of scientific picture of the relation of speakers to their environment, and of the role of language ... as *internal realism*.[7] (Putnam [1976c], p. 123)

By December of 1976, addressing the American Philosophical Association in Boston, he has adopted a dramatic two-level position that he traces to Kant.[8] As usual, the first level encompasses ordinary scientific explanations; given Putnam's focus on realism, which for him means a focus on truth, this is just the internal realism described above. At the second level, again as usual, we want something more; in particular, we want a perfectly general account of how theories relate to the world, not one tied to our parochial science. One such account is 'metaphysical realism':

It is, or purports to be, a model of the relation of *any* correct theory to all or part of THE WORLD. ... What makes this picture different from *internal* realism (which employs a similar picture *within* a theory) is that (1) the picture is supposed to apply to *all* correct theories at once ... and (2) THE WORLD is supposed to be *independent* of any particular representation we have of it—indeed, it is held that we might be *unable* to represent THE WORLD correctly at all ... *truth* is supposed to be radically non-epistemic. (Putnam [1976c], pp. 123–125)

On the basis of a sharp, multifaceted critique,[9] Putnam rejects this second-level account—much as Kant rejects transcendental realism—in favor of what he calls 'internalism':

[7] Sometimes 'internal realism' seems to include the 'internalist' second-level account described below—e.g., see Putnam [1976c], pp. 135–138—but I stick to the more limited understanding explicated in the text.

[8] See Putnam [1981a], pp. 60–64, 74.

[9] This includes the much-discussed 'model theoretic argument' (Putnam [1976c], [1977]).

It is characteristic of this view to hold that *what objects does the world consist of?* is a question that only makes sense to ask *within* a theory or description. ...'Truth', in an internalist view, is some sort of (idealized) rational acceptability—some sort of ideal coherence of our beliefs with each other and with our experiences *as those experiences are themselves represented in our belief system*—and not correspondence with mind-independent or discourse-independent 'states of affairs'. (Putnam [1981a], pp. 49–50)

The parallel with Kant is immediate: our ordinary knowledge at the first level is really, when viewed from the second level, only knowledge of the world as described or conceptualized; at the second level, we see that the metaphysical realist is wrong, that no knowledge of 'something totally uncontaminated by conceptualization' is possible (Putnam [1981a], p. 54).

The Second Philosopher's reaction to this picture is by now predictable. The first-level inquiry into the function of language in human undertakings is entirely comprehensible to her. Indeed, she agrees with Putnam (in his first-level mode) that epistemic accounts of truth are not accurate characterizations of the role that notion plays in her discussions of language and the world—she sees important distinctions between having certain experiences and there being a tree outside her window, between meters reading certain ways and the existence of particles, between a theory having true observational consequences and its being true—but she also withholds judgment, pending more detailed investigation, before she concludes that truth is 'correspondence' (see Part II). She is surprised, then, when Putnam goes on to insist that truth is *really* ideal rational acceptability: didn't we agree, she asks, that a theory's being true is different from its being reasonable for us to believe it?! When she asks Putnam, as she did Kant and Carnap before him, why he makes this new claim, where he takes her to have gone wrong, he tells her, as they did, that there's nothing wrong with her proceedings and conclusions, for her purposes, but that he has different goals. And, as before, she responds by asking: what are those goals, and how are they to be pursued?

Putnam's higher purpose is fairly clear: for all the attractions of a lower-level empirical account, he wants a theory of truth that isn't restricted to the confines of our current science.[10] In his description of metaphysical

[10] When Putnam writes: '*Internal realism is all the realism we want or need*' (Putnam [1976c], p. 130) he doesn't mean internal realism is all we need, period, he means it's all the *realism* we need.

realism, he expresses this as a desire for a model of the relation between language and the world—or as he puts it 'THE WORLD'—that applies to all theories at once, not just our current science, and that takes THE WORLD to be independent of our representation of it. This will not enlighten the Second Philosopher because she too seeks an account of the relation between language and the world that holds for any discourse, and she too believes that the world is as it is independently of our ways of representing or describing it. Unless the capital letters are doing some work, she doesn't understand how the metaphysical realist's 'higher' purpose differs from hers.

But the capital letters aren't just for emphasis: THE WORLD is different from the world. To see this, consider Putnam's formulation of a venerable criticism of correspondence theories of truth:

> One cannot think of truth as correspondence to facts (or 'reality') because ... thinking of truth in this way would require one to be able to compare concepts directly with unconceptualized reality—and philosophers [are] fond ... of pointing out the absurdity of such a comparison. (Putnam [1976b], p. 110)

Putnam is surely right about the popularity of this charge! But this 'Comparison Problem'[11]—how can we compare language with raw reality?—doesn't apply to the Second Philosopher's project: she understands the challenge as explaining the relation between human language use—as described in linguistics, psychology, sociology, etc.—and the world—as described in physics, chemistry, biology, astronomy, botany, etc.; she will never be involved in the search for 'a "correspondence" to noumenal things' (Putnam [1981a], p. 73) because she is deaf and blind to the lure of the Kantian transcendental project in the first place (see I.4). Putnam's use of 'THE WORLD' signals the contrast: his metaphysical realist requires a correspondence between language and 'something totally uncontaminated by conceptualization' (Putnam [1981a], p. 54).[12]

[11] I borrow the term from Marino [2006]. See II.2.

[12] Though Putnam clearly associates THE WORLD with Kant's noumenon, there seems to be one important difference. On Putnam's understanding of the latter, the operative notion is '*the whole noumenal world* ... you must *not* think that because there are chairs and horses and sensations in our representation [the world as experienced], that there are correspondingly noumenal chairs and noumenal horses and noumenal sensations' (Putnam [1981a], p. 63). THE WORLD, on the other hand, does seem to come portioned into objects or 'pieces' (see, e.g., Putnam [1976c], p. 124). THE WORLD, despite its unconceptualized character, appears to have at least some KF-structure, in the sense of III.3.

Following Putnam's typographical lead, we might say that his second-level realist wants a 'Correspondence Theory', while the Second Philosopher's first-level inquiry can at best deliver a 'correspondence theory'. The motivation Putnam provides for this capitalized pursuit in answer to the Second Philosopher, the higher purpose he commends to her, is the hope of a conception of truth that stands outside our best understanding of the world, that sees our current science as just one among many possible 'theories or descriptions'. From this lofty vantage point, he recognizes the hopelessness of concocting a Correspondence Theory, and so opts for internalism, the view that truth, within any 'theory or description' is just idealized rational acceptability, 'a sort of ideal coherence with' that particular system of belief.[13] If the Second Philosopher were pushed this far, she might well ask on what grounds Putnam draws any conclusions in this higher context, where all her ordinary methods have been set aside, but she will not be pushed this far. When Kant pressed the desire for an account of a priori knowledge as motivating his transcendental inquiry, she found no shortcoming in her own empirical approach to the a priori; faced with Putnam's truth theoretic motivations, she finds no deficiency in her characteristic ways of investigating word–world relations.

Assuming, then, that Putnam's two-level position holds no more second-philosophical attractions than Kant's or Carnap's, let's turn now turn to his direct attacks on naturalism itself, in the University of California Howison Lectures of 1981 (Putnam [1981b] and [1981c]). The two together purport to describe

> the failure of contemporary attempts to 'naturalize' metaphysics [and] attempts to naturalize the fundamental notions of the theory of knowledge. (Putnam [1981c], p. 229)

[13] It isn't entirely clear whether Putnam already required this 'higher' sort of theory, in addition to internal realism, in May of 1976, or something in the transition to December of 1976 drove him to it. His own description of his change of mind (Putnam [1976c], pp. 129–130) suggests the former, as he admits that 'metaphysical realism [was] a picture I was wedded to' and all his energies seem focused on rejecting it, not on arguing that internal realism, by itself, is explanatorily inadequate. If this is right, then he holds a two-level position even in May—internal realism plus metaphysical realism—though only one of these is mentioned. This might begin to explain why the Comparison Problem turns up on p. 110 of Putnam [1976b], though it's still hard to see why it's in any way relevant to internal realism, the view explicitly under discussion at that point.

One example of such an attempt is what Putnam calls 'materialism',[14] the view that science is the best source of information about what the world is like. Putnam continues:

We don't *need* intellectual intuition to do *his* sort of metaphysics: his metaphysics, he says, is as open-ended, as infinitely revisable and fallible, as science itself. In fact, it *is* science itself! ... The appeal of materialism lies precisely in this, in the claim to be *natural* metaphysics, metaphysics within the bounds of science. (Putnam [1981b], p. 210)

This view Putnam now considers not only false, but pernicious:

Metaphysical materialism has replaced positivism and pragmatism as the dominant contemporary form of scientism. Since scientism is, in my opinion, one of the most dangerous contemporary intellectual tendencies, a critique of its most influential contemporary form is a duty for a philosopher who views his enterprise as more than a purely technical discipline. (Putnam [1981b], p. 211)

'Scientism' is a pejorative, of course,[15] but it seems to me not entirely unfair as applied to many forms of naturalism, and to Second Philosophy in particular.[16] It would take some considerable effort to sort out precisely which position or positions Putnam means to be attacking in these papers, but fortunately, for our purposes, we need only ask to what extent his critique might be effective against the Second Philosopher. The answer to this question has no obvious bearing on its cogency as addressed to whichever opponents Putnam himself has in mind.

Perhaps the central theme of Putnam's critique is the charge that the naturalist has not learned the lesson of Kant, that she continues to pursue 'a coherent theory of the noumena':

The approach to which I have devoted this paper is an approach which claims that there *is* a 'transcendental' reality in Kant's sense, one absolutely independent of our minds ... *but* (and this is what makes it 'natural' metaphysics) we need no *intellektuelle Anschauung* ... the 'scientific method' will do ... 'Metaphysics within the bounds of science alone' might be its slogan. (Putnam [1981b], p. 226)

[14] This usage seems non-standard (see e.g., Blackburn [1994], p. 233), as science doesn't tell us that everything is material, but the following quotation certainly has a naturalistic ring.

[15] See Blackburn [1994], p. 344.

[16] In the contemporary spirit of co-opting slurs, see Wilson [2006], p. 614, footnote 19: 'I yield the Lamp of Scientism to no one!'

Now Putnam himself, as we've seen, 'can sympathize with the urge behind this view'; he feels a strong pull to 'talk about the transcendent' and even speculates that 'one's attitude toward it must, perhaps, be the concern of religion rather than of rational philosophy' (Putnam [1981b], p. 226). Still, however Putnam views such a project, we've also seen that the Second Philosopher feels no impulse to undertake any higher-level inquiry, nor is she motivated to do so by the allurements of Kant, Carnap, or Putnam. She does believe that the world she studies is independent of us, but this is only what Kant would endorse as 'empirical realism', what Carnap would characterize as true internal to the scientific framework, what Putnam would call 'internal realism'. To aspire to anything else presupposes a step the Second Philosopher doesn't take in the first place.

So the Second Philosopher is no 'transcendental realist', but perhaps she has offended against what Putnam calls Kant's 'corollary':

> The corollary Kant drew from all this is that even experiences are in part constructions of the mind ... the idea that all experience involves mental construction, and the idea that the dependence of physical object concepts and experience concepts goes *both* ways, continue to be of great importance in contemporary philosophy. (Putnam [1981b], pp. 209–210)

Of course, the claim that human cognizers perform some processing on raw sensory stimulations is a commonplace of contemporary psychology; there is a concerted scientific effort to determine how this is done, to describe the mechanisms involved. But Putnam has more than this in mind: it is

> Silly [to think] that we can have knowledge of objects that goes beyond experience. (Putnam [1981b], p. 210)

> One idea ... definitely sunk by Kant ... is that we can think and talk about things as they are, independently of our minds. (Putnam [1981b], p. 205)

The 'Kantian corollary' seems to be that we cannot hope to know what the world is like independent of our perceptual and conceptual processors or independent of our scientific theorizing.

To the Second Philosopher, this sounds either false or unproblematic. When empirical psychology tells us that we are prone to certain sorts of perceptual or cognitive mistakes, it is telling us that the world is not as our basic processors tend to see it. Likewise, scientific progress sometimes takes the form of discovering that the way the world appears to us is not the way it actually is: as Einstein showed that our perception of the world as Euclidean is actually a parochial, small-scale take on a large-scale non-Euclidean universe, or as quantum mechanics suggests that our everyday ideas about causation aren't applicable in the micro-world. In all such cases, careful application of the scientific method allows us to 'see around' our most basic forms of perception and conceptualization, to better understand how the world is independently of us. And it is clearly possible to 'see around' any particular scientific theory; this is how science progresses, by replacing one theory with another. So the only complaint that remains is that we can't find out about the world without using our scientific methods—something the Second Philosopher would hardly contest!

It appears, then, that the Kantian core of Putnam's attack actually involves a dispute at his second, transcendental level of inquiry—can we or can we not achieve knowledge of THE WORLD?—a dispute to which the Second Philosopher is not party. Still, his overall attack on naturalism includes various independent lines of thought that may help illuminate the distinctive character of Second Philosophy. Let me take up three of these in order: one metaphysical, one epistemological, and one truth theoretic.

Putnam's metaphysical complaint is that the naturalist believes there is

> One true theory, the true and complete description of the furniture of the world ... this belief in one true theory requires a *ready-made* world ... : the world itself has to have a 'built-in' structure. (Putnam [1981b], pp. 210–211)

Certainly the Second Philosopher does believe the world has a 'built-in' structure, meaning that the world is as it is independently of our perception, cognition, theorizing, and so on; this is the structure we're trying to capture in our scientific efforts to 'see around' our common ways of thinking, to set aside our prejudices and reveal the world as it is. This much is straightforward, but Putnam adds something more: the assumption that there is one and only one theory that correctly describes this 'built-in' structure. Otherwise,

> Any sentence that changes truth-value upon passing from one correct theory to another correct theory ... will express only a *theory-relative* property of THE WORLD. And the more such sentences there are, the more properties of THE WORLD will turn out to be theory-relative. (Putnam [1976c], p. 132)

> truth would lose its absolute (non-perspectival) character. (Putnam [1981b], p. 211)

The idea seems to be that if there were more than one correct theory, then each such theory would impose its properties on the world; the world would have those properties only relative to, from the perspective of, the relevant theory. Perhaps, even more radically, the world would have no structure of its own and could be imposed on in any way we choose!

The Second Philosopher rejects all of this. For comparison, suppose the world consisted of a deck of cards: one true theory describes it as made up of fifty-two card-like objects, another as four suit-like clump-objects, still another as a single complex whole; it seems reasonable to say that all these theories are correct, that each describes aspects of the way the world is, that each of them picks out properties that are 'built in'. Similarly, the Second Philosopher holds that the world our science studies has a built-in structure, that our methods are designed to help us get at this structure, but she needn't insist that there is only one correct way to do this, and she needn't deny that which built-in structures we tend to pick up on is at least partly a function of our cognitive structures and our interests. To say that there might be several correct ways of describing the world is not to say that the features each attributes to it are theory-relative, merely perspectival. And it certainly isn't to say that every way of describing the world is equally good: the history of science is littered with theories that didn't work![17]

In truth I think the possibility of more than 'one true theory' troubles Putnam because Quine's example has led him to expect that metaphysics naturalized should be a more straightforward undertaking that it is. For the Quinean naturalist, all ontological questions are settled by our best scientific theory, a vast, intertwined web of beliefs well matched to experience at its edges.[18] Viewed in this light, Putnam's concern is entirely understandable: if there isn't one unique correct theory, how can we answer our ontological questions? But we've already seen that Quine's picture is too simple for

[17] Discussions of scientific realism often employ the grisly image of the scientist 'carving nature at the joints'. In these terms, the Second Philosopher allows that nature probably has joints she hasn't noticed, joints that don't interest her, joints forever hidden from her, but this doesn't mean that it has no joints (or, equivalently, that it has joints everywhere).

[18] Actually, Quine requires that the theory be regimented into first-order logic before ontological determinations can be reached, but his attitude toward this regimentation differs sharply from Carnap's. For Carnap, study of the logic of science begins as a pure discipline; for Quine, in Ricketts' words, 'Our language, our speech habits, encompasses the use of logical notation for the regimentation of some statements into others for various purposes' (Ricketts [2003], p. 275). See Ricketts's [2003] for further discussion (especially pp. 269–276).

many reasons—atomic theory was part of our best theory before the existence of atoms was established; some parts of our best theory are mere conventions, adopted for want of something better—and others can be added—for example, the common use of explicit idealizations (see IV.2.i). Pace Quine, determining what our successful theories tell us about what there is cannot be a simple matter of reading off their existential claims. I come back to the prospects for post-Quinean metaphysics naturalized in IV.5, by which point more Second Philosophy will have collected on the table, but even then, I can only sketch the barest beginnings. Still, the point that bears emphasis for now is that none of this undermines the core of metaphysics naturalized: the conviction that the Second Philosopher's methods are the best way we have of finding out what the world is like. We must face the fact that this 'finding out' is a more difficult task than Quine once led us to believe, but the Second Philosopher has no reason to suppose that the project is doomed or that there is any better way to pursue it.

Turning to epistemology, Putnam's critique centers on two bold claims: that naturalism tends to either cultural relativism or cultural imperialism, and that both these alternatives are self-refuting.[19] To see how this charge might apply to the Second Philosopher, suppose she is engaged in her scientific study of science: she calls on her physiological, psychological, neurological accounts of human perception and conceptualization, her linguistic, psycholinguistic, cognitive scientific theories of the workings of human language, her physical, chemical, astronomical, biological, botanical, geological descriptions of the world in which these humans live; she uses these and any other relevant scientific findings to explain how these humans, by these means, come to know about this world. Now suppose that along the way she encounters various human linguistic practices that differ markedly from hers. Some of these—chanting or story-telling—don't seem to play the characteristic roles of claims about the world, but others—astrology or theology—apparently do. The Second Philosopher recognizes that the evidential standards and methodological norms of these

[19] See Putnam [1981c], pp. 234–240. Characteristically, Putnam describes his 'cultural relativist' and 'cultural imperialist' in terms of truth: the former defines truth as 'correct according to the norms of my culture' and adds that someone in another culture defines truth analogously in terms of his own cultural norms; the latter defines truth as the former does but holds this to be the only notion of truth. In what follows, I use the terms 'relativist' and 'imperialist' without modifiers to distinguish attitudes toward alternative methods of inquiry.

practices are not the methods she uses in her inquiries. What does she say about these oddities?

One answer, the relativist's answer, goes like this: 'Clearly their standards and norms are different from mine. I think mine are justified—I continually examine and re-examine their efficacy and attempt to refine and extend them as I go along—but I must recognize that my justifications themselves rely on my standards and norms. I can't expect the participants in these other practices to be any more impressed by my justification of my methods by means of my methods than I am by their justification of their methods by means of their methods. Given the symmetry of the situation, I must concede that their practice is as good as mine.' Putnam objects that when the relativist says 'they see that my methods are justified by my methods', she's actually asserting this claim on the basis of her norms and standards, not theirs. In other words, the cultural relativist refutes herself, because her own position implies that she can't make a claim of symmetry that conveys what it ought to convey.

Now it may be that some professed naturalists would react to the discovery of other practices in this way, but the Second Philosopher does not. In some cases, she might conclude that the seemingly assertive practice is actually pursued for other reasons: perhaps to produce a certain spiritual state, in the case of theology; or as a tool in a type of quasi-psychoanalytic process, in the case of astrology. But her analysis—based on psychology, sociology, anthropology, etc.—may well determine that the odd practice *is* aimed, just as her second-philosophical practice is aimed, at telling us what the world is like: the astrologer may insist that human behavior can be predicted from the positions of the stars; the theologian may insist that certain phenomena are supernatural miracles. Faced with this sort of claim, the Second Philosopher doesn't fall back on any relativism; she simply believes that the standards and norms of these practices are incorrect, that what the practitioners of these alternative practices offer as evidence does not in fact support their stated opinions.

Once again, the Second Philosopher's reaction here doesn't take the form: this is unscientific! Her objections are ordinary objections, like those voiced in this passage from Richard Feynman (quoted in part in I.1):

> Astrologists say that there are days when it's better to go to the dentist than other days. There are days when it's better to fly in an airplane, for you, if you are

born on such a day and such and such an hour. And it's all calculated by very careful rules in terms of the position of the stars. If it were true it would be very interesting. Insurance people would be very interested to change the insurance rates on people if they follow the astrological rules, because they have a better chance when they are in the airplane. Tests to determine whether people who go on the day that they are not supposed to go are worse off or not have never been made by the astrologers...

Maybe it's still true, yes. On the other hand, there's an awful lot of information that indicates that it isn't true. We have a lot of knowledge about how things work, what people are, what the world is, what those stars are, what the planets are that you are looking at, what makes them go around more or less... so what are you going to do? Disbelieve it. There's no evidence at all for it.... The only way you can believe it is to have a general lack of information about the stars and the world and what the rest of the things look like. If such a phenomenon existed it would be most remarkable, in the face of all the other phenomena that exist, and unless someone can demonstrate it to you with a real experiment, with a real test, took people who believe and people who didn't believe and made a test, and so on, then there's no point in listening to them.

Tests of this kind, incidentally, have been made in the early days of science. It's rather interesting. I found out that in the early days, like in the time when they were discovering oxygen and so on, people made such experimental attempts to find out, for example, whether missionaries—it sounds silly; it only sounds silly because you're afraid to test it—whether good people like missionaries who pray and so on were less likely to be in a shipwreck than others. And so when missionaries were going to far countries, they checked in the shipwrecks whether the missionaries were less likely to drown than other people. And it turned out that there was no difference. (Feynman [1998], pp. 91–93)

Her opponents might protest that she reaches these conclusions by means of her own standards and norms, her own methods, but to this she happily agrees. She admits, of course, that her methods are open to criticism and modification—she is vigilant in striving to improve them—but neither the astrologer nor the theologian has presented any grounds for concern about their efficacy.[20]

Clearly the Second Philosopher is far more susceptible to the charge of imperialism than she is to the charge of relativism: she holds that her methods are correct and that the astrologer's, insofar as they differ

[20] Note the similarity to the second-philosophical response to Stroud's pseudo-Cartesian in I.2. Such cases come up again in IV.1.

from hers, are not. This gives added significance to Putnam's claim that cultural imperialism is self-refuting—which leads in turn to his third style of objection to naturalism, namely, to his doubts about the naturalist's conception of truth. Addressed to the cultural imperialist, this line of thought begins:

> [The imperialist] can say, 'Well then, truth—the only notion of truth I understand—is defined by the norms of *my* culture.' ('After all,' he can add, 'which other norms should I rely on? The norms of *somebody else's* culture?') (Putnam [1981c], p. 238)

Thus the imperialist's notion of truth 'cannot go beyond right assertibility' (Putnam [1981c], p. 239). The trouble, according to Putnam, is that our culture does not include a norm of the form

> (*) A statement is true (rightly assertible) only if it is assertible according to the norms of modern European and American culture. (Putnam [1981c], p. 239)

As our culture doesn't endorse (*), it follows, by our imperialist notion of truth, that (*) isn't true. Or, to put it the other way round: 'If this statement [(*)] is true, it follows that it is not true ... Hence it is not true QED' (Putnam [1981c], p. 239). Thus imperialism, for us, is self-refuting. Ironically, it wouldn't be for 'a totalitarian culture which erected its own cultural imperialism into a required dogma, a culturally normative belief' (Putnam [1981c], p. 239), but this is cold comfort to the naturalist!

Applied to the Second Philosopher, the argument would go something like this. To determine whether or not a statement is true, the Second Philosopher applies her current standards and norms, so her notion of truth is 'right assertibility by those standards and norms'. But why should this be so? When the Second Philosopher is asked to settle a question of truth, she does indeed apply her methods, but she needn't view this connection as a *definition* of truth. In fact, we've seen that epistemic accounts are unlikely to emerge from her investigation of the notion of truth. Defining truth as 'right assertibility' would convert one of the most important challenges of her study of her own methods—the task of assessing the reliability of her norms and standards—into an analytic certainty: right assertability by her methods would be what truth *is*! Any theory of truth that trivializes this important job of self-assessment should certainly be rejected. So I think my imperialistic Second Philosopher is not committed to a 'right assertibility' account of truth.

We've seen, then, that the Second Philosopher adopts neither an epistemic notion of truth nor (in the typography introduced earlier) a Correspondence Theory. Indeed, nothing in our characterization of Second Philosophy commits her to any particular theory of truth; she is bound only to follow her inquiry wherever it leads. Still, at some point she must face the question Putnam raised in the first place—how does our use of language function in our interactions with each other and the world? what word–world connections are involved?—and a theory of truth may or may not end up playing a role in her answers.

My plan now is to reorient the project of illustrating the practice of Second Philosophy. So far, I've been describing the Second Philosopher's approaches to various questions and reactions to various challenges, comparing and contrasting them with the approaches and reactions of various other broadly naturalistic inquirers. What I propose to do next is take up one contemporary philosophical debate—this very question of word–world connections and the theory of truth—and investigate the extent to which it is and isn't a naturally occurring piece of Second Philosophy. In the process, I hope to make some second-philosophical progress on the issues themselves.

PART II

The Second Philosopher at Work

II.1

What's left to do?

No doubt the resistance of some philosophers to forms of naturalism as austere as Second Philosophy springs from the concern—stated or implied, conscious or unconscious—that if all becomes science, nothing is left for the philosopher to do. I hope the sheer bulk of this volume helps debunk that idea, but perhaps it's worth addressing the issue explicitly for a paragraph or two, before getting down to work.

The primary reason there remain so many jobs for the Second Philosopher is quite simple: there are important questions (typically classified as philosophical) that don't fit within a single scientific discipline.[1] We've already touched repeatedly on one example in the second-philosophical inquiry into the reliability of perception (see I.1, I.2, and I.6): to determine when, why, and how our perceptual beliefs tend to be accurate, it isn't enough to investigate the workings of the retina and visual cortex with the vision specialist; we must also investigate the sources of the inputs, the nature of the objects perceived, and the behavior of light as it is absorbed, reflected, and detected, which takes us (at least) into various branches of physics. The scientific study of the methods of science itself is another example (see I.6), and the central questions addressed in Parts III and IV below—what is the ground of logical and mathematical truth? How do we come to know these things? What role do they play in our investigation of the world?—these questions too require attention to such diverse fields as psychology, physics, and the methodologies of mathematics and the natural sciences. Questions like these tend not to be asked by scientists, or to be asked and answered with an unsatisfying narrowness of vision.[2]

[1] Cf. Quine [1995], p. 16: naturalistic 'inquiry proceeds in disregard of disciplinary boundaries but with respect for the disciplines themselves and appetite for their input'.

[2] e.g., the invaluable work of psychologists on the nature of mathematical cognition is often marred by a seemingly myopic inference from 'we cognize mathematically' to 'mathematics is purely a matter

But it isn't just her distinctive questions or her ability to draw together insights from various scientific disciplines that set the Second Philosopher apart; there is also her training in the history and contemporary practice of philosophy. Philosophical writings of various sorts have provided inspiration to scientists themselves, as Einstein demonstrates in his remarks on

> a paradox upon which I had already hit at the age of sixteen: If I pursue a beam of light with the velocity *c* (velocity of light in a vacuum), I should observe such a beam of light as a spatially oscillatory electromagnetic field at rest. However, there seems to be no such thing, whether on the basis of experience or according to Maxwell's equations ... Today everyone knows, of course, that all attempts to clarify this paradox satisfactorily were condemned to failure as long as the axiom of the absolute character of time, viz., of simultaneity, unrecognizedly was anchored in the unconscious. Clearly to recognize this axiom and its arbitrary character really implies already the solution of the problem. The type of critical reasoning which was required for the discovery of this central point was decisively furthered, in my case, especially by the reading of David Hume's and Ernst Mach's philosophical writings. (Einstein [1949], p. 53)

This isn't to say that Einstein adopted the views of these empiricist philosophers—rather he describes himself as an 'unscrupulous opportunist' drawing on whichever philosophical position suits his purposes in a given context[3]—but there is no doubt that first philosophies can play this sort of inspirational or heuristic role, for the disciplinary scientist or the Second Philosopher. We see an example of the latter in the influence of Kantian thought on the second-philosophical thinking of Part III.

What's more, an awareness of the philosophical tradition can provide prophylactic as well as inspirational benefits. Many appealing ideas that occur to the untutored have been tried out once or more than once over the centuries; knowing which of these are clear dead ends and why, and the strengths and weaknesses of the rest, can vastly improve the Second Philosopher's instincts. Furthermore, Mark Wilson uses an example from the field of folklore preservation to illustrate how seemingly innocent, quasi-philosophical modes of thought—'ur-philosophy' in his terminology—can

of cognition'. Close attention to matters outside the field of psychology—from the structure of the world to the role of mathematics in science—serves to temper such overreaching.

[3] The quotation comes from Einstein [1949], p. 684. See my [1997], pp. 188–190, for further discussion.

distort ordinary practices in disturbingly counterproductive ways.[4] So the Second Philosopher's intellectual experience protects her against missteps to which others might more easily succumb.

In sum, then, the Second Philosopher is uniquely placed—without strict disciplinary allegiance—and peculiarly well trained—in philosophy—to address scientific versions of traditionally philosophical questions. I should remark that she is also fully capable of the sort of meta-philosophical ruminations that have filled this book up to now—the Second Philosopher is as free as anyone else to take an interest in the human practice of philosophy, its methods and its motivations—though to have made this explicit would have complicated the presentation. Nothing I've said about her relies on methods that aren't easily classified as roughly 'scientific' or on any definitive separation of science from non-science; in discussions of the two-level positions of Kant, Carnap, and Putnam, the separation of levels was their assumption, not mine. So I hope that what's gone before and what's to come will dispel the concern that there is no such pursuit as Second Philosophy.

If this is right, then we should expect that Second Philosophy is already practiced on occasion, if not by that name. In fact, I think it is, but rarely in pure form (as I.3 and I.6 suggest). I hope to illustrate this by returning to the question of how human language use functions in our interactions with each other and the world, of how our words relate to things, and examining one contemporary debate for its second-philosophical content. The notions of truth and reference tend to play a prominent role in discussions of word–world connections, so our story begins with one strand of debate in the theory of truth, in particular, with the writings of Hartry Field and Stephen Leeds. I then turn for contrast to Crispin Wright, whose approach is unabashedly first philosophical, and finally to Mark Wilson, whose work embodies the particularly austere naturalism of our Second Philosopher ('embodies' rather than 'embraces', because Second Philosophy, as a way of addressing questions rather than a doctrine, can be followed unselfconsciously, without talking or perhaps even thinking

[4] See Wilson [2006], chapter 2. The ur-philosophy in this case concerns what philosophers would call the distinction between primary and secondary qualities, namely, the theory that primary qualities (size, shape) are objective features of an object, but secondary qualities (color, taste) are subjective, 'in us'.

much about it[5]). In the end, I hope to have isolated a second-philosophical core in these discussions and to have provided the beginnings of a second-philosophical take on truth, reference, and our linguistic interactions with the world, but my main goal is to illuminate the nature of Second Philosophy by means complementary to those of Part I.

[5] I'm fond of the term 'natural-born naturalist'.

II.2

An illustration: truth and reference

The modern study of truth begins with Alfred Tarski's semantic theory, developed in 1929 and published in the early 1930s (Tarski [1933]). Tarski aims to give a 'materially adequate' definition of truth, by which he means a definition that implies, for each sentence '...', the T-sentence

'...' is true if and only if... ,

one example being the famous

'snow is white' is true if and only if snow is white.

The definition is called 'semantic' because it attempts to capture a notion of truth that belongs to semantics, along with such relations as 'denotes' (as in 'George Washington' denotes George Washington) and 'applies to' (as in the predicate 'was the first president of the United States' applies to George Washington).[1]

Unfortunately, such a definition for ordinary English, applied to the sentence I'm about to type and call '*c*'—'*c* is false'—would produce the unfortunate T-sentence

'*c* is false' is true if and only *c* is false,

and thus (as *c* is the sentence '*c* is false')

c is true if and only if *c* is false,

the infamous Paradox of the Liar.[2] In light of this difficulty, Tarski turns his attention from natural to formal languages—with variables and logical operators, interpreted names and predicate letters[3]—which he thinks of

[1] Tarski also calls it a 'correspondence theory' (Tarski [1933], p. 153), but this doesn't seem to square with contemporary terminology, as we'll see below.

[2] See Tarski [1933], pp. 157–158.

[3] These are interpreted symbols, not mere syntax: 'we are not interested here in "formal" languages... to the signs and expressions of which no meaning is attached' (Tarski [1933], p. 166). Cf. Field [1972], p. 4, footnote 2.

as representing some 'wider or narrower' portion of ordinary language.[4] Truth for such an 'object language' (that is, the formal language under investigation) is defined in a meta-language (that is, the language in which the investigation is carried out). If this is to work, the T-sentences must be proved in the meta-language, so it must contain (a translation of) ... on the right hand side of each T-sentence, and it must be able to construct a name for ... , like '...' on the left-hand side.[5] (I use ' "*e*" ' for the meta-language's name for the object language expression *e*, and '*e**' for its translation of *e*.[6])

Defining truth for a given object language depends on a prior notion of what it is for an assignment (of things to the variables) to satisfy an open sentence[7] of that language. If *s* is such an assignment and "*x*" is a variable of *L*, we say that *s* assigns it *s*(*x*); if "*c*" is a name of *L*, we say *s* assigns it *c**.[8] Then the definition of satisfaction (for the simplified case of a language with only one-place predicates) goes like this:

An open sentence *S* of *L* is *satisfied by an assignment s* if and only if one of the following:[9]

(i) *S* consists of a predicate "*P*" followed by a variable "*x*", and $P^*(s(x))$.[10]

(ii) *S* consists of a predicate "*P*" followed by a name "*c*", and $P^*(c^*)$.

(iii) *S* consists of a "not" followed by an expression "*U*" and "*U*" is not satisfied by *s*.

(iv) *S* consists of an expression "*U*" followed by "and" followed by an expression "*W*", and "*U*" and "*W*" are satisfied by *s*.

[4] Cf. Tarski [1933], p. 165, footnote 2: 'The results obtained for formalized language also have a certain validity for colloquial language ... if we translate into colloquial language any definition of a true sentence which has been constructed for some formalized language, we obtain a fragmentary definition of truth which embraces a wider or narrower category of sentences.'

[5] The meta-language must also be 'essentially stronger' than the object language: e.g., Tarski shows that arithmetical truth can be defined in set theory but not in arithmetic itself. See Enderton [1972], pp. 228–229, for a textbook treatment.

[6] e.g., suppose the object language is a portion of French used for chemistry and the meta-language is some suitably equipped portion of English. Then "platine est un métal" is the meta-linguistic name for the object language sentence whose translation into the meta-language is: platinum is a metal.

[7] An open sentence may have unbound variables: e.g., '*x* is wise' or 'everything with *P* bears *R* to *x* and *y*'.

[8] To continue in the spirit of footnote 6, if "Angleterre" is a name in the object language, then *s* assigns it England.

[9] Imagine the other connectives are defined as usual in terms of 'not', 'and', and 'for all'.

[10] e.g., if *S* consists of the predicate "is white" followed by the variable "*x*", and *s*(*x*) is white.

(v) S consists of "for all x" followed by an expression "U", and "U" is satisfied by all assignments s', where s' is just like s, except possibly in its assignment to the variable x.[11]

It turns out that for an open sentence with no unbound variables—that is, for 'sentence' proper—the particular assignment s is irrelevant—if it's satisfied by one, it's satisfied by all—so the definition of truth comes out like this:

A sentence S of L is *true* if and only if it's satisfied by some assignment s (or equivalently by all assignments s).

The definition of satisfaction is recursive—notice that 'satisfied by' occurs in the definition itself—but it can be converted into a standard definition—S is satisfied by s if and only if so-and-so, where so-and-so doesn't contain 'satisfied by'—using well-known methods.[12] The meta-language can now express a T-sentence for each sentence of the object language:

'...' is true if and only if... *.

In the special case where the object language is a portion of English and the meta-language is a richer portion of English, this becomes the T-sentence at the opening of this section:

'...' is true if and only if....

In cases where the meta-language is also strong enough to prove the T-sentences, Tarski has achieved his materially adequate definition.

So much for preliminaries. The story I want to tell here begins with Hartry Field's critique of Tarski's theory (Field [1972]). On Field's reading, Tarski's contribution is best understood against the backdrop of the rise of scientific philosophies of logical positivism and logical empiricism. So, for example, speaking of the period just before Tarski, Carnap reports that

> in our philosophical discussions we had, of course, always talked about [truth and word–world relations], but we had no exact systematized language for this purpose. (Carnap [1963], p. 60)

Even when Tarski's work first appeared, before it was fully assimilated, Carnap reports the belief of Reichenbach and Otto Neurath that truth

[11] For simplicity, I've given this clause for the special case with object and meta-languages as portions of English. For the general case (with English as the meta-language), (v) would begin: S consists of the object language expression e such that e^* is 'for all x', and....

[12] Assuming, again, that the meta-language is 'essentially stronger' than the object language (see footnote 5 above).

'could not be reconciled with a strictly empiricist and anti-metaphysical point of view' (Carnap [1963], p. 61). Carnap considered this a serious misunderstanding, arguing that Tarski had 'provided for the first time the means for precisely explicating' the concept of truth (Carnap [1963], p. 61). And, as Field points out, Tarski himself occasionally speaks in similar terms: he expresses concern that his methods be in 'harmony with the postulates of the unity of science and of physicalism' and declares that with his work 'the problem of establishing semantics on a scientific basis is completely solved' (Tarski [1936], pp. 406, 407).

Field contends that this conclusion is unwarranted, that Tarski hasn't reduced truth to straightforward scientific terms, but only to other troublesome semantic terms. To follow Field's line of thought, consider clause (ii) of the definition of 'satisfied by'. The assignment *s* makes no difference here, so we get

"*Pc*" is true if and only if $P^*(c^*)$,

where c^* and P^* are translations of the object language "*c*" and "*P*" into the meta-language. Field's plan is to illuminate Tarski's method here by contrasting (ii) with the more straightforward and intuitive (ii)′:

"*Pc*" is true if and only if "*P*" applies to the object "*c*" denotes.

This second proposal obviously won't do the job Tarski has set for himself—employing, as it does, the semantic terms 'denotes' and 'applies to'—but the question is how Tarski has managed to avoid these.

The answer is that he's used translation. For example, Field argues that the illicit semantic term

N denotes *a*

has been replaced by a long disjunction with one disjunct for every name in the object language[13]

(*N* is "*c*" and *a* is c^*) or (*N* is "*d*" and *a* is d^*) ... or (*N* is "*z*" and *a* is z^*).

This definition, Field notes, satisfies something very like Tarski's adequacy condition for his definition of truth, namely, it implies every D-sentence

"*e*" denotes *a* if and only if e^* is *a*

[13] If French is the object language and English the meta-language, the illicit semantic term '*N* refers to *a*' is replaced by a long disjunction with one disjunct for every French name: *N* is "Angleterre" and *a* is England or ... or *N* is "Allemagne" and *a* is Germany.

or, when some portion of English is the object language, and some richer portion of English is the meta-language:

"*e*" denotes *a* if and only if *e* is *a*

which is to say:

"*e*" denotes *e*.

'Applies to' can be given a similar treatment: 'X applies to a' is replaced with

(X is "P" and $P^*(a)$) or (X is "Q" and $Q^*(a)$) or ... or (X is "Z" and $Z^*(a)$)

where there is one disjunct for every predicate of the object language. Much as before, this yields the *A*-sentences

"P" applies to *a* if and only if $P^*(a)$

or, in the special case

"P" applies to *a* if and only if $P(a)$.

None of the long disjunctions contains semantic terms, and they are all that's needed to convert the likes of Field's (ii)′ into the likes of Tarski's (ii), so it seems Tarski has in fact effected the promised scientific reduction of truth.

To show this semblance to be illusory, Field invites us to compare this treatment of 'denotes' with the chemist's treatment of 'valence'. The latter term, introduced in the mid-nineteenth century, describes an element's powers of combination in terms of a single number. This notion is 'physicalistically important', and thus

> If physicalism is correct it ought to be possible to explicate this concept in physical terms—e.g., it ought to be possible to find structural properties of the atoms of each element that determine what the valence of that element will be. (Field [1972], pp. 15–16)

This was eventually accomplished, and thus, Field writes, 'The notion of valence was ... shown to be a physicalistically acceptable notion' (Field [1972], p. 16). But, Field asks, what if a late nineteenth-century chemist had proposed the following 'reduction' of valence, with one disjunct for each element?

E has valence *n* if and only if *E* is potassium and *n* is $+1$, or ... or *E* is sulfur and *n* is -2.

Obviously chemists would not have regarded this as satisfactory—they single-mindedly pursued a reduction in terms of atomic structure—and Field maintains that Tarski's analogous 'reduction' of denotation is no better:

> It seems clear that [the above 'reduction' of denotation does] not really reduce denotation to non-semantical terms, any more than [the disjunctive 'reduction' of valence] reduces valence to nonchemical terms. (Field [1972], p. 18)

What's needed, in the present case, is a robust account of how names 'latch onto their denotations' (Field [1972], p. 19), most likely, Field thinks, something along the lines of a causal account of reference.[14]

What's come to be called a causal theory of reference has its origins in the writings of Saul Kripke and Hilary Putnam,[15] as an alternative to the reigning description theory of names. On that account, a proper name—'Kurt Gödel'—is associated with a description—say, 'the man who proved the incompleteness theorem'—(or perhaps a cluster of such descriptions) and the name then refers to the individual who satisfies the associated description (or perhaps some sufficient, possibly weighted subset of the cluster). Kripke points out that in fact we sometimes don't know enough about a person to have a uniquely identifying description:

> Consider Richard Feynman, to whom many of us are able to refer. He is a leading contemporary theoretical physicist. Everyone *here* (I'm sure!) can state the contents of one of Feynman's theories so as to differentiate him from Gell-Mann. However, the man in the street, not possessing these abilities, may still use the name 'Feynman'. When asked, he will say: well he's a physicist or something. He may not think that this picks out anyone uniquely. I still think he uses the name 'Feynman' as a name for Feynman. (Kripke [1972], p. 292)

For that matter, the descriptions we associate with the name may well be false:

> Very often we use a name on the basis of considerable misinformation. ... [For example,] what do we know about Peano? What many people in this room may 'know' about Peano is that he was the discoverer of certain axioms which

[14] 'Refers' here is intended to cover both 'denotes' and 'applies to'.

[15] See Kripke [1972], and, e.g., Putnam [1975a]. This is the naturalistic Putnam whose passing is noted in I.7 above.

characterize the sequence of natural numbers, the so-called 'Peano axioms'. Probably some people can even state them. I have been told that these axioms are not actually due to Peano but to Dedekind. Peano was of course not a dishonest man. He includes them in his book with an accompanying credit in his footnotes. Somehow the footnote has been ignored. So, on the [description theory] the term 'Peano', as we use it, really refers to—now that you've heard it you see that you were really all the time talking about—Dedekind. But you were not. (Kripke [1972], pp. 294–295)

The counter-proposal—the so-called 'causal theory'—goes like this:

Someone, let's say, a baby, is born; his parents call him by a certain name. They talk about him to their friends. Other people meet him. Through various sorts of talk the name is spread from link to link as if by a chain. A speaker who is on the far end of this chain, who has heard about, say Richard Feynman, in the market place or elsewhere, may be referring to Richard Feynman even though he can't remember from whom he first heard of Feynman or from whom he ever heard of Feynman. He knows Feynman was a famous physicist. A certain passage of communication reaching ultimately to the man himself does reach the speaker. He then is referring to Feynman even though he can't identify him uniquely. (Kripke [1972], pp. 298–299)

Such a chain begins with an 'initial baptism' and then is passed link to link by word of mouth, by written records, by scholarly investigation, and so on. The result is that 'Peano discovered the Peano Axioms' is a false claim about Peano, not a true claim about Dedekind.

Kripke and Putnam make a similar proposal for kind natural words like 'water' and 'gold':

If we imagine a hypothetical (admittedly somewhat artificial) baptism of the substance, we must imagine it picked out as by some such 'definition' as, 'Gold is the substance instantiated by the items over there, or at any rate, by almost all of them'.... I believe that in general, terms for natural kinds (e.g., animal, vegetable, and chemical kinds) get their reference fixed in this way; the substance is defined as the kind instantiated by (almost all) of a given sample. The 'almost all' qualification allows that some fools' gold may be present in the sample. (Kripke [1972], p. 328)

So, 'water' is the substance instantiated by the stuff in this glass, in lakes and streams, in rainfall. We associate with the term a range of descriptions—gold is a yellow metal, water is a colorless tasteless liquid—but the referent is determined by the samples, and extends to everything with the same internal structure. In the course of scientific investigation, we often discover better descriptions of the internal structure of the stuff—gold has atomic number 79, water is H_2O[16]—with the result that some of the old descriptions are dropped—some gold is white, some water is solid—and even some of the original samples might be reclassified—a bit of iron pyrites or heavy water. And so on. Alas, for all the attractions of this picture,[17] many serious efforts over the decades since these passages were written have demonstrated just how devilishly hard it is to fill in the details of this causal account of reference.[18]

Notice that when Field advocates such a reduction of 'denotation' and 'application', he's not thinking of a traditional correspondence theory of truth (that is, a Correspondence Theory in the typography of I.7). He imagines Quine objecting:

To ask for more than these schemas [schematic versions of the T- and R-sentences]—to ask for causal theories of reference to nail language to reality—is to fail to recognize that we are at sea on Neurath's boat: we have to work *within* our conceptual scheme, we can't glue it to reality from the outside. (Field [1972], p. 24)

Quine is here rejecting the sort of thing the Putnam of I.7 also derides: a theory that involves a correspondence between language and a noumenal, unconceptualized reality, a theory susceptible to the Comparison Problem.[19] Field replies:

In looking for a theory of truth and a theory of primitive reference we *are* trying to explain the connection between language and (extralinguistic) reality, but we are *not* trying to step outside our theories of the world in order to do so. Our accounts of primitive reference and of truth are not to be thought of as something that could be given by philosophical reflection prior to scientific information—on the contrary, it seems likely that such things as psychological models of human beings and investigations of neurophysiology will be very relevant to discovering the mechanisms involved in reference. (Field [1972], p. 24)

[16] See I.1, footnote 9.

[17] Cf. Kripke [1972], p. 303: 'I may not have presented a theory, but I do think that I have presented a better picture than that given by description theorists.'

[18] For a sampling, see Devitt [1981] for an early, book-length effort, or Stanford and Kitcher [2000] for a more recent proposal.

[19] The difficulty of 'gluing language to reality from the outside' is what we called the Comparison Problem in I.7: how can we compare our sentences with raw reality?

Presumably a causal theory of the sort Field imagines adding to Tarski's inductive clauses could in principle result from the Second Philosopher's straightforward scientific investigation of the relations between human language—as understood by linguistics, psychology, neuroscience, and so on—and the world—as understood by physics, chemistry, geology, biology, botany and so on—much as the naturalistic Putnam once conjectured (in I.7).

Field's interpretation of Tarski is controversial—we get a sense of why in a moment—but what matters for our purposes isn't so much its historical accuracy as the debate that it launched. In reaction to Field [1972], Stephen Leeds (in Leeds [1978]) calls attention to one particular turn of the argument: if the analogy between 'truth' and 'valence' is to carry its required weight, then 'truth', like 'valence', must play a substantial role in our scientific account of the world; science isn't expected to give a full treatment of any odd notion whatsoever! This is not a point Field himself missed:

> The clarity of 'valence' ... before reduction—and even more, [its] *utility* before reduction—did provide physicalists with substantial reason to think that a reduction ... was possible, and, as I remarked earlier, a great deal of fruitful work in physical chemistry ... was motivated by the fact. Similarly, insofar as semantic notions like 'true' are useful, we have every reason to suspect that they will be reducible to non-semantic terms, and it is likely that progress in linguistic theory will come by looking for such reductions. (Field [1972], p. 25)

Indeed, he expressed some concern:

> Of course, this sort of argument ... is only as powerful as our arguments for the utility ... of the term 'true'—the purposes it serves, and the extent to which those purposes could be served by less pretentious notions such as warranted assertibility—needs much closer investigation. (Field, [1972], p. 25)

This is where Leeds makes his entrance, to argue that a less pretentious notion can do the job.

What Leeds has in mind isn't warranted assertibility or any such epistemic option,[20] but the simple expedient of resting content with what Tarski has given us: the T-sentences and the R-sentences.[21] The central question is

[20] See I.7 for more on epistemic accounts of truth.

[21] Perhaps with some device for collecting them. See footnotes 27 and 28.

whether or not these modest resources can serve the scientific purposes of truth. So what scientific purposes *are* served by truth? Here Leeds follows Quine:

> Truth hinges on reality ... Where the truth predicate has its utility is in just those places where, though still concerned with reality, we are impelled by certain technical complications to mention sentences. Here the truth predicate serves, as it were, to point through the sentence to the reality; it serves as a reminder that though sentences are mentioned, reality is still the whole point. (Quine [1970b/1986], p. 11)

For single sentences, then, truth is just a device of disquotation:

'snow is white' is true if and only if snow is white.

We do no more by asserting its truth than by asserting the sentence itself.

Truth does more work when we wish to affirm without repeating, saying for example 'What John said is true', and even more when we wish to affirm what we can't repeat, 'Whatever John said while I was out of the room was true'. Handier still is its role in allowing us to affirm infinitely many things at once—any sentence of the form '*p* or not-*p*' is true[22]—and it's hard to see any other way of denying an infinite collection of sentences when we aren't in a position to deny any particular member—'not every consequence of this theory is true'.[23] In such cases, truth is playing the role of an infinite conjunction or disjunction: 'John said "snow is white" and snow is white, or John said "grass is green" and grass is green, or ... '(with one clause for every sentence of English); snow is white or snow isn't white, and grass is green or grass isn't green, and ... '(with one clause for every sentence of English); Consequence 1 is false, or Consequence 2 is false, or Consequence 3 is false, or ... (with one clause for every consequence of the theory in question).

In addition to using the truth predicate for such purposes, we also make general claims about truth: 'we aim to assert only truths', or 'there are truths we will never know'. These can also be understood as infinite conjunctions—'we aim to assert "snow is white" only if snow is white, and we aim to assert "grass is green" only if grass is green, and ... '—or

[22] See Quine [1970b/1986], p. 11.

[23] See Field [1986a], p. 57, or [1986b], pp. 485–486.

disjunctions—'snow is white but we'll never know it, or grass is green but we'll never know it, or…'. Putnam's use of 'true' to characterize realism (see I.7) is similar:

(1) the sentences of [the theory] are true or false; and
(2) what makes them true or false is something *external*—that is to say, it is not (in general) our sense data, actual or potential, or the structure of our minds, or our language, etc. (Putnam [1975b], p. 69)

would be short for an infinite conjunction of the form—' "snow is white or snow is not white, and what makes it one or the other is something external…" and "grass is green or grass is not green, and what makes it one or the other is something external…" and…'.[24] Examples could be multiplied, but the idea should be clear: truth is a device for semantic ascent (from 'snow is white' to ' "snow is white" is true') and descent (from ' "snow is white" is true' to 'snow is white'); it serves as a device of infinite conjunction and disjunction. In Quine's phrase: 'Truth is disquotation' (Quine [1990], p. 80).

According to the disquotationalist,[25] then, the notion of truth is given by the T-sentences. Similarly, reference is given by the R-sentences for names and predicates:

"*c*" refers to a iff *c* is *a*, and
"*P*" refers to *a* iff *P*(*a*).

With this much in place, we can reproduce instances of the Tarskian inductive clauses;[26] for example

"*Pc*" is true iff *Pc* iff there is something that "*P*" applies to and "*c*" denotes.
"*R* or *S*" is true iff *R* or *S* iff "*R*" is true or "*S*" is true.

Field points out that if we also allow the disquotationalist some means of generalization, as he is inclined to do, then truth and reference can be defined outright. Obviously, infinite conjunctions and disjunctions would do the trick:

[24] Thus a truth theoretic notion of realism requires the rejection of epistemic theories of truth, but not necessarily the adoption of a correspondence theory (much less a Correspondence Theory).

[25] 'Deflationism' is a general term for theories that 'deflate' truth, of which disquotationalism is one variety (for others, see Field [1986a] or [1986b] or the 'minimalism' of II.5). The theory sketched here and in II.4 is a variation on what Field calls 'pure disquotational truth'.

[26] See Field [1994a], pp. 114–115, 124–125.

S is true iff *S* is 'snow is white' and snow is white
or *S* is 'grass is green' and grass is green
... or *S* is 'roses are red' and roses are red

where there's one disjunct for every sentence of our language, but more palatable options—like a substitutional quantifier over sentences[27] or a role for schematic letters[28]—would also suffice. With any of these devices at hand, the fully general Tarskian clauses can be proved; for example,

for all sentences *R* and *S*, *R* and *S* is true iff *R* is true and *S* is true

But where Tarski takes these clauses to be part of a definition of truth, the disquotationalist regards them as simple consequences of his own analysis. It follows, then, that disquotational truth will apply everywhere, even if some areas of language aren't compositional.[29]

So far, we have a preliminary sketch of disquotationalism and a mere gesture at one variety of correspondence theory (a causal theory of reference plus the Tarskian inductive clauses), but I think this is enough to generate two broad diagnostics for separating accounts of the first sort from accounts of the second sort. Though commentators sometimes write as if there were disquotational truth conditions and correspondence truth conditions, in fact theorists of both sorts agree that the truth conditions of 'snow is white' are snow's being white.[30] The difference between them comes out rather in their assessment of the status of the shared assertion that

'snow is white' is true iff snow is white.

For the disquotationalist, this is an obvious fact about truth, sometimes described as trivial or conceptual or analytic or necessary or whatever, but however that may be, always perfectly straightforward and superficial. In contrast, the correspondence theorist sees it as an important fact about 'snow is white', in need of analysis and explanation: what makes it the case

[27] Ordinary quantification is objectual: 'for all *x*, *x* is red' is true iff every object *x* is red. What we want to say is: for all sentences *S*, *S* is true iff *S*, but this is ill formed, because '*S*' is a name of a sentence and doesn't make sense in the third slot. For the substitutional quantification Π*S* ('*S*' is true iff *S*), we're to imagine substituting the string of symbols *S* into both slots of the biconditional, generating a name for *S* on the left-hand side, and *S* itself in the second.

[28] Here we add a new sort of variable, a schematic variable, *X*, and a rule that allows the replacement of such a variable by any sentence. We can then assert: '*X*' is true iff *X*. (Tarski's T-schema is somewhat different: '"*X* is true if, and only if, p" ... with "*p*" replaced by any sentence of the language ... and "X" replaced by a name of that sentence' (Tarski [1944], pp. 119–120).)

[29] See Field [1994a], p. 125.

[30] See Field [1994a], p. 111, footnote 9, Leeds [1995], p. 33, footnote 5.

that 'snow is white' has precisely these truth conditions? (On the version of correspondence theory suggested above, the answer would involve causal relations between our use of 'snow' and instances of certain kinds of precipitation, and so on.) So the first diagnostic is this: are the T- and R-sentences simply obvious (or trivial or ...) facts about truth and reference, or do they stand in need of analysis and explanation?[31]

As a test case, let's apply the proposed diagnostic to Tarski himself. Commenting on a particular T-sentence, he writes:

> It is clear that from the point of view of our basic conception of truth these sentences [the two sides of the biconditional] are equivalent. (Tarski [1944], p 119)

It certainly sounds as if Tarski is siding here with the disquotationalist, understanding the T-sentence as an obvious conceptual fact of some sort. Perhaps, then, it is no surprise that he goes on to suggest each T-sentence

> may be considered a partial definition of truth ... The general definition has to be, in a certain sense, the logical conjunction of all these partial definitions ... (Tarski [1944], p. 120)

thus apparently embracing disquotationalism more or less as we understand it.

The second diagnostic emerges when we say that our friend Pierre, who speaks French, has said something true. For the correspondence theorist, this is a fact about Pierre, the sentence he uttered, various causal word–world relations, and the world itself: Pierre said truly 'neige est blanc', because 'neige' denotes a certain form of precipitation, 'blanc' applies to things that reflect certain wavelengths, and so on. Most importantly, Pierre's having said something true has nothing to do with us. The non-French-speaking disquotationalist, on the other hand, seems unable to say that Pierre has spoken truly, because he has in his language no T-sentence for 'neige est blanc'. The natural move is for the disquotationalist to translate what Pierre said into his own language, in other words, to hold that: Pierre said truly 'neige est blanc', because 'neige est blanc' translates as 'snow is white' and snow is white.[32]

[31] Cf. Leeds [1995], pp. 3, 7, 9–10.

[32] Cf. Leeds [1995], p. 6, Field [1994a], pp. 127–130. If we don't know precisely what Pierre said, we appeal to the same generalizing function as in our own language: Pierre said something true because

The disquotationalist's appeal to translation in this context invites objections of two very different sorts. On one extreme, it might be argued that a proper translation must preserve reference in the correspondence theorist's sense. Considering the difficulties encountered by the causal theorists, much more needs to be said to make this a serious concern; until the various factors that lead the makers of French–English dictionaries to pair 'neige' with 'snow' are spelled out and catalogued, and a robust theory of reference constructed from them, the burden of proof here must remain with the objector.

At the other extreme lie concerns arising from Quine's notorious discussions of translation:

> A rabbit scurries by, the native says 'Gavagai', and the linguist notes down ... 'Rabbit' (or 'Lo, a rabbit') as tentative translation, subject to testing in further cases. ... Who knows but what the objects to which this term applies are not rabbits after all, but mere stages, or brief temporal segments, of rabbits? ... Or perhaps the objects to which 'gavagai' applies are all and sundry undetached parts of rabbits ... A further alternative ... is to take 'gavagai' as ... naming the fusion ... of all rabbits: that single though discontinuous portion of the spatiotemporal world that consists of rabbits. ... a still further alternative ... is to take it as ... naming a recurring universal, rabbithood. (Quine [1960], pp. 29, 51–52)

At first blush, it seems we could readily distinguish these, but

> Consider, then, how. Point to a rabbit and you have pointed to a stage of a rabbit, to an integral part of a rabbit, to the rabbit fusion, and to where rabbithood is manifested. Point to an integral part of a rabbit and you have pointed again to the remaining four sorts of things; and so on around. (Quine [1960], pp. 52–53)

When we have more of the native's language in hand, we can ask specific questions:

> If ... we take 'are the same' as translation of some construction in the jungle language, we may proceed on that basis to question our informant about sameness

Pierre said 'neige est blanc', which translates to 'snow is white', and snow is white, or Pierre said 'herbe est verte', which translates to 'grass is green', and grass is green or ...

Obviously this won't work if Pierre's sentence has no English equivalent; Field considers various moves to cover such cases (Field [1994a], pp. 147–151). If the utterance in question cannot be made comprehensible to me in any way—by translation, by expansion of my language, by my learning the other language—it seems the disquotationalist should come to doubt that it makes sense to call it true (cf. Leeds [1995], p. 7).

of gavagais from occasion to occasion and so conclude that gavagais are rabbits and not stages. But if instead we take 'are stages of the same animal' as translation of that jungle construction, we will conclude from the same subsequent questioning of our informant that gavagais are rabbit stages. (Quine [1960], p. 72)

Quine concludes that

There can be no doubt that rival [translations] can fit the totality of speech behavior to perfection, and can fit the totality of dispositions to speech behavior as well, and still specify mutually incompatible translations of countless sentences. (Quine [1960], p. 72)

This is Quine's 'indeterminacy of translation'.[33] If translation is so thoroughly indeterminate, perhaps it can't carry the weight the disquotationalist assigns it.

In fact, I think even Quine himself, at least in some of his moods,[34] would find this concern to be exaggerated:

Translation between kindred languages, e.g., Frisian and English, is aided by resemblance of cognate word forms. Translation between unrelated languages, e.g., Hungarian and English, may be aided by traditional equations that have evolved in step with a shared culture ... a chain of interpreters of a sort can be recruited of marginal persons across the darkest archipelago. (Quine [1960], p. 28)

What Quine has in mind for his 'gavagai' example is radical translation, 'i.e., translation of the language of a hitherto untouched people' (Quine [1960], p. 28). But even here, he admits that the problem, which he sometimes calls 'artificial' and 'contrived',[35] would not arise in practice:

[33] This is one of the less-naturalistic Quinean themes alluded to in I.6, footnote 3.

[34] Despite the reasonable-sounding passages cited below, the line of thought beginning in 'indeterminacy of translation' continues to 'ontological relativity', which often sounds close to Putnam's 'culturally relative' naturalism: ontology is relative to our background theory (see Quine [1968], pp. 45–51). I prefer to understand these Quinean passages as arguments that it makes no sense to ask for a theory of Reference (in the typology of I.7): 'It is meaningless to ask whether, in general, our terms "rabbit", "rabbit part" ... really refer respectively to rabbits, rabbit parts ... rather than to some ingeniously permuted denotations. It is meaningless to ask this absolutely; we can meaningfully ask it only relative to some background language ... And in practice we end the regress of background languages, in discussions of reference, by acquiescing in our mother tongue and taking its words at face value' (Quine [1968], pp. 48–49). If 'our mother tongue' is the one the Second Philosopher speaks, maybe this reading can be maintained, but I haven't the scholarship to know.

[35] See, e.g., Quine [1968], pp. 30, 33, 35.

An actual field linguist would of course be sensible enough to equate 'gavagai' with 'rabbit', dismissing such perverse alternatives as 'undetached rabbit part' and 'rabbit stage' out of hand. ... The implicit maxim guiding his choice of 'rabbit' ... is that an enduring and relatively homogeneous object, moving as a whole against a contrasting background, is a likely reference for a short expression. (Quine [1968], p. 34)[36]

Quine counts the linguist's maxim as an 'imposition', but the Second Philosopher would argue that he's forgetting the common perceptual and cognitive machinery we humans inherit from our shared evolutionary past.

Quine also considers actual cases: for example, he reports that in Japanese, an expression like 'five pencils' involves a number word, a pencil word, and a third expression that can be understood either as converting the number word into a form suitable to counting stick-like objects, or as converting the pencil word from a mass term like 'pencil stuff' to an individuating word like 'pencil'. In such cases, there is genuine indeterminacy of translation, and it's only reasonable to grant that many real-life instances of translation are subject to such vagaries. But this doesn't keep the ordinary, practical notion of translation from doing the job the disquotationalist assigns it in determining the reference of Pierre's word 'neige', where the correct translation is at least as unproblematic as it is for 'gavagai'.

A tamer worry, also derived from Quine, is that translation rests on a notion of synonymy, which comes under attack in the course of Quine's critique of the analytic/synthetic distinction. As indicated above (in I.6), the Second Philosopher departs from Quine in allowing for the possibility of a respectable notion of meaning, arising perhaps in psycholinguistics, but nothing like this is needed to establish that human languages can actually be translated. It seems Leeds has something similar in mind when he reminds us that there are standard translations:

I suspect ... there are all sorts of special circumstances that hold in the case of human languages that make the standard translations salient—roughly, our common origins and common psychology, and the fact that, after a point, all human languages have evolved in interaction with each other. (Leeds [1995], p. 33)

[36] See also: '[the translator] is much influenced by his natural expectation that any people in rabbit country would have *some* brief expression that could in the long run be best translated simply as "Rabbit"' (Quine [1960], p. 40).

Field also denies that any notion of synonymy is needed to support 'standards of translation' suitable to the disquotationalist's purposes (Field [1994a], p. 129).[37]

Assuming, then, that the disquotationalist can help himself to this mundane notion of translation, we have our second diagnostic question: when I claim that what Pierre said is true, is what I'm claiming exclusively about Pierre, his language, and the world, or does it also involve me and my language, via a translation of Pierre's utterance? For the disquotationalist, my referring to snow when I say 'snow' is not a matter of a complex causal connection between me and a certain form of precipitation, but an obvious fact about reference, and my attribution of reference to Pierre rests on a combination of that same obvious fact with my translation of his 'neige' as my 'snow'. Where the correspondence theorist sees an objective fact about Pierre, his language, and the world, the disquotationalist sees a determination that rests on how it's best to translate Pierre's language into ours (plus the trivial fact that 'snow' refers to snow). This is a second stark difference between the two.

Applying this second diagnostic, Tarski once again seems to fall on the disquotational side. In the case of Pierre, my English is the meta-language and Pierre's French is the object-language; I'm making assertions in English about the semantic properties of Pierre's utterances in French. Then

'neige est blanc' is true iff snow is white

is a straightforward instance of Tarski's

'...' is true iff... *.

So it's not hard to see why commentators objected to Field's characterization of Tarski as a correspondence theorist.[38]

In any case, a disquotational theory like the one we've been exploring here is the 'less pretentious notion' that Leeds advocates as a scientifically

[37] I'm not sure that the second-philosophical line on translation coincides perfectly with what Field and Leeds have in mind; e.g., Field writes that 'the skeptic about interpersonal synonymy holds that though we of course have standards of translation, they are a matter of being better or worse, not of right and wrong, and also are highly context-dependent (since the purposes for which they are better or worse might vary from one context to the next)' and then declines 'to discuss whether this rejection of interpersonal synonymy is a reasonable position' (Field [1994a], p. 129). The Second Philosopher, I think, should be willing to say that the translation of 'gavagai' as 'rabbit' or 'neige' as 'snow' is correct.

[38] See, e.g., Etchemendy [1988] and Soames [1984]. Field replies to Soames in the 2001 postscript to the reprinting of his [1972] (Field [2001], pp. 27–28). See also Field's comments on Soames position in his [1986a], p. 66, or [1986b], pp. 497–498. (Field maintains that Soames's theory is actually a correspondence theory, if I understand correctly, on both the diagnostics discussed above.)

adequate alternative to Field's correspondence theory.[39] He proposes that science has no need for a substantive property of 'truth' or relation of 'reference' that needs explicating, that all it uses or requires are accounts that amount to those given above:

S is true if and only if
S is 'snow is white' and snow is white, or
S is 'grass is green' and grass is green, or ...

N denotes *a* if and only if
N is 'France' and *a* is France, or
N is 'Germany' and *a* is Germany, or ...

P applies to *a* if and only if
P is 'white' and *a* is white, or
P is 'green' and *a* is green, or ...

An entirely understandable first reaction to this suggestion is the sense that something has gone badly wrong, that the most vital component—the word–world connections—has somehow evaporated. Field writes of 'the commonly voiced worry that deflationism cuts language off from the world' (Field [1994a], p. 126), and Leeds of 'the kind of linguistic Kantianism that continues to haunt' the subject (Leeds [1995], p. 10). The question is: can '"Hume" refers to Hume' really be all there is to the connection between the word 'Hume' and the man?

In fact, the disquotationalist allows that much can and should be said about the relation of our linguistic utterances to the world:

You probably believe quite a few sentences that involve the name 'Hume', and a large proportion of them are probably disquotationally true ... Surely this correlation between your 'Hume' beliefs and the Hume-facts cries out for explanation ... the general lines of the explanation [are suggested by these causal facts]: you acquired your 'Hume' beliefs largely through interactions with others, who in turn acquired theirs from others, and so on until we reach believers with a fairly immediate causal access to Hume or his writings or whatever ... the causal

[39] Field notes that even the correspondence theorist would have use for a disquotational notion of truth, in addition to his correspondence notion, because 'if I want to deny Euclidean geometry, I don't want to express my denial by saying that not all axioms of Euclidean geometry are true in a correspondence sense. For what I want to say is something about the structure of space only, not involving the linguistic practices of English speakers' (Field [1986a], p. 58, or [1986b], pp. 487–488). See also Field [1994a], p. 123.

network has multiple independent chains, and contains historical experts who have investigated these independent chains systematically, so the chance of large errors remaining isn't that high. (Field [1994a], pp. 118–119)

We have here an explanation of 'the otherwise mysterious correlation between a knowledgeable person's beliefs involving the name "Hume" and the facts about Hume' (Field [1994a], p. 118). Leeds gives a similar account of how it is that

when there is an outbreak of flu virus in my vicinity, I will tend—via an elaborate causal chain—to come to believe 'there is an outbreak of flu'. And we know—at least in broad outline—why ... this takes place: ... researchers long ago found out tests for flu virus, and there are people whose business it is to carry out the tests, and tell the rest of us the results. Nor is it an inexplicable accident or coincidence that these causal relations are in place; rather ... we can explain how [they] came to hold by tracing the history of research into viruses. Such a history—already available in the local library—will show us, among other things, how it came about that, at a certain point, the causal connections between 'virus' and viruses were fairly firmly set up: so that from that point onward it was nearly guaranteed, given our theory of viruses and given our inductive procedures—and given also how viruses actually work—that new beliefs about viruses that won general acceptance would tend to be [disquotationally true]. (Leeds [1995], pp. 10–11).

Clearly what's being described here is connections between words and ordinary things in the world, not mere relations between one bit of language and another.

Field calls causal connections of this sort 'indication relations': I tend to believe 'It's raining' when it's raining, and when I believe 'It's raining', it tends to be raining, so my belief state[40] is a good indicator of whether or not it's raining. 'This is simply a correlation, there to be observed; and a deflationist is as free to take note of it as is anyone else' (Field [1994a], p. 109). What's happened here is that the very sorts of causal connections that the causal theorist will point to as constituting the robust relation of reference are now being understood as part of our account of the reliability of our beliefs (about Hume, viruses, or the weather).[41] Thus the

[40] I'm assuming here, with Field, that 'we can speak of a language-user as believing ... sentences of his or her own language'. See Field [1994a], p. 109, also footnote 7.

[41] Field makes this explicit in his [1994a], pp. 118–119. Leeds seems to make a similar suggestion in his [1995], pp. 10–11. 'Indication relations' appear to constitute more (or) less what epistemologists call 'tracking' (see Roush [2006]).

disquotationalist agrees that the investigation of robust connections between words and extra-linguistic things is unavoidable, indeed crucial, to our understanding of the world. Word–world connections aren't eliminated, their description just isn't to be found under the heading of truth or reference; we might say they turn up in the local epistemology.[42]

At this point, it's natural to wonder if the disquotationalist isn't simply involved in a project of relabeling:[43]

> This [raises] the possibility... that by the time the deflationist is finished explaining [why our 'Hume' beliefs are largely disquotationally true] and similar facts, he will have reconstructed the inflationist's relation 'S has truth conditions p', in fact if not in name. (Field [1994a], p. 119)

In fact, Field himself once argued in this way.[44] This style of worry plays a central role in the subsequent debate between correspondence theorists and disquotationalists, as we'll see in II.4, but first let's pause and address the question we started with: to what extent can the debate so far described be understood as purely second-philosophical line of thought?[45]

[42] General epistemology might determine that knowledge is generated by reliable belief-forming processes or that knowledge tracks the world, but what we're interested in here—what I've just called 'local epistemology'—are the actual circumstances that form the beliefs or mediate the tracking in particular cases (Hume, viruses). Incidentally, the tendency toward epistemological externalism on display in this remark may not be shared by Field and/or Leeds. (An externalist assesses the quality of our beliefs at least partly in terms of the worldly context in which they're generated; the contrasting internalist holds that only our conscious justifications are relevant.)

[43] This is a central theme of Marino [2002].

[44] See Field [1986a].

[45] Individual points at which the story as told here may take a more second-philosophical turn than is to be found in Field and/or Leeds have already been touched on above (see footnotes 37 and 42), and there may be more, but the issues raised in the following section concern the overall architecture of the debate.

II.3
Reconfiguring the Debate

An aura of scientific philosophizing certainly pervades the debate as described so far (in II.2): Tarski hopes to give a scientifically acceptable account of truth; Field and Leeds come to focus on the scientific functions of 'true' and 'refers'. Still, on closer inspection it appears that some of its underlying assumptions and motivations differ from those of the Second Philosopher.[1] The key to uncovering these differences is a re-examination of Field's central analogy between truth and reference on the one hand and valence on the other. My goal here is a second-philosophical reorganization of the dispute between correspondence and disquotation.

As we've seen, the Field's case against Tarski rests on this analogy: the disjunctive analysis of valence—*E* has valence *n* iff *x* is *E* is potassium and *n* is 1 or... —is obviously unsatisfactory; if Tarski's disjunctive account of reference runs parallel, it should also be rejected. Furthermore, the two cases do run parallel on Field's reading, because both are fueled by a single, well-supported principle of scientific methodology, namely Physicalism. Let's begin then with a closer look at this crucial methodological maxim.

The term 'physicalism', which Field takes over from Tarski, may strike the philosophical ear as suggesting a substantive metaphysical doctrine, perhaps a piece of a priori first philosophy to the effect that there are no abstract objects or minds or whatever.[2] But however others may understand the term, it's clear that this is not what Field has in mind. Let me quote at some length, to give the flavor of his analysis:

The doctrine of physicalism functions as a high-level empirical hypothesis, a hypothesis no small number of experiments can force us to give up. It functions, in other words, in much the same way as the doctrine of mechanism... once

[1] See Wilson [2006], pp. 634–635, for a different take on how Field's concerns differ from mine. See also II.2, footnotes 37 and 42.

[2] See Blackburn [1994], p. 287.

functioned: this latter doctrine has now been universally rejected, but it was given up only by the development of a well-accepted theory (Maxwell's) which described phenomena (electromagnetic radiation and the electromagnetic field) that were very difficult to account for mechanically, and by amassing a great deal of experiment and theory that together made it quite conclusive that mechanical explanations of these phenomena (e.g., by positing 'the ether') would never get off the ground. Mechanism has been empirically refuted; its heir is physicalism, which allows as 'basic' not only facts about mechanics, but facts about other branches of physics as well. I believe that physicists a hundred years ago were justified in accepting mechanism, and that, similarly, physicalism should be accepted until we have convincing evidence that there is a realm of phenomena it leaves out of account. Even if there *does* turn out to be such a realm of phenomena, the only way we'll ever come to know that there is, is by repeated efforts and repeated failures to explain these phenomena in physical terms. (Field [1972], pp. 11–12)

Field continues with an example from biology:

Suppose, for instance, that a certain woman has two sons, one hemophilic and one not. Then, according to standard genetic accounts of hemophilia, the ovum from which one of these sons was produced must have contained a gene for hemophilia, and the ovum from which the other son was produced must not have contained such a gene. But now the doctrine of physicalism tells us that there must have been a *physical* difference between the two ova that explains why the first son had hemophilia and the second one didn't, if the standard genetic account is to be accepted. We should not rest content with the special biological predicate 'has-a-hemophilic-gene'—rather, we should look for nonbiological facts (chemical facts; and ultimately physical facts) that underlie the correct application of this predicate. That at least is what the principle of physicalism tells us, and it can hardly be doubted that this principle has motivated a great deal of very profitable research into the chemical foundations of genetics. (Field [1972], p. 12)

So Field's criticism of Tarski's definition of truth purportedly rests on the legitimate methodological demands of our best current science.

This sounds for all the world like a characteristically second-philosophical argument, so let's try to fill in a few of the details. In the first paragraph quoted above, Physicalism is understood as the successor to Mechanism.[3] In their popular history of physics, Einstein and Infeld begin their description of Mechanism with a discussion of simple forces of attraction and repulsion

[3] The following discussion of Mechanism is (more or less) a shorter rehash of my [1997], pp. 111–116 (see also pp. 137–141 for discussion of the kinetic theory).

acting on a line between two objects, with magnitudes depending only on the distance between them—after which they pose the question, 'is it possible to describe all physical phenomena by forces of this kind alone?'

> The great achievements of mechanics in all its branches, its striking success in the development of astronomy, the application of its ideas to problems apparently different and non-mechanical in character, all these things contributed to the belief that it *is* possible ... Throughout the two centuries following Galileo's time such an endeavor, conscious or unconscious, is apparent in nearly all scientific creation. ... this so-called *mechanical view*, most clearly formulated by Helmholtz [in the mid-nineteenth century] played an important role in its time. The development of the kinetic theory of matter is one of the greatest achievements directly influenced by the mechanical view. (Einstein and Infeld [1938], pp. 53–55)

Mechanism, then, is the methodological admonition always to seek explanations in terms of these simple forces. As Field notes, it rested on a well-confirmed empirical hypothesis of great power and range, and its adoption was entirely justified in its day.

Alas, difficulties arose: the intricacies of applying Mechanism in electrostatics and magnetism left 'no reason to be particularly proud or pleased' (Einstein and Infeld [1938], p. 84); a moving electric charge was found to produce a force that acts in a direction perpendicular to the line between it and a stationary magnet, and that varies with the velocity of the charge; and most famously, the light ether stubbornly resisted mechanical treatment. Meanwhile, the electrodynamic anomalies eventually led to the discovery that light is an electromagnetic wave:

> This great result is due to the field theory. Two apparently unrelated branches of science are covered by the same theory ... Slowly and by a struggle the field concept established for itself a leading place in physics and has remained one of the basic physical concepts. The electromagnetic field is, for the modern physicist, as real as the chair on which he sits.[4] (Einstein and Infeld [1938], p. 150–151)

Thus Mechanism was eventually rejected, on sound empirical grounds:

> The field was at first considered as something which might later be interpreted mechanically with the help of the ether. By the time it was realized that this program could not be carried out, the achievements of the field theory had already

[4] See IV.5 for more on the reality of the electromagnetic field.

become too striking and important for it to be exchanged for a mechanical dogma. (Einstein and Infeld [1938], pp. 152–153)

This is the story Field alludes to the first quotation, if we understand his 'Physicalism' as the expansion of Mechanism to allow for explanations in terms of fields.

What's crucial for our purposes is that this methodological history concerns the structure of physical explanations: should they be purely mechanical or are other types permissible? No mention is made here of the relation of physics to chemistry or biology or any other subject matter, or of reduction in particular. Yet reduction *is* the topic of Field's second quotation, where he puts forward an illustration of the doctrine of Physicalism, and his later writings continue in this vein (e.g., Field [1992]). I think we must conclude that the Physicalism at work in Field's valence analogy is not in fact the Mechanism + Fields that emerged as described above in the history of physics. Field's Physicalism is actually a modification of what he calls 'classical reductionism', a position familiar from the philosophy of mind:

The classical reductionist proposes ... that for each sentence in the language of a successful special science like ... psychology, there is a sentence in the language of a lower level science—ultimately, in the language of physics—that in some intuitive sense 'expresses the same facts' ... [and] higher-level laws and generalizations should themselves, when physically transcribed, admit physical explanation. (Field [1992], pp. 272–273)

Field softens this view in various ways, but whatever its details, this is clearly not the successor to Mechanism described by Einstein and Infeld.

Though the Second Philosopher might, in the course of her investigations, come to adopt Field's Physicalism in addition the successor to Mechanism (Mechanism + Fields), I confess to doubts on this point. This is no place for a full analysis of reductionism, nor am I capable of delivering one, but perhaps the flavor of my concern is conveyed by a recent discussion of color by Howard Stein.[5] Consider, for example, a psychological

[5] On the controversies surrounding color, let me note Stein's demonstration that science quite ably answers ordinary questions about the colors of things (the sky, the sun, etc.) without any worry about whether colors are or aren't 'really in things' (see Stein [2004], pp. 155–156). Wilson ([2006], pp. 74–76, 104–106, 289, 392–394, 437–438, 453–467), relying on his own range of real-world considerations, reaches the complementary conclusion that the question 'are colors really in things?' is itself based on mistaken presuppositions.

('special science') explanation of how Joe came to step on the gas when he saw a green traffic signal. Stein gives a persuasive scientific and historical case that our efforts to understand what it is to be aware of a color will most likely transform our understanding of the question itself; he concludes:

> If there is to be an explanation of color experience, it will have to take the form of (a) an account of the effect of bodies upon the light incident on them—an account we *do now have*; (b) an account of the effect of light that reaches the retina upon the central nervous system—and this account we have *in part*; and (c) an account—a *theory* of some sort—of how neural processes produce 'experience'. The last part, (c), we do not have today ... We *know* that living organisms, and therefore in particular sentient organisms (including ourselves), have developed in a world in which there once were none. *Prima facie*, it is a very striking question how such a capacity as that of 'seeing green' can have developed out of an inanimate world. ... I believe that the question of *what it is*, from the point of view of the basic principles that govern the physical world, for a system to be 'aware'—'conscious'—is at present both poignant and ... entirely unclear. I do not know whether that question can ever, possibly, acquire clarity—or *develop* into a *related* question that is clear (it is exactly this latter kind of thing that happened to the question about the motion of the earth) ... I find it hard to believe that an advance of science that ... explains 'someone's inclination to utter the word "green" when visually confronted with something' ... will not transform our understanding of [the question of awareness] ... *Today* we do not even have a set of concepts that allow us to *envisage* [a theory for (c) above]. [Some philosophers are] inclined to think ... that such a theory is in principle impossible. *I* think ... that the fate of Locke's view that scientific physics is impossible; of Kant's view that scientific chemistry is impossible; of Compte's view that knowledge of the chemistry of the *stars* is impossible; should all conduce to skepticism about that kind of philosophical skepticism. (Stein [2004], pp. 154, 165–166)

Field would presumably share Stein's optimism,[6] but I fear his Physicalism represents an implausible attempt to foresee the structure of the theory this inquiry might ultimately produce.

[6] One intriguing line of investigation involves the study of 'brain-damaged subjects who retain the very capacities they think they have lost ... [e.g.] a "blind-sight" patient may claim not to "see" the very visual stimuli that he or she can be shown to be able to discriminate by "guessing" ... in every major class of defect in which patients apparently lose some particular cognitive ability through brain damage, examples of preserved capacities can be found of which the patient is unaware. ... These facts ... offer both a challenge and an opportunity to consider the brain mechanisms of conscious awareness, and what its functional status might be' (Weiskrantz [1997], p. 1; see also Weiskrantz [1986]). For discussion of this and related avenues of research, see Baars et al. [2003]; for critical analysis, see Block [1995].

Fortunately, we can investigate the viability of Field's analogy without resolving the general issue of reduction for the special sciences. What's at issue is the viability of the analogy between valence and reference—do the two cases run parallel? in particular are the two disjunctive definitions unsatisfactory for parallel reasons?—and these questions can be pursued without appeal to an overarching methodology of reduction. Let's begin, then, with a look at the valence case, with special attention to the possibility of a disjunctive definition and the factors that make such a move unsatisfactory, but with an eye also to any reductionist motivations that might turn up along the way.

To get a sense of the context in which the theory of valence arose, consider the electrochemical theory of chemical bonding proposed by Berzelius in the early nineteenth century.[7] Impressed by the decomposition of salts by electrolysis, Berzelius suggested, in 1811, that compounds are held together by the very electrical polarities that draw their components to opposite poles when they are exposed to electric current.[8] So, for example, the acids and bases that make up salts are themselves electronegative and electropositive, respectively, and the component atoms themselves range from the most electronegative (oxygen) to the most electropositive (potassium). As the historian J. R. Partington remarks, here 'combination is maintained by polarity and not by a superadded special chemical force of affinity' (Partington [1957], p. 199).

It could be that some vaguely reductionist impulse motivated Berzelius' theory, but if so, as far as I know, this wasn't made explicit. There was, however, a serious methodological controversy very much discussed at the time over the status of organic compounds. Some, like Lavoisier, held that organic chemistry is simply a branch of ordinary chemistry; others, like Berzelius—troubled by various difficulties in the analysis of organic compounds—were uncertain whether or not organics should be set apart from inorganics; still others, a significant proportion, recommended two separate studies, believing that

[7] As will become clear in the following footnotes, my sources here are the histories Partington [1957] and Ihde [1964].

[8] See Partington [1957], pp. 196–200, or Ihde [1964], pp. 132–133.

Organic compounds were produced through the agency of a vital force that was present only in living plants and animals ... they could not be synthesized from the elements. (Ihde [1964], p. 163)

This controversy was resolved in 1828, when Wöhler, a former student of Berzelius, synthesized urea. He wrote at once to his teacher, 'I must tell you that I can make urea without the use of kidneys, either man or dog' (quoted by Ihde [1964], p. 165)! Thus we see that both Vitalism and Mechanism were specific methodological guidelines that arose explicitly in the practice of their respective disciplines; both were sharply debated and tested, and eventually overturned as those practices progressed. It's not clear that Field's Physicalism is playing any such role.

In any case, Berzelius' electrochemical theory came under pressure after the so-called 'incident of the smoky candles':

At a ball held in the Tuileries in Paris, fumes given off by the candles made the guests start coughing. The problem of finding the cause was referred to Brongniart, whom the king frequently consulted. Brongniart passed it on to his son-in-law, Dumas ... (Ihde [1964], p. 193, see also Partingon [1957], p. 240)

Dumas soon determined that the candles had been bleached with chlorine, that when the wax was heated, one part hydrogen had been replaced by one part chlorine, after which the hydrogen freed from the wax had combined with the remaining chlorine to form hydrochloric acid. This inspired Dumas to study the effects of chlorine on various substances: for example, the action of chlorine (Cl_2) on acetic acid ($C_4H_4O_2$) produces trichloric acid ($C_4HCl_3O_2$)and hydrochloric acid (HCl); three Hs are replaced by three Cls. Thus, in the course of the 1830s, Dumas was led to his theory of substitution: for example,

When a substance containing hydrogen is exposed to the dehydrogenising action of chlorine, bromine, or iodine, for every volume of hydrogen that it loses, it takes up an equal volume of chlorine, bromine, etc. (quoted by Ihde [1964], p. 193, by Partington [1957], p. 241)

Acetic acid and trichloric acid are so similar that the challenge to electrochemical theory was immediate: how could electronegative chlorine be substituted for electropositive hydrogen without producing a fundamental chemical change?

While Berzelius struggled with this problem, Dumas and his followers rejected electrochemical theory in favor of a theory of general chemical types that were assumed to persist despite such substitutions as chlorine for hydrogen.[9] Type theories were also called 'unitary theories', because a complex molecule was looked upon as a fundamental unit, into which substitutions could be made. The most elaborate version was formulated by Gerhardt in the early 1840s, classifying organic substances in terms of four basic types.[10] Gerhardt regarded the formulas he generated—for alcohols, ethers and acids (water type), hydrocarbons, ketones and aldehydes (hydrogen type), chlorides (hydrogen chloride type), and amines and amides (ammonia type)—as classificatory tools only, not as representations of real atomic structures.[11] Here the contrast with Berzelius' thinking is stark: for example, when (in 1827) Berzelius was first to (name and) explain the phenomenon of isomerism—compounds with the same composition, but different chemical properties—and he did so by taking seriously the spatial arrangement of atoms in their molecules.[12] Partington remarks, 'this is clearly the beginning of the theory of structure' (Partington [1957], p. 204).

It was in this context that the theory of valence arose, in the work of Frankland and Kolbe in the early 1850s.[13] Despite the ascendance of type theory, both were followers of Berzelius (Kolbe was in fact a student of Wöhler):

> Although these two men were young enough to realize the strength of the unitary concept of chemical compounds, they still preferred to regard radicals as stable atomic groupings rather than as arbitrary arrangements made solely for classification purposes. (Ihde [1964], p. 216)

[9] In the midst of this controversy, Wöhler wrote a letter under the name 'S. C. H. Windler', describing the gradual replacement of every atom of manganese acetate by chlorine, resulting in pure chlorine that nevertheless behaved true to type, as manganese acetate. Originally sent to Berzelius, the joke eventually found its way to Liegen, who published it in his journal *Annalen* for 1840 with an editorial footnote remarking that the English had produced bleached cotton cloth that consisted entirely of chlorine and was 'much sought after' by fashionable Londoners (see Partington [1957], p. 251, see also Ihde [1964], p. 197).

[10] See Partington [1957], pp. 258–268, Ihde [1964], pp. 209–216.

[11] See Partington [1957], p. 274: 'For Gerhardt and his school ... the radicals were mere phantoms, "the ghosts of departed reactions", incapable of isolation and existing only in the imagination.' See also Partington [1957], p. 267, and Ihde [1964], p. 216.

[12] 'It would seem as if the simple atoms of which substances are composed may be united with each other in different ways' (quoted by Partington [1957], p. 204).

[13] See Partington [1957], chapter XII; Ihde [1964], pp. 216–225.

Their researches were guided mainly by the older teachings of Berzelius, and when they were completed the type theory as such had ceased to have any interest for chemists. (Partington [1957], p. 274)

Dalton's atoms had been hypothesized in the first decade of the nineteenth century to explain the Laws of Definite Proportion—compounds break down in fixed proportions[14]—and the Law of Multiple Proportions—the definite proportions in which substances can combine are simple multiples.[15] Frankland focused also on substitutions—for example, Dumas's observation that two atoms of chlorine replace two atoms of hydrogen, but only one atom of oxygen—and what he called 'symmetries'—for example, his own observation that nitrogen tends to combine with either three or five atoms of other elements. In 1852, he writes:

When the formulae of inorganic chemical compounds are considered, even a superficial observer is impressed with the general symmetry of their construction ... Without offering any hypothesis regarding the cause of this symmetrical grouping of atoms, it is sufficiently evident from the examples just given, that such a tendency or law prevails, and that, no matter what the character of the uniting atoms may be, the *combining power* of the attracting element, if I may be allowed the term, is always satisfied by the same number of these atoms. (quoted Ihde [1964], p. 220, italics mine; see also Partington [1957], p. 285).

Comparing his view to type theories, he notes that the nature of a compound

is essentially dependent upon the electrochemical character of its single atoms, and not merely on the relative positions of these atoms. (quoted in Partington [1957], p. 285)

The key idea is that the atoms of a given element have a fixed and characteristic 'combining power', or valence, of electrochemical origin.

The theory of valences developed steadily during the late 1850s: for example, Kekulé and Couper determined that carbon is quadravalent and that it enters into combination with itself (so that the apparent valence of two carbon atoms is six, not eight).[16] Around the same time, Cannizzaro reinstated Avogadro's hypotheses (a fixed volume of any gas contains the

[14] e.g., 9 grams of water always decomposes into 1 gram of hydrogen and 8 grams of oxygen.

[15] e.g., 3 grams of carbon combines with 4 grams of oxygen or with 8 grams of oxygen.

[16] See Partington [1957], pp. 288–290; Ihde [1964], pp. 224–225.

same number of molecules and some molecules are diatomic) and achieved a stable calculation of atomic weights. By 1860, these discoveries had unified chemistry, set the stage for the classification of elements in the periodic table, and underwritten the rise of structural chemistry.[17]

But understanding of the underlying electrochemical mechanism of chemical bonding—first posited by Berzelius decades before—had to wait until the discovery of the electron (by Thompson in 1897) and the proton (by Rutherford and Geiger in 1908), the development of the Rutherford model of the atom as a positive central nucleus with orbiting electrons (in 1911), and finally, the advent of the Bohr atom with its discrete energy levels, based on quantum mechanical ideas (in 1913).[18] Once these tools were in place, the light dawned quickly: by 1916, Lewis proposed that atoms of the inert gases have eight electrons in their outer shell and that other atoms tend toward this stable eight either by gaining or losing electrons to form charged ions[19] or by sharing electron pairs with other atoms to form compounds.[20] This crude model gives electrochemical accounts of ionic and covalent bonding that match well with empirical determinations of valence. In the years since, elaborations and descendants of Lewis's system have been developed to deal with anomalies and to provide additional information about the shapes of the molecules described. Meanwhile, the rise of quantum mechanics has produced deeper explanations of electron pairing and electron distribution generally, as well as quantitative analyses of lengths, angles, and strengths of individual bonds. By now, Frankland's simple idea of valence as the 'combining power' of an atom has given way to a complex and varied theory of inter-atomic forces and an array of related notions: ionic valence, covalence, oxidation number, metallic valence, and so on.

[17] See Partington [1957], pp. 256–258; Ihde [1964], pp. 226–230.

[18] For the developments of this paragraph, see Partington [1957], chapter 16, Ihde [1964], 499–507, 536–545.

[19] e.g., sodium loses the one electron in its outer shell to expose a stable shell of eight (like neon) and become a positive ion. Similarly, chlorine adds one electron to its outer shell to make eight (like argon) and become an ion with a single negative charge. Thus, sodium chloride.

[20] e.g., a carbon atom has four outer electrons, an oxygen atom has six, so a single carbon atom can share two of its electrons with one oxygen atom, bringing that oxygen atom to its full eight, and its other two electrons another oxygen atom, bringing both that oxygen atom and the carbon atom up to eight. Thus, carbon dioxide.

Hydrogen is a special case because losing its one electron doesn't expose a stable shell, but a bare nucleus. As its corresponding inert gas is helium, it bonds by sharing its one electron in a pair with another atom, as in diatomic hydrogen gas.

With this sketchy history in place, we can return to the question of Field's analogy, but notice first that no general reductionism or Fieldian Physicalism seems to have played a role in the search for an understanding of valence in terms of atomic structure: valence was explicitly introduced as an electrochemical feature of the atom, not as a purely chemical notion that might or might not be 'reducible' to underlying physical notions; the physical terms of the theory had already been set by Berzelius decades earlier. As noted above, it might be argued that Berzelius himself was originally moved by a reductionist impulse to replace the chemical notion of 'affinity' with something structural, but if he was so guided, he apparently didn't mention it and no explicit debate on the topic appears to have taken place. Furthermore, there is a much simpler reading of Berzelius' motive: he wanted to provide an explanation of chemical bonding where there had been none before.

In any case, our central interest here is in the purported analogy between valence and reference; in particular, we're keen to isolate what it is that makes the disjunctive account of 'valence' unacceptable. Oddly enough, it now appears that in one sense at least, there's nothing at all wrong with the definition 'E has valence n iff E is E is potassium and n is $+1$, or... or E is sulfur and n is -2'. It amounts to a table of the 'combining powers' of the elements, and in the days of Frankland and Kolbe, there was no guarantee that such a thing was possible—no guarantee that chemical bonds could be understood in terms of atoms at all (remember type theory), or that the atoms of a given element would have stable powers that could be represented by simple whole numbers, or that electrical polarity would be involved. In other words, the disjunctive definition or table of values represents a rich empirical achievement, the result of considerable theoretical imagination and experimental effort; we shouldn't forget that Kekulé and Couper made an important contribution when they identified the valence of carbon as 4! Insofar as that empirical effort was successful, it supported the theory of valence, but of course it wasn't entirely so, as the need for further elaborations and the eventual fragmentation of 'valence' demonstrates.

Obviously, the disjunctive definition was just the first step toward understanding the chemical bond in terms of electrical features of the atoms involved. As an explanation, the table of values is clearly incomplete: what electrochemical features of the atoms are responsible for their valences? No progress could be made on this chemist's question until the physics of the atom caught up, but as we've seen, the moment it did, the chemists were there to take advantage. The development of our understanding of the chemical bond has continued to this day, and will no doubt progress from here, even as the term 'valence' itself is perhaps superseded.[21] In sum, then, the disjunctive definition was an important advance, but not a satisfactory stopping place in the ongoing project of explaining the chemical bond.

So, finally, what of the analogy? Does the case of reference run parallel to the case of valence? One stark disanalogy should strike us immediately: the disjunctive definition of 'refers', unlike the table of valences, is hardly an empirical achievement.[22] The disquotationalist will say that the R-sentences are trivial, the correspondence theorist will insist that they require explanation, but neither will claim that serious investigation on the part of many researchers, past and present, was required to come up with them. This observation points to another: while the definition of valence occurs in the course of a specific explanatory project—that of elucidating the chemical bond—Field's criticism of Tarksi's definition of truth seems to arise when a general methodological principle, Physicalism, is applied in isolation, without a clear ongoing context of empirical investigation. This leaves a crucial gap in the analogy, because the case for the inadequacy of the disjunctive definition of 'valence' as a viable stopping point rests entirely on the structure of the inquiry in which it is embedded; without an analogous sense of the explanatory project at hand or the role 'true' or 'refers' is supposed to play in it, a parallel case for the inadequacy of Tarski's definitions cannot be made. In sum, then, the analogy fails on two points: on the one hand, the disjunctive definition of valence is not a trivial matter like the disjunctive definition of reference, but an important advance in the investigation of the chemical bond; on the other, the inadequacy of disjunctive definition of valence as a stopping point is dictated by specific

[21] Physicalism, as Field describes it, only requires that we reduce the terms of our 'approximately true' theories; other terms are rather 'candidates for elimination' (Field [1992], p. 283). This would seem to be a difficult distinction to draw, even in retrospect, as the case of 'valence' illustrates (which is it?); it's no doubt much more difficult to draw in advance, at the very point when Physicalism should be providing methodological guidance. We should get a better sense of the roots of this murkiness in II.4.

[22] One could insist that the disjunctive definition of 'valence' be understood as stipulative, as introducing the term by pure convention, but this overlooks the role the term was designed to play in the theory of the chemical bond.

features of that context of investigation, while the debate over reference appears to take place in an explanatory vacuum.

Eventually, of course, Leeds posed his challenge to Field—what scientific role does 'truth' play?—and the debate between correspondence and disquotation focused thereafter on various uses of 'true' and whether or not disquotational truth is adequate for those uses. But notice that the Second Philosopher's interest in the correspondence/disquotation dispute arose quite differently: at the end of I.7, she felt the force of the naturalistic Putnam's challenge to explain how language functions in our interactions with each other and the world, to explain the word–world relations involved; this was the explanatory context in which the notions of 'truth' and 'reference' were potentially involved. As it happens, this challenge of Putnam's evolved out of his early responses to Leeds[23]—he was suggesting an empirical, explanatory role that correspondence truth, but not disquotational truth might play—so following the course of the actual philosophical debate eventually returns us to the Second Philosopher's point of entry. The second-philosophical reconfiguring I have in mind then is simply to view the correspondence/disquotation debate as arising, not from the application in isolation of a questionable reductionist principle, but in the course of this ongoing empirical project of investigating the function of human language. What we've learned so far (in II.2) is that there is a disquotational notion of truth that's useful for semantic ascent and descent and for various generalizing tasks, but we've made little progress on our motivating project and we don't know if a notion like correspondence truth will be required in order to do so. We now return to the actual debate-in-progress.

[23] Putnam's first discussion of Field and Leeds actually appeared in his [1976a], before Leeds [1978] was published.

II.4
Disquotation

If 'true' and 'refers' play a central role in our explanation of how our language functions, of how our words relate to the world, then a Second Philosopher embedded in that inquiry will be able to assess what sort of account of truth and reference is needed, much as the chemist embedded in the inquiry into chemical bonding knew what sort of account of valence was needed for the job at hand. Field and Leeds debate a more general question—is there any explanatory role for truth and reference that requires a correspondence-style explication of the T- and R-sentences?—but we've seen that Putnam's reply rests on the role of truth in the ongoing empirical investigation of the functioning of human language. Let's pick up this thread.

Field frames Putnam's challenge to the disquotationalist in this way:

> It is perfectly obvious that in explaining how a pilot manages to land a plane safely with some regularity, one will appeal to the fact that she has a good many true beliefs: beliefs about her airspeed at any moment, about whether she is above or below the glidescope, about her altitude with respect to the ground, about which runway is in use, and so forth. ... The deflationist obviously needs to grant this explanatory role for the truth of her beliefs, and will have to say that it is somehow licensed by the generalizing role of 'true'. Can this deflationist strategy be maintained? (Field [2001], p. 153, from the 2001 postscript to Field [1994a])

In the course of his debate with Leeds over this question, Field became convinced that his earlier conclusion—that a case for correspondence truth could be mounted on these grounds[1]—was incorrect, that the disquotationalist can handle cases of this sort.[2] After 'vacillat[ing] between

[1] See Field [1986a], to which Leeds [1995] offers a reply.

[2] In Field [1994a]. In the 2001 postscript to the reprinting of this paper (Field [2001], pp. 153–156), Field explicitly agrees with Leeds that his earlier argument was incorrect.

the two pictures', Field reports that he 'came down fully in favor of the deflationary viewpoint' (Field [2001], p. viii in his [1994a]).[3]

Unfortunately, the published record of this exchange is well-nigh impenetrable to the outsider, but it seems to me that the overarching issues are less arcane than the wealth of detail there would suggest. In essence, our correspondence theorist thinks the pilot succeeds in making a safe landing because many of her relevant beliefs are true, where this involves appropriate causal connections that constitute a relation of reference between her words and the world; this referential relation plays the same role in many other such explanations, so we're naturally led to seek a general theory of reference and of how it manages to contribute to our success. The disquotationalist, in contrast, thinks the pilot succeeds because she believes 'my current airspeed is so-and-so', and she tends to believe 'my current airspeed is so-and-so' when and only when her current airspeed is so-and-so; she believes 'my altitude is such-and-such' and she tends to believe 'my altitude is such-and-such' when and only when her altitude is such-and-such; and she believes 'when my airspeed is so-and-so and my altitude is such-and-such, I should do this in order to facilitate a safe landing', and she tends to believe 'when my airspeed is so-and-so and my altitude is such-and-such, I should do this in order to facilitate a safe landing' when and only when the safest thing to do with airspeed so-and-so and altitude such-and-such is indeed this—in other words, because her relevant beliefs are good indicators of the relevant worldly conditions.[4]

Here, once again, where the correspondence theorist sees robust referential relations, the disquotationist sees local epistemology,[5] which returns us to the charge of relabeling that arose at the end of II.2: what the correspondence theorist calls reference, the disquotational theorist calls indication

[3] Though Field so describes himself in the preface to his collection [2001], the text of [1994a] actually advocates 'methodological deflationism': 'we should assume full-fledged deflationism as a working hypothesis. That way, if full-fledged deflationism turns out to be inadequate, we will at least have a clearer sense than we have now of just where it is that inflationist assumptions about truth conditions are needed' (Field [1994a], p. 140, see also p. 119).

[4] If we don't know enough about the pilot's beliefs or about flying to give this sort of detailed explanation, then (according to the disquotationalist) we may well fall back on the formulation—she lands successfully because her beliefs are true—but the 'true' is functioning in its disjunctive role: she believes '...' and ..., or she believes '---' and ---, etc. Or perhaps more accurately: she believes something that translates as '... ' and ..., etc. See Field's 2001 postscript to his [1994a] (Field [2001], pp. 153–156).

[5] See II.2, footnote 42.

relations; the disquotationalist is appealing to a reduction of non-trivial reference relations without using the name. The disquotationalist might well respond that no one has in fact succeeded in formulating a robust account of reference based on the causal connections available in the local epistemology,[6] indeed that one might doubt the causal connections involved in various cases have anything substantive in common.[7] Both these seem to me to be offshoots of a more direct response: one comes to imagine that indication relations might add up to a reduction of reference relations or truth conditions only because one's sampling of indication relations is badly skewed. Let me explain.

Consider first a whimsical example of Field's: an ancient Greek has various beliefs about the gods, as a result of which he tends to believe 'Zeus is throwing thunderbolts' when and only when there is thunder in his vicinity.[8] In this case, the Greek's belief state is a good indicator of thunder. What it's not is a good indicator of its own truth conditions: he doesn't tend to believe 'Zeus is throwing thunderbolts' when and only when Zeus is throwing thunderbolts, because there is no Zeus in the first place. But this defect in the indication relations of the Greek's belief state doesn't keep them from figuring in the explanation of his success in coming indoors before the thunderstorm starts: he succeeded in keeping dry because his belief state is a good indicator of certain facts about his environment. Indeed, in a slight adjustment of the pilot's case, her beliefs about airspeed might actually be inaccurate (say her gauge is off), but she succeeds anyway, because she's naturally impatient and keeps her speed (she thinks) higher than it needs to be. Cases like these suggest that what explains our success isn't that our beliefs are true, or that they are good indicators of their truth conditions, but that they are good indicators of something or other that's important in helping us get along.

[6] See Field ([1994a], p. 110): 'the project of giving anything close to a believable reduction of talk of truth conditions to talk of indication relations is at best a gleam in the eye of some theorists.'

[7] See Leeds [1995], p. 11: 'the deflationist will want to deny that the class of causal connections between words and their R-referents is a natural class in a way that requires uniform explanation.' He suggests that this class appears natural only because it's confused with another—for each of our words, our belief that it is causally connected to its R-referent—and that the explanation of this other class is superficial, not substantive (Leeds [1995], pp. 11–12). Finally, he goes on to argue that a correspondence theory is impossible, for reasons that strike me as inadequate (see IV.4, footnote 29).

[8] Field [1994a], p. 110. See also Field [2001], p. 154.

To see that this isn't just a matter of artificial examples, consider a case much discussed in the philosophy of science literature: Priestley's beliefs about dephlogisticated air.[9] Priestley and his fellow phlogistonites were out to explain combustion and related phenomena. They held that combustion involves the transfer of phlogiston from the burning substance to the air, that heated metals release phlogiston to produce a 'calx', that combustion in a closed space will cease when the air has absorbed all the phlogiston it can, that heating calx of mercury in air produces mercury and air from which phlogiston has been removed (because the phlogiston released when the calx was formed is now returned to convert the calx back to mercury), and that this dephlogisticated air supports combustion and respiration better than ordinary air (because it has greater capacity to absorb phlogiston).[10]

Imagine, then, Priestley in his lab, hoping to ease his sick friend's breathing.[11] He heats up some calx of mercury, believing 'this process will produce dephlogisticated air' and 'dephlogisticated air is good for breathing', and these beliefs lead him to administer the product to his friend, who is indeed helped. We explain Priestley's success by pointing out that his first belief is a good indicator that he's undertaking a process that will produce air rich in oxygen, and his second belief is a good indicator that air rich in oxygen is good for breathing—that is, our detailed story of his investigations, of how he came to these beliefs, shows that he tends to believe 'this process produces dephlogisticated air' when and only when the process produces air rich in oxygen and that he tends to believe 'dephlogisticated air is good for breathing' when and only when air rich in oxygen is good for breathing. But neither of these beliefs of Priestley's indicates its own truth conditions,[12] because there is no such

[9] Here I draw on Kitcher [1978] and [1993], pp. 97–105, Partington [1957], pp. 85–89, and Ihde [1964], pp. 40–50.

[10] On this last, Kitcher ([1993], p. 100) quotes the following passage: 'My reader will not wonder, that, after having ascertained the superior goodness of dephlogisticated air by mice living in it, and other tests above mentioned, I should have had the curiosity to taste it myself. I have gratified that curiosity by breathing it... The feeling of it to my lungs was not sensibly different from that of common air; but I fancied that my breast felt peculiarly light and easy for some time afterwards.'

[11] Partington notes that Priestley 'recommended its [dephlogisticated air's] use in medicine', and that he went on to speculate: 'Who can tell but that, in time, this pure air may become a fashionable article in luxury. Hitherto, only two mice and myself have had the privilege of breathing it' (Partington, [1957], p. 118).

[12] This formulation in terms of truth conditions will need some adjustment below. See footnote 25.

thing as phlogiston.[13] Again, this and similar cases suggest that it is the indication relations of our beliefs, not their truth or falsity, that explains our success. Given the state of our current fundamental physics, it seems overwhelmingly likely that the correct explanation of the success of the many advanced technologies we now enjoy does not run through the truth of our theories!

These considerations undercut the Success Argument—that is, the claim that the disquotationalist can't explain successful uses of language—but opponents of disquotationalism have also questioned its ability to account for various other aspects of human language use. It's been argued, for example, that the ever-vexing problem of vague language is especially problematic for the disquotationalist,[14] or that the disquotationalist can't make sense of the claim that a particular area of language isn't in the business of stating facts.[15] I'll have something to say about vagueness in III.7 (see especially footnote 21) and about non-factual interpretations of mathematics in IV.4, so I leave those aside here. Both these objections fall in the general category of 'Critical Stance Problems':[16] how can we question the proper semantic functioning of (a part of) our language while assuming that its semantics is trivial? For purposes of illustration, let me take up another of these, one that lies closer to the issues we've been examining so far, namely: can the disquotationalist coherently raise the question of whether or not our terms have determinate reference?

The question is inspired by the history of science; Field's central example is drawn from physics. He argues that, given special relativity, there are two equally good candidates for the referent of Newton's word 'mass': rest mass and relativistic mass. Some of Newton's assertions involving 'mass' come out true or approximately true (at low velocities) on one or the

[13] Partington ([1957], p. 121) remarks: 'Although most of Priestley's experimental results were accurate, he was led astray in explaining them by his use of the phlogiston theory. In this respect, his own candid self-criticism is ... true: "I have a tolerably good habit of circumspection with regard to *facts*; but as to conclusions from them, I am not apt to be very confident".'

[14] Ever-vexing because of sorites paradoxes: if a man with no hairs is bald, and a man with $n + 1$ hairs is bald whenever a man with n hairs is bald, then all men are bald. For discussion, see, e.g., Field [1994b], section 1, Leeds [1997], section IV, Field [2000]. Apparently vagueness is supposed to pose a special problem for disquotationalists—over and above the problem it poses for everyone—because various of the standard solutions, like supervaluationism, are believed to be unavailable to the disquotationalist.

[15] e.g., ethics as it's understood by expressivists (i.e., ethical statements are expressions of attitudes, not beliefs). See Field [1994b], section 4.

[16] The term, and this way of classifying objections to disquotationalism, is due to Marino [2002]. See also Marino [forthcoming].

other of these readings; some are true on both. In his [1973], where this example is introduced, Field is still in the grip of the correspondence theory, so he proposes a modification of its usual referential apparatus to allow for 'partial denotation': Newton's 'mass' partially refers to rest mass and partially refers to relativistic mass; such assertions as are true (false) for both referents are true (false);[17] the rest lack truth value. He concludes that we have every reason to believe that our current language also harbors terms of indeterminate reference, even if we can't be entirely sure which they are.[18] This concern survives his conversion to disquotationalism, hence the question: can the disquotationalist make sense of the claim that our own terms are referentially indeterminate, given that reference, for him, comes down to the R-sentences?; if the triviality—'...' refers to ... —is all there is to reference, what can it mean to say that '...' is referentially indeterminate?

The example of 'mass' has proved controversial,[19] so to avoid complexities in the philosophy of physics, let's switch from Newton back to Priestley. In a recent analysis, Kyle Stanford and Philip Kitcher suggest that Priestley's various utterances involving 'dephlogisticated air' can be divided into a number of categories.[20] Some occur in the course of theoretical discussions of phlogiston theory, where it's clear that phlogiston is taken to be the stuff emitted during combustion and that dephlogisticated air is air without much phlogiston, which is clearly problematic because there is no phlogiston. In experimental contexts, however—for example, when he comes to his friend's aid in the laboratory—his use of the term is causally based in the stuff, whatever it is, that eases breathing, or the stuff, whatever it is, generated by heating calx of mercury. Stanford and Kitcher's subtle refinement of the causal theory of reference issues the judgment that when Priestley uses the term 'dephlogisticated air', he sometimes refers to oxygen, sometimes refers to air rich in oxygen, sometimes partially refers to

[17] This oversimplification sounds circular. Field actually defines a 'structure' M to be an ordinary model that assigns a term one of the things it partially denotes. Then a sentence is true if it comes out true-in-M for every structure. (See Field [1973], p. 190.)

[18] Early on, Field seems to have drawn this conclusion from a pessimistic induction on the deficiencies of past scientific theories, but later (e.g., in Field [1994a], p. 236), he prefers to base it on the visible shortcomings of our current theories: 'quantum theory and general relativity... don't fit together coherently, and indeed the individual theories themselves often harbor inconsistencies.'

[19] e.g., see Earman and Fine [1977], where it is argued that Newton's term 'mass' referred to rest mass, and Field's reply in his [1994a], p. 235. The discussion of II.3 suggests that the mid-19th-century term 'valence' might make a nice case study, but I stick with our more familiar example in the text.

[20] See Stanford and Kitcher [2000], p. 125.

both, and sometimes refers to nothing at all. Telling which cases are which involves a careful analysis of the causal genesis of each utterance, its function in the context of that utterance, and so on. Can the disquotationalist capture the spirit of this analysis?

Field suggests the following approach:

> A natural deflationist move is to say that what the 'mass' example shows is simply that there is no best translation of Newton's word into our current scientific language. (Field [1994b], p. 237)

For the disquotationalist, reference is given by the R-sentences in my language, so Newton's term doesn't refer at all except insofar as it's translated. Describing the case from the point of view of Quine's stance on translation (see II.2), Field writes:

> On the most radical of Quine's views, when we ask what Newton was referring to there is nothing to be right or wrong about, there is only the question of whether translating his term one way or the other would meet our goal-driven standards, standards which may vary from one context of translation to the next. In some contexts, the translation 'relativistic mass' may be preferred, in other contexts, the translation 'rest mass' may be preferred; in other contexts we may prefer to translate it by a made up term 'Newtonian mass', which we recognize to be true of nothing.... The idea isn't that one of these translations is correct for some of Newton's uses of 'mass', a second for others, and a third for still others; rather, it is that even when translating a given use of 'mass' there are multiple acceptable standards of translation, depending on the translator's needs of the moment, with talk of 'correctness' not straightforwardly applicable. (Field [1973], from the 2001 postscript, pp. 195–196)

As we've seen, the Second Philosopher denies that translation is as unconstrained as Quine suggests: she thinks 'gavagai' is correctly translated as 'rabbit' and 'neige' as 'snow'. She acknowledges, of course, that there are more troublesome cases, such as Priestley's 'dephlogisticated air' (if not Newton's 'mass'), but even here, for example, she sees no reason to deny that some uses, for some purposes, are correctly translated as 'oxygen', while other uses, perhaps for other purposes, are correctly translated otherwise. There is genuine indeterminacy—for example, in the case of Quine's Japanese modifier (see II.2)—but it isn't ubiquitous.[21]

[21] Again, I won't try to assess Field's views on translation. See II.2, footnote 37.

This approach could incorporate much of what Stanford and Kitcher describe: the careful investigation of what Priestley was doing in a given context, what causal factors supported his particular utterances, what worldly facts were actually responsible for his successes and failures, and so on—all this would be taken into account by any conscientious translator, or we might just as well say 'interpreter', as Priestley was writing in English, and this well-informed translation/interpretation of Priestley's terms would vary from one context of usage to another, as do Stanford and Kitcher's reference relations. In addition, the translator adds one factor not present in Stanford/Kitcher's story: namely, the context of translation, the translator's interests and goals.[22] If the translation is intended for chemistry students, the translator might want to highlight the ways Priestley was interacting with and manipulating the world as we now understand it; preferring 'oxygen' or 'air rich in oxygen' wherever possible, pushing the collateral wording in those directions, would be appropriate.[23] If the translation is intended instead for historians, the translator will stay as true to the texts as possible, to give a more 'literal' translation that would preserve as closely as possible the thought processes of the man himself.[24] The correspondence theorist thinks these fine points of translation are irrelevant to the objective facts of Priestley's reference; the disquotationalist thinks they precede the attribution of reference to Priestley.[25] But prior to this disagreement, both theorists tell the same story about the details of Priestley's interaction with his environment—what the disquotationalist calls the indication relations of his utterances—and both will go on to judge that his reference is indeterminate.

[22] For a real-life example, see Avigad's remarks on his own policy in translating Dedekind: 'Since my goals are not primarily historical, I will generally use contemporary terms to describe the mathematical substance of the developments. For example, where I speak of the ring of integers in a finite extension of the rationals, Dedekind refers to the "system" of integers in such a field. Readers interested in terminological nuances and historical context should consult the sources cited above' (Avigad [2006], footnote 4).

[23] e.g., such a translator might say: when Priestley writes 'heating a metal in air yields calx of the metal and dephlogisticated air', he really means that heating a metal in air produces metal oxide and air poor in oxygen. See Kitcher [1978], p. 530.

[24] Such a translator would retain the word 'phlogiston', much as Field imagines the same sort of translator inventing a term 'Newtonian mass'.

[25] The disquotationalist thinks our judgments on whether or not Priestley's claims are true will also come only after translation/interpretation. This is the added complication alluded to in footnote 12. What the disquotationalist should say there is that Priestley's beliefs don't indicate the truth conditions we would assign them on our more literal translations/interpretations of his language.

Assuming, then, that this provides the disquotationalist with a reasonable rendition of the claim that the terms of past science have sometimes suffered from referential indeterminacy, can he say the same of his own current language? The correspondence theorist notes that we have reason to wonder about various aspects of our current best scientific theories; to take two familiar, but needlessly esoteric examples,[26] we have reason to doubt that our description of spacetime as a continuous manifold in general relativity will pan out in the long run,[27] and reason to suspect that, despite its fantastic predictive success, there's something going wrong in quantum mechanics as well. In other words, we have no reason to think that we're better off than Priestley, no reason to think that the reference of all our scientific terms is determinate. How can the disquotationalist make sense of this claim when all he has to say about reference is 'spacetime' refers to spacetime and 'collapse of the wave packet' refers to the collapse of the wave packet? Field's first thought is to call once again on translation:

> For many of our terms, there are possible improvements of our present best theory involving that term, such that the term is translatable into each of these improved theories in more than one way... The gap between this conclusion and the conclusion that some of our terms have no definite extension does not seem so large, so I tentatively think that the deflationist should reach the same conclusions as the inflationist on the issue of indeterminacy in current scientific terms. (Field [1994b], p. 237)

In subsequent discussions (Field [1998a] and [2000]), an example from mathematics[28] apparently persuades Field to reject this translation-based

[26] See II.6 for more down-to-earth cases.

[27] See my [1997], pp. 146–152. See also Field [2000], p. 279.

[28] See Field [1998a], pp. 271–273, [2001], p. 276, and [2000], pp. 280–282, 303. The terms '*i*' and '$-i$' (that is, the square root of -1 and its additive inverse) are distinct but entirely interchangeable. Assuming mathematical objects exist, as Field does for the purposes of this discussion, it's tempting to say that both '*i*' and '$-i$' are referentially indeterminate: each could refer to either value, as long as they don't both refer to the same one. But in this case it seems odd to suggest that a future improved theory might uncover more information about complex numbers and allow for two distinguishable translations of '*i*'. I take this to be Field's worry, but I don't share it.

For one thing, I'm disinclined to set much store by an example that rests on a robust mathematical Platonism that I (and presumably Field) regard as problematic (see IV.4). But beyond that, I don't think the mathematics supports the reading it's given here. When 'i' is introduced, it's most often with an indefinite article: e.g., 'denote a solution [of $x^2+1 = 0$] by i' (Ahlfors [1979], p. 5) or 'i is a square root of -1' (Bak and Newman [1997], p. 2). In such cases, we're fully aware that there are two square roots and that we aren't specifying which is which; '*i*' isn't functioning as a name—it's more like 'let *n* be a prime number between 3 and 11 and let n^* be the other'—and we know we can get away with this

approach in favor of his now preferred account in terms of non-standard probability functions, a suggestion that involves considerable technical and metaphysical footwork.[29] I won't pursue this proposal here, because it seems to me that Field has drifted off course.[30] I hope to sketch a simpler and more straightforward disquotational path through these thickets, one perhaps more salient and more congenial in the Second Philosopher's modified context of inquiry (see II.3). Imagine, then, that our disquotationalist is a Second Philosopher engaged in an empirical inquiry into the functioning of human language, and that she is examining the possibility that a disquotational theory of truth is adequate to her purposes.

We now return to Priestley. Recall that the Stanford/Kitcher correspondence theorist and our disquotationalist agree on the importance of a sensitive understanding of what Priestley thought he was doing in particular contexts, what facts in the world account for his successes and failures, what causal networks produced his various utterances of 'dephlogisticated air', and so on. From the disquotationalist's point of view, it seems fair to describe this inquiry as a delineation of the various indication relations that held between Priestley's beliefs states and the world—some of his belief states were good indicators of the presence of oxygen, some didn't indicate anything at all, and so on—and the explanation of Priestley's successes and failures, as we've seen, will rest on such things. Both the correspondence theorist and the disquotationalist seek these explanations of success and failure, and their explanations will appeal to the same chemical facts, the same causal networks, and so on. The difference between the two only comes later, when the correspondence theorist wants, in addition, to determine which among this welter of causal relations constitutes an objective relation of reference between Priestley's words and his environment. To the disquotationalist, for whom reference doesn't enter the

because of the symmetry. When we do eventually get nervous about not really understanding what we're doing, as mathematicians did historically, we replace $x + yi$ with (x,y), and then we can easily distinguish (0,1) from (0,−1). Indeed, textbooks often begin their official treatment with ordered pairs of reals, then define 'i' as (0,1) (e.g., see Apostol [1967], pp. 358–362).

[29] For the latter, I have in mind Field's efforts to explain away the impression that he's characterizing what it is to regard a question of truth or reference as indeterminate, as opposed to what it is for such a question to be indeterminate. See Field [2000], p. 302, [2001], p. 309.

[30] For what it's worth, Leeds's position on Critical Stance questions strikes me as even less attractive (see Leeds [1997]).

picture until translation/interpretation has taken place, the correspondence theorist's search for causal relation of reference between Priestley and his surroundings alone is a fool's errand; from this point of view, the correspondence theorist is exercising his considerable ingenuity in search of a will-o'-the-wisp, a fact of the matter that doesn't exist.

So what should the disquotationalist say about our own language? Given the aforementioned worries about our best scientific theories, the disquotationalist, like the correspondence theorist, has every reason to believe that we're in more-or-less Priestley's position in an unknown number of ways. For the disquotationalist, this means she thinks that our theories are overwhelmingly likely to be good indicators of something or other that's relevant—just as Priestley's were—because they're so successful—at least as much so as Priestley's were. But it's also overwhelmingly likely that they often accomplish this by means other than indicating their own truth conditions. 'The wave packet has collapsed' is a good indicator of something, but probably not quite the collapse of a wave packet. Likewise, the belief that spacetime is continuous.

Saying all this is unproblematic for the disquotationalist—what more do we want from her? What Field wants is for the disquotationalist to say that in such cases, our own language is most likely referentially indeterminate, to say that our terms most likely have indeterminate reference just as Priestley's did.[31] But for our disquotationalist, the indeterminacy of Priestley's reference is secondary, a derivative fact, dependent on our translational or interpretive goals; the real story about Priestley himself is the one about his indication relations, the one our disquotationalist and the correspondence theorist share. Granted, the disquotationalist is within her rights to go on to claim that Priestley's reference is indeterminate, meaning that there are several good translations/interpretations into our language; and in the same way, following Field's first suggestion, she might go on to predict that our reference will turn out to have been indeterminate, meaning that, if all goes well, future scientists will have better theories and good reasons to see the translation of our theories into theirs as indeterminate. My point is that these judgments of reference

[31] See, e.g., the introductory section to Field [2000].

are really a side issue, arising from our perspective on Priestley or future scientists' perspective on us, plus the trivial R-sentences. The important facts about Priestley's word–world connections or about our word–world connections lie in the indication relations, in the real causal connections that held in Priestley's time and now hold in ours. The effort to figure out what those relations and connections are, in our own time, is exactly the effort to figure out what it is about our theories that makes them work and not work, which is what sets us on the road to improving them.

My suggestion, then, is that Field's emphasis on the question of reference, for Priestley and for us, distorts the true disquotational impulse; after all, for the disquotationalist reference is trivial. My disquotationalist agrees with Field that there is an important and widespread phenomenon present in Priestley's case and almost certainly in ours as well. Field thinks this phenomenon involves reference (as in partial denotation) and he worries that the disquotationalist can't accommodate it. In contrast, my disquotationalist thinks that what went wrong with Priestley wasn't primarily a matter of reference, but of indication relations; though it can be described in terms of reference, doing so introduces the extraneous element of our interest in translating/interpreting him. Furthermore, what's going wrong with us is in all likelihood just the same sort of thing that went wrong with Priestley—an anomaly in our indication relations—and again, it doesn't need to and probably shouldn't be described in terms of reference: our term 'collapse of the wave packet' refers to the collapse of the wave packet, yes, if there is such a thing, but what we need to focus on is what our beliefs about 'wave packets' are actually indicating.

The upshot of all this is that the disquotationalist I'm describing here is no less concerned about word–world connections than the correspondence theorist. In fact, as the case of Stanford and Kitcher shows, the two may well be focused on precisely the same word–world connections. The difference is that the disquotationalist wants to understand them as what (she thinks) they are, as indication relations; that is, she wants to focus attention on understanding how our complex linguistic, scientific interactions with the world manage to work as often as they do. From this perspective, the correspondence theorist's quest for the real relations of reference appears not only as a lost cause, but also as a distraction from the real cause. We would do better, says the disquotationalist, to study indication relations for themselves, in their full richness and complexity, without trying to shoehorn them into a probably non-existent theory of truth and reference.[32]

Let me stop here. We've considered here two illustrative examples of arguments against disquotationalism, in particular, two examples of arguments that our explanations of word–world connections require something more than a disquotational account of truth and reference: the first, the Success Argument, asked us to explain the role of language in our successful undertakings; the second, a Critical Stance Argument, asked us to explain what's going on in cases of purportedly indeterminate reference. We've suggested that a second-philosophical variety of post-Fieldian disquotationalist can meet both these challenges. This amounts to no more than the beginnings of a complete argument that no stronger notion of truth or reference is ever needed in explaining the functioning of human language, but I hope to have brought some plausibility to the claim that the answer is no.

In the Field/Leeds version of the correspondence/disquotation debate, this would bring us near the end of the story: if disquotational truth can play all the required scientific roles, then there is no need for a correspondence theory, after all. For the Second Philosopher, what's been accomplished is different: if a correspondence theory of truth need not be part of her explanation of word–world relations, her task is clarified and various red herrings removed, but much more must be done to complete a positive account of how our language functions. This difference of perspective is why the second-philosophical disquotationalist expends more energy than Field and Leeds on the structure and uses of indication relations, and thus on her shared interests with the likes of Stanford and Kitcher. She sees more at stake here than the theory of truth.

Suppose, then, that the Second Philosopher tentatively adopts the disquotational account of truth sketched here as necessary for various expressive purposes and as sufficient for truth's role in her explanatory project, and

[32] I've put the case, e.g., in terms of deciding whether or not Priestley is referring to oxygen, but the same considerations carry over to deciding whether or not Priestley's claims should count as true. See footnote 25.

that she tentatively resolves to approach the as-yet unanswered questions of word–world relations from that perspective. This point of view will come under further pressure in II.6, but first let me pause to contrast the Second Philosopher's position with a very different approach to truth.

II.5

Minimalism

Minimalism about truth comes in deflationary and inflationary varieties,[1] neither of which, as we'll see, is particularly Second Philosopher friendly. For the sharpest contrast, I focus here on the inflationary version, advocated by Crispin Wright,[2] with indications here and there of how it differs from deflationary minimalism, advocated by Paul Horwich.[3] As in Part I, I hope to further illuminate the Second Philosopher's methods by tracing out her reactions to these uncongenial positions,[4] but I have also a more specific motive in this case. Even the deflationary version of minimalism is quite distinct from the position sketched in II.4, and any understanding of this second-philosophical strain of disquotationalism will be severely hampered if the two aren't sharply distinguished.[5]

Though Wright roundly rejects deflationism,[6] he sees minimalism as following in its footsteps: 'Minimalism about truth ... is spiritually akin to but supersedes the ... deflationary conception of truth' (Wright [1996], p. 5).[7] He describes the central insight this way: deflationism

[1] The terms 'deflationary' and 'inflationary' are used to roughly distinguish views that 'deflate' truth, of which disquotationalism is a particularly austere example, from those that 'inflate' it, like correspondence theories (see Field [1994a]). Cf. II.2, footnote 25.

[2] See, e.g., Wright [1992], [1996], [1999], [2001].

[3] See, e.g., Horwich [1990/1998].

[4] Obviously, the Second Philosopher's sometimes negative reactions to minimalism do not amount to criticisms of the view for anyone who doesn't share her viewpoint, which Wright, Horwich, and others attracted to this line of thought clearly do not.

[5] Cf. Marino's related division of a wide range of deflationists into two groups, with Wright and Horwich in one and Field in the other (Marino [2002], [2005]). In Marino [2002], she advocates a correspondence theory partly on the grounds that adopting a deflationary theory of the second sort risks confusion with those of the first sort. This section is largely motivated by the hope of avoiding this danger for the Second Philosopher's brand of disquotationalism.

[6] Wright's argument for this rejection can be found in his [1992], pp. 12–24, [1999], pp. 209–219, and [2001], pp. 754–759; it eventually ends in a standoff. (See Wright [2001], footnote 12, pp. 783–784, which begins 'This is, of course, by no means the end of the dialectic.' The reader is referred to Wright [1999], for more, but even there it isn't clear that the deflationist's position cannot be sustained.)

[7] In Wright [1992], p. 24, minimalism is described as 'a species of deflationism'.

is sound in its instinct that it does not, metaphysically, take very much for a discourse to qualify as truth-apt, nor for us to be entitled to claim that many of its statements are true. (Wright [1996], p. 5)

This stark characterization of deflationism should bring us up short. Nothing in the discussion of II.4 would suggest that truth-aptness, much less truth itself are easy.

To see what Wright is getting at here, we need to look at the structure of his minimalism. Beginning from the disquotational idea that the T-sentences more or less exhaust the concept of truth, the minimalist disagrees only by adding a few more of what Wright calls 'platitudes':

The minimalist view about truth ... is that it is necessary and sufficient, in order for a predicate to qualify as a truth predicate, that it satisfy each of a basic set of platitudes about truth ... for instance, that to assert a statement is to present it as true; that 'S' is true if and only if S (the Disquotational Scheme); that statements which are apt for truth have negations that are likewise; that truth is one thing, justification another; that to be true is to correspond to the facts; and so on. (Wright [1996], p. 4; see also Wright [1992], pp. 24–29)

From either our disquotationalist's or the minimalist's perspective, one might say that it's easy to be a truth predicate, but what Wright claimed a moment ago is that it's easy for a discourse to be truth-apt.[8] For this conclusion, we need the second plank of minimalism: a discourse counts as truth-apt, or assertoric,

just in case its ingredient sentences are subject to certain minimal constraints of syntax—embeddability within negation, the conditional, contexts of propositional attitude, and so on—and of discipline: their use must be governed by agreed standards of warrant. (Wright [2003], pp. 4–5)

The phrase 'standards of warrant' may seem to court circularity—for a discourse to be truth-apt is for it to have standards that warrant attributions of truth—but Wright makes clear that he means only 'firmly acknowledged standards of proper and improper use' (Wright [1992], p. 29). He goes on to argue that a minimalist truth predicate can be defined on any truth-apt or assertoric discourse.[9]

[8] As opposed to discourses with some goal other than asserting truths about the world, e.g., moral discourse on the expressivist's reading (see II.4, footnote 15).

[9] In the terms used below to discuss Wright's pluralism, the idea is that a legitimate truth property—that is, a property satisfying the minimalist truth concept (the Platitudes)—can be defined directly

Wright's central example is discourse about the comic. When I call a scene in a movie funny, I might wonder if I'm attributing an objective property to the scene before me; I might wonder whether my claim should be taken literally, realistically, as a description of the world. Wright contends that this worry of mine is entirely independent of truth.[10] The syntax of comic discourse is entirely in order,[11] and claims about the comic are governed by standards of proper and improper use:

> When I claim that something is funny, I'm not simply reporting my own reaction to it; for I can readily conceive that my reaction might be wrong—insensitive or misplaced. Nor am I reporting a majority reaction, or conjecturing the direction it might take; for the same point applies—the comic sensibilities of the majority may be blunted, 'off the wall' or debased. Regarding something as funny incorporates a judgment about the *fittingness* of the comic response ... [there is] reasoned appreciation and debate of what is funny, and ... criticism of others' opinions about it. (Wright [1992], pp. 8, 9)

It follows that comic discourse is truth-apt. If we are warranted, by our standards, in asserting that Buster Keaton taking an elaborate fall is funny, then 'Buster Keaton's fall is funny' is true.[12]

On this picture, all it takes for comic discourse to be truth-apt is some fairly superficial features—flexible syntax, standards of proper use—and after that, one of its claims is true if it meets those standards.[13] Hence, Wright's remark that 'it doesn't take much, metaphysically speaking ... for us to be entitled to claim that many of its statements are true'. But, again, this conclusion rests on a minimalism about truth-aptness that doesn't follow simply from the deflationary view that the concept of truth is (more or less) exhausted by the T-sentences, even if that analysis is supplemented by a few more platitudes. What's at work here is the further conviction

from the standards of warrant for any such discourse (namely, what Wright calls 'superassertability' by those standards). See Wright [1992], chapter 2.

[10] Or perhaps better: that the issues involved in my worry about realism will be clearer if we adopt a theory of truth that makes the worry independent of truth. Horwich also holds that minimalism disentangles issues of realism from truth (Horwich [1990/1998], pp. 7–8). The contrast is with Dummett and Putnam's truth theoretic characterization of realism (see I.7).

[11] We can say, 'that's not funny', 'if that's funny, I'll eat my hat', 'I hope the movie will be funny', and so on.

[12] More accurately, Wright takes this minimalist line of thought to constitute only a prima facie case for the truth of my claim. See Wright [1992], pp. 86–87, [1996], p. 5. I come back to this caveat below.

[13] See previous footnote.

that whether or not a sentence makes a claim about the world is fully determined by the surface structure of the sentence and the discourse in which it occurs:

> ... if things are in all these surface respects as if assertions are being made, then so they are. ... assertoric content [is not] a *potentially covert* characteristic of discourse ... (Wright [1992], pp. 29, 35)

So, given its syntax and discipline, comic discourse is automatically descriptive, assertoric, truth-apt.[14]

In fact, just as the descriptive character of a discourse is settled by its surface syntax and discipline, so too the content of its individual statements is fully revealed in their superficial form:

> no more can be asked of a purportedly conditional statement, for instance, than that it should *overtly* behave like a conditional statement. ... there is simply nothing to achieve by ... syntactic-reconstructive manoeuvres. (Wright [1992], p. 36)

> There is ... no *deep* notion of singular reference such that an expression which has all the surface syntactic features of a [referring term], and features in, say, true contexts of (by surface criteria) predication and identity, may nevertheless fail to be in the market for genuine—'deep'—reference. (Wright [1992], pp. 28–29)

(Of course, establishing truth, in this second passage, doesn't require any independent determination of whether or not the purported objects exist; for the minimalist, establishing truth is just a matter of checking that the conditions of proper use are satisfied.[15]) The upshot is that it would be pointless to try to reconstrue 'Buster Keaton's fall is funny'—which seems to attribute a property to an event—as disguised conditional 'if I see it, I laugh'. In fact, the very worry that might lead me to attempt such a reconstrual—the worry that comic discourse isn't properly descriptive as it stands—is misplaced from the minimalist's point of view.

Apart from the extra platitudes, what's been described so far is a version of deflationism that differs from the Second Philosopher's largely in its striking faith in surface features of a discourse: its syntax and its accepted standards of proper use. Pondering the reliability of syntax, the Second

[14] Horwich would seem to agree with this line of thought—see his [1990/1998], pp. 84, 146—though his example is moral discourse.

[15] Again, subject to the caveat in footnote 12.

Philosopher considers, for example, the case of talk about rainbows.[16] Under certain conditions, she agrees that there is a rainbow over there, and hence that 'there's a rainbow over there' is true, but she doesn't regard the surface syntax—which attributes a location to an object—as matching the underlying nature of the phenomenon. Surely, 'there's a rainbow over there' is true if and only if there's a rainbow over there, but further investigation reveals that there being a rainbow over there is actually an optical phenomenon generated by sunlight on water droplets in the air, whose perceived 'location' depends on the angle of observation, and so on.[17] And as far as accepted standards of use are concerned, the Second Philosopher has her carefully examined and finely tuned methods, and any proposed new ones will be evaluated on the same terms. The fact of their being used by some linguistic community doesn't by itself give them any warrant, even prima facie.

We might put the difference this way. For the Second Philosopher, truth and reference are built up from below. By her scientific means, she determines what there is and what it is like. This in turn tells her which of her words refer and which fail to do so—'cat' refers to cats, and there are cats, so 'cat' refers; 'witch' refers to witches, but there aren't any, so 'witch' does not refer—and truth follows in train (e.g., as in Tarski's definition, see II.2). For the minimalist, in contrast, surface features of the discourse come first, and if they are in order, then a prima facie case for reference and truth has been made. In a phrase: for the Second Philosopher, truth and reference bubble up from below, from the world to the syntax; for the minimalist, they trickle down from above, from the syntax to the world.[18]

Where Wright's minimalism differs most emphatically from the deflationary variety is in his insistence that truth is a property. In its pure form, the difference between truth's being and not being a property can be difficult to pin down,[19] but Wright's idea is bold enough to swamp the

[16] A favorite example of Wilson, whose views are discussed in II.6.

[17] It's sometimes argued that no disquotationalist can question surface syntax, because the truth conditions of '...' must simply be .... What's suggested in the text is that nothing stops the disquotationalist from noting that ... if and only if ________, or from regarding ________ as a more penetrating analysis of the phenomenon. This would only conflict with disquotationalism if she went on to say that the truth conditions of '...' are really ________, not ..., but I see no motivation for her to say that.

[18] This is presumably an example of the 'linguistic Kantism' from which Leeds was keen to dissociate his version of disquotationalism (Leeds [1995], p. 10).

[19] See, e.g., Wright [2001], pp. 753–754, Horwich [1990/1998], pp. 141–144.

subtleties: he holds that the minimalist's concept of truth can be instantiated by different properties in different discourses. Thus there might be one kind of truth in science, another in comic discourse, yet another in moral discourse, and so on. The debate between realism and anti-realism about a given discourse, then, is not about whether or not it is truth-apt (assuming it has appropriate syntax and standards of proper use), not even about whether or not its assertions are true (assuming they're reached by exercise of its recognized standards), but over the nature of the truth property that instantiates its merely minimal truth predicate.

On this approach, anti-realism about the discourse is the claim that

> nothing further [beyond the Platitudes] is true of the local truth [property] which can serve somehow to fill out and substantiate an intuitively realistic view of its subject matter. (Wright [1992], p. 174)

So, for example, we might ask: does this truth property guarantee that 'differences of opinion—where not within the tolerances permitted by various relevant kinds of vagueness—have to involve some form of cognitive shortcoming' (e.g., ignorance, inattention, distraction, oversight, prejudice, etc.)?[20] If so, then we might say the local truth property involves a robust correspondence to the world. We might also ask: given that the states of affairs the discourse deals with explain our beliefs about them—this much is guaranteed by the platitudes—'*what else* [is there that] citation of them can contribute towards explaining' (Wright [2003], p. 8)? The wider the variation in such explananda, the more the discourse 'deals in facts which are independent and substantial' (ibid.).[21] If a local truth property passed both these tests, it would qualify as a form of correspondence more substantial than that guaranteed by the Correspondence Platitude alone.[22] Presumably the truth property for comic discourse fails both these tests of realism, while the truth property for natural science passes them.

Of course the Second Philosopher—with her relentlessly bottom-up understanding—requires the reality of the subject matter and the reliability of its methods to be secured before matters of truth can be determined, so any purported 'truth property' beyond what Wright's minimalist would

[20] The quotation comes from Wright [2003], p. 8. The list comes from Wright [1992], p. 93. Wright calls this condition 'Cognitive Command'.

[21] This condition Wright calls 'Wide Cosmological Role'.

[22] That is, 'to be true is to correspond to the facts' (Wright [1996], p. 4).

call 'scientific truth' would be hard pressed to meet her standards.[23] Even in the case of 'scientific truth' itself, she remains unconvinced that there is any unified 'truth property' to be found.[24] So the Second Philosopher will be unsympathetic both to Wright's pluralism and to his inflationism.

But perhaps we've unfairly oversimplified Wright's position by ignoring his insistence that the top-down considerations rehearsed here give only *prima facie* support for truth and reference in a given discourse. To see how this goes, consider the case of arithmetic.[25] Given that there can hardly be a more well-behaved syntax than that of number theory, and that proof from, say, the Peano Axioms[26] serves admirably as a standard of proper use, we would expect the minimalist to grant the truth-aptness of the discourse and the truth of '$2+2=4$'. So far, the truth involved is merely minimal, but it should be possible to argue further that differences of opinion must surely trace to cognitive shortcomings, and that arithmetic facts are used to explain a wide range of phenomena,[27] thus guaranteeing a robust correspondence in the local truth property. There would be no room (by the minimalist's lights) for reinterpreting arithmetic truths—for saying, for example, that '$2+2=4$' really means 'the string "$2+2=4$" can be generated from these strings (corresponding to the Peano Axioms) by these rules of combination)' or some such thing—because the surface syntax with its names ('2' and '4'), its function ('+') and its relation ('=') must be trusted. Nor is there room for an Error Theory, which would deny that proof from the Peano Axioms is enough to establish the truth of '$2+2=4$' and go on to count standard mathematical claims as false. Thus Wright's minimalism would seem to deliver his Neo-Fregean Platonism.[28]

But all is not as it first appears. According to Wright, this minimalist line of thought provides only 'a strong prima-facie case for admitting numbers as

[23] Notice that here, as in Part I, it is her opponent, not the Second Philosopher, who imagines he has a criterion for separating the scientific from the rest.

[24] Presumably this is the sort of thing Stanford and Kitcher seek in II.4.

[25] Serious consideration of mathematics will have to wait until Part IV, but perhaps it isn't too early to introduce the Benacerrafian worries that figure in the background of so much contemporary philosophy of mathematics. The classic source is Benacerraf [1973]; see also Field [1989], pp. 25–30, or my [1990], pp. 36–48.

[26] See Enderton [1977], p. 70, or Blackburn [1994], pp. 279–280.

[27] Wright ([1992], p. 199) gives a nice example: 'it is because a prime number of tiles have been delivered... that a contractor has trouble in using them to cover, without remainder, a rectangular bathroom floor, even if he has never heard of prime numbers and never thought about how the area of a rectangle is determined.'

[28] See Wright [1983], or more recently, Hale and Wright [2001].

objects' (Hale and Wright [2001], p. 8). It turns out there is still room for an Error Theory of arithmetic: 'For all that's so far been said ... mathematical discourse may be made out to fail even of minimal truth' (Wright [1992], p. 35). It could be, for example, that we have consistently failed to correctly apply the standards of proper use—that is, we may be entirely mistaken about what we take to be proofs—or that proof is not, in fact, a coherent standard. Neither route is plausible. But another option for the Error Theorist remains:

> It might be conceded that the argument makes a prima-facie case, but argued that the relevant statements cannot really be seen as involving reference to numbers, on the ground that there are (allegedly) insuperable obstacles in the way of making sense of the very idea that we are able to engage in identifying reference to, or thought about any (abstract) objects to which we stand in no spatial, causal or other natural relations, however remote or indirect. A closely related objection—originating with Paul Benacerraf and subsequently pressed in a revised form by Hartry Field—has it that a platonist account of the truth conditions of mathematical statements puts them beyond the reach of humanly possible knowledge or reliable belief, and must therefore be rejected. (Hale and Wright [2001], p. 9)

Wright traces his susceptibility to this venerable objection[29] to his disagreement with deflationism: his minimalism

> does after all allow, in contrast to deflationism, that truth is a genuine property—to possess it is to meet a normative constraint distinct from assertoric warrant—which warranted assertions are therefore not guaranteed to possess. (Wright [1992], p. 35)

So, Wright's minimalist cannot insist that arithmetic truth is inseparably linked to the relevant standards of proper use (that is, to proof), but must allow that there is a gap here, that it is possible to argue that the truth property of a discourse is not straightforwardly tied to its standards.[30]

So the Error Theorist has not been vanquished by minimalism alone, and Wright concedes that he 'must therefore seek to answer such challenges'

[29] See footnote 25.

[30] e.g., that the truth property is distinct from superassertability. In Wright [1992] (p. 87), Wright wonders how we could ever have reason to attribute a local truth property according to which the standards of a discourse are entirely unreliable, as the Error Theorist does, but in Wright [2001] (pp. 9–10), he takes this possibility more seriously.

(Hale and Wright [2001], p. 10). The response appeals to Hume's Principle: the number of Fs = the number of Gs if and only if there is a one-to-one correspondence between the Fs and the Gs.

> Provided that facts about one–one correlation of concepts ... are, as we may reasonably presume, unproblematically accessible, we gain access, via Hume's Principle and without any need to postulate any mysterious extrasensory faculties or so-called mathematical intuition, to corresponding truths whose formulation involves reference to numbers. (Hale and Wright [2001], pp. 10–11)[31]

This suggestion naturally generated considerable disagreement between Wright and the prominent Error Theorist Hartry Field,[32] but what's of interest to us here isn't the ins and outs of the debate, but the fact of its taking place at all. What Field and Wright are debating is whether or not we talk about or know about numbers—which is just the issue of realism vs. anti-realism in arithmetic—and what's at stake is the cogency of the minimalist's argument for arithmetic truth. But minimalism was supposed to recast the realism/anti-realism debate, to place it *after* the issue of truth had already been settled by minimalist considerations, to locate it in the nature of the local truth property. Alas, it seems the old debate has re-emerged, in its old terms, before the minimalist has secured his route to truth. It seems it isn't so easy, after all, 'for us to be entitled to claim that many of [a truth-apt discourse's] statements are true' (Wright [1996], p. 5).[33]

But this concession to bottom-up thinking isn't nearly enough for the Second Philosopher. By her lights, the top-down form of argument doesn't provide even a prima facie case for truth and reference; all standards, all truth must be laboriously built up from below. She is left with her familiar intra-scientific notions of truth and reference, and those notions, for all she's seen so far, might be thoroughly disquotational. Let's now look at a recent apparently inflationary notion, this time (I suggest) from a fully second-philosophical source.

[31] Wright first proposed this solution in his [1983], §xi.

[32] Field's defense of the view that there are no mathematical objects and (thus) that many mathematical claims are false appears in Field [1980] and [1989]. Chapter 5 of Field [1989], 'Platonism for Cheap?', is Field's critique of Wright [1983].

[33] For what it's worth, it seems to me that Wright's minimalism might have taken a different turn: he might have maintained that the argument from minimalism to his neo-Fregean Platonism is conclusive, that those considerations do yield minimal truth for arithmetical claims and existence for numbers. Field's objection would then arise after these minimal gains were in place, in a subsequent realism/anti-realism debate centering on epistemic access, Cognitive Command, and so forth.

II.6
Correlation

In his book *Wandering Significance*,[1] Mark Wilson addresses one central strand in the Second Philosopher's larger investigation of how human language functions—roughly, a study of the role of concept words, including 'concept' itself—and his interest in truth and reference, like hers, arises only in the course of this project. Though he never avows any version of naturalism—he believes that science should be used, not mentioned, as a natural-born naturalist would[2]—he does allow himself the occasional joking reference:

I yield the Lamp of Scientism to no one! (Wilson [2006], p. 614, footnote 19)

I have never understood clearly in what the sin of scientism consists, unless it merely connotes an eagerness to talk about scientific fact beyond tasteful limits. (Wilson [2006], p. 137)

As remarked in I.7, the pejorative 'scientism' in some of its meanings can serve as a rough-and-ready equivalent of 'Second Philosophy'. It seems to me that Wilson does in fact behave consistently as a Second Philosopher, and I hope that what follows will lend plausibility to this claim.

Wilson's focus is on predicates, rather than names or sentences, so his concerns most often lie in the vicinity of A-sentences like:

'is red' applies to *a* if and only if *a* is red.

As in the cases of reference and truth, what's at issue is the status of such claims, which deflationists will regard as trivial in one way or another. Switching to truth, Wilson aptly characterizes deflationism this way:

So what form of evaluation do we offer when we declare a sentence to be 'true'? Deflationists provide a novel answer: adding 'is true' after a claim S is essentially

[1] Wilson [2006].

[2] Fittingly, in the section of Wilson [2006] with the subject heading 'science should be used but not mentioned' (1.iv), science is hardly mentioned.

equipollent to decorating S within supplementary filigree or rephrasing an active assertion as passive. Such notational bric-a-brac is not entirely useless because it can be employed for various purposes. If I have branded all my cows with numerals, I can instruct the foreman to round up only the dogies that bear prime numbers. Just so, I can exploit the 'is true' filigree to make assertions like 'Everything Nixon believes is true', for the implausible contention that I agree with Nixon on everything is not conveniently rendered otherwise. From this point of view, the so-called Tarski biconditionals ... seem as if they must be largely constitutive of 'true' 's utility, for they tell us when to add the filigree and when to take it away. (Wilson [2006], p. 626)

This idea Wilson staunchly opposes:

We may swaddle infant 'true' in comforting truisms, but, sooner or later, it must face the harsh adult world of correlational complication. (Wilson [2006], p. 634)

Given the attractions of the second-philosophical version of disquotationalism sketched in II.4, we wonder: what are these 'correlational complications'? Will attending to them revive some version of the correspondence theory?

First it's worth noting that Wilson's discussions of deflationism most often focus on those who 'regard talk of word/world correlation as inherently incoherent' (Wilson [2006], p. 634). He has in mind here those, like Gary Ebbs, who deny

that we can conceive of the entities and substances and species of the 'external' world independently of any of the empirical beliefs and theories we hold or might hold in the future. To accept this picture, we must conceive of the relationships between our words and the 'external' world from an 'external' perspective. We must imagine that we can completely distinguish between what we believe and think about the things to which we refer, on the one hand, and the pure truth about these things, on the other. (Ebbs [1997], p. 203, quoted in Wilson [2006], p. 79)

Such 'anti-correlationalist' (Wilson [2006], p. 79) or 'veil of predication' (Wilson [2006], p. 634) deflationists are precisely those inspired by the Comparison Problem: how can we compare our terms with raw, unconceptualized reality?[3]

[3] See also Wilson [2006], pp. 263–264. Wilson also opposes top-down deflationists like Wright (Wilson [2006], pp. 659–660), but he is more interested in the Comparison Problem motivation than in Wright's more ambitious goals.

Recall how the early Field, replying to a Quinean version of this objection (in II.2), rejected this characterization of the word–world connections:

We *are* trying to explain the connection between language and (extra-linguistic) reality, but we are *not* trying to step outside our theories of the world in order to do so. Our accounts of primitive reference and of truth are not to be thought of as something that could be given by philosophical reflection prior to scientific information. (Field [1972], p. 24)

Resolutely using rather than mentioning science (in typically second-philosophical fashion), Wilson instead reacts by considering a particular case of Ebbs's concern, imagining him to deny

that we can conceive of rabbits and their liking for carrots independently of any of the empirical beliefs and theories we hold or might hold about such mammals and their vegetative preferences in the future. To accept this picture, we must conceive of the relationships between our words and rabbits from an 'external perspective'. We must imagine that we can completely distinguish between what we believe and think about rabbits and their favorite foods, on the one hand, and the pure truth about these issues, on the other. (Wilson [2006], p. 79)

But this, Wilson notes,

winds up expressing little beyond the banal observation that rabbits (at least in the wild) pretty much go about their own business, independently of how we happen to think about them. (Wilson [2006], p. 80)

Of course the Second Philosopher heartily agrees. So we're left wondering whether Wilson's 'correlational perspective' in fact clashes with our second-philosophical disquotationalism.

To address this question, we turn to the substance of Wilson's account of how predicates link up with the world.[4] The story begins with the word's so-called directivities:

I employ 'directivity' as a non-technical means for capturing the loose bundle of considerations that we might reasonably cite, at various moments in a predicate's career, in deciding how the term should be *rightly applied*. (Wilson [2006], p. 95)[5]

[4] Obviously, I can't do justice to the many subtleties of Wilson's position in a few pages. Wilson [2006] is highly recommended.

[5] Here Wilson uses 'directivities' in place of the more common terms—'intension', 'conceptual norm', 'content'—to avoid seeming to endorse various presuppositions: that a predicate's directivities are unified, that they are stable throughout its usage, that they are open to conscious inspection, etc.

So, for example, to decide whether or not 'is red' should be applied to a stone, we might look at it in good light; to decide whether or not 'is under high pressure' should be applied to a portion of fluid, we might measure the pressure with a pitot tube, or perhaps, in other circumstances, 'calculate its value from the boundary conditions using finite differences' (Wilson [2006], p. 95). The world's contribution is uncovered by attending to

> The *physical information* that is captured when [the predicate] is fruitfully employed ... the physical environment in which the usage achieves its practical objectives. (Wilson [2006], p. 135)

The worldly features that underlie the predicate's successful use Wilson calls its 'support'.[6]

In the simplest sort of case, this is a familiar picture: the predicate 'is a dog' is useful in our interactions with animals because the objective attribute of being a dog involves various developmental patterns, adult behaviors, beneficial and unbeneficial food preferences, and so on:

> A simple 'is a dog'/***being a dog*** association does seem ... to genuinely capture the true center of what is involved in canine-oriented talk. (Wilson [2006], p. 61)

Despite Quine's skepticism—in Wilson's typography, about 'gavagai' and ***being a rabbit*** as opposed to ***being an undetached rabbit part***, etc. (see II.2)—we are able to uncover such correlations:

> Consider a sorting machine that distinguishes cans of peaches from cans of pears. Insofar as I can determine, Quine's somewhat hazy methodological strictures require us to say that 'there is not fact of the matter' ... whether our device sorts the cans by weight rather than through the patterns on their labels. But such doubts are plainly excessive—weight and label sorters operate with dramatically different mechanisms and it won't require lengthy investigation to determine what we have before us. (Wilson [2006], p. 265)

Similarly, 'is a rabbit':

> A few gestures at relevant specimens are likely, *pace* Quine, to lead to an employment that is properly described in terms of a genuine correlation between predicates and physical traits. (Wilson [2006], p. 269)

[6] Here again, Wilson avoids such familiar terms as 'extension', because he doesn't presuppose that 'any of the physical attributes involved in [a predicate's supports] will map onto the term ... in any regular or fixed way' or that they 'correlate neatly with any genuine physical grouping' (Wilson [2006], p. 135).

Such correlations seem of a piece with our disquotationalist's 'indication relations': our beliefs involving 'is a dog' are good indicators of the facts about dogs, just as the machine's 'accept' is a good indicator that it's scanning a can of peaches.

Of course, Wilson's central interest is in cases more complicated than this:

> The associated directivities of a predicate commonly come in a wide variety of grades, some of which are quite *easy to follow* and some of which border on the totally *opaque*. (Wilson [2006], p. 112)

Consider an example from applied mathematics, namely, light reflecting off a razor blade.[7] As Wilson tells the story (Wilson [2006], pp. 319–327), the problem was solved by Sommerfeld in 1894, in terms of a series of Bessel functions, but the series is so slow to converge that computation of actual values is entirely impractical.[8] Sommerfeld circumvented this problem by dividing the area to be described, the area around the razor blade, into three sectors, and providing a simpler exponential term for each. These replacements not only radically reduce the computational complexities, they also provide a more perspicuous picture or understanding of the phenomena involved—all at the minor expense of ignoring the behavior on the borders between regions. In such a case:

> We want our descriptive vocabulary to prove useful in dealing with the material goods around us, but the manipulative acts that we can readily perform as users of language ... are unlikely to suit Nature's patterns very well in their own right. ... Accordingly, if our usage is to suit the real world's properties, our easy-to-follow directivities must be cut and pasted together according to the strategic

[7] Attention to the obstacles faced by applied mathematicians, manufacturers, designers, etc. and the various ingenious methods they use to overcome them supplies a central theme of Wilson's book, what he sometimes calls 'the lesson of applied mathematics': 'Why do predicates behave so perversely? ... I believe the answer rests largely at the unwelcoming door of Mother Nature. The universe in which we have been deposited seems disinclined to render the practical description of the macroscopic bodies around us especially easy. ... Insofar as we are capable of achieving *descriptive successes* within a workable language ... we are frequently forced to rely upon unexpectedly roundabout strategies to achieve these objectives. It is as if the great house of science stands before us, but mathematics can't find the keys to its front door, so if we are to enter the edifice at all, we must scramble up backyard trellises, crawl through shuttered attic windows and stumble along half-lighted halls and stairwells' (Wilson [2006], p. 26; see also p. 452).

[8] Apparently even computers can't overcome the computational load in the related case of diffraction off water droplets in a rainbow (see Wilson [2006], pp. 319–320).

dictates of an organizational plan derived from a less transparent directive center. (Wilson [2006], pp. 114–115)

In Wilson's terminology, the series of Bessel functions form the 'core' directivity for 'light intensity' in this case, while the more workable exponentials are 'the satellite directivities it spawns' (Wilson [2006], p. 115).[9]

A case like this is fairly straightforward—the satellite directivities give handy approximations to underlying values determined by the core directivities; applied mathematicians can even justify the claim that the approximation is a good one—but in other cases, the core directivities are actually hidden from us. A colorful example comes from the history of metallurgy, in particular, from the development of effective methods for preparing steel for swords. Wilson informs us that

Extracting a desirable cutting tool from what was formerly hematite or native iron is no mean accomplishment, for it requires the unnatural trapping of unstable phases with the material matrix. ... All the traditional arsenal of the smithy—quenching, cold working, annealing, etc.—serves to install a very refined polycrystalline structure, delicately sensitive to impurities, within the steel, although virtually none of its mechanics was understood until well into the twentieth century. (Wilson [2006], p. 229)

The story goes that the ancient Japanese would quench a sword by stabbing it into the body of a prisoner! We have here a grisly directivity—part of determining that 'is a well-made sword' applied to a given sample involved verifying that it had been 'plunged into the belly of a noble foe'—whose support is now easy to find: it serves as a satellite for the core directivity 'quickly lower outer temperature to lock in ferrite grain' (Wilson [2006], p. 229). But of course, the Japanese swordsmith knew nothing of this; he may well have mistaken his satellite for a core.

What Wilson is giving us, it seems to me, is a more wide-ranging, detailed, and subtle discussion of a phenomenon already known to our disquotationalist: a piece of language serving as a good indicator, but of something other than its truth conditions.[10] Sommerfeld's exponential

[9] In these two quotations from Wilson [2006], pp. 114–115, Wilson is actually discussing the underlying core directivities and the satellite computational techniques for the vibrations of a conga drum (see also Wilson [2006], pp. 254–256), but for our purposes the structure of the two cases is the same.

[10] As noted in II.4, for our disquotationalist, assessment of truth depends on how we translate or interpret the stretch of language in question, but it seems unlikely there would be any context in

calculations function as good indicators of the local light intensity. The Japanese swordsmith's 'I must now plunge the metal into a prisoner' is a good indicator that he must now quickly lower the outer temperature, much as the ancient Greek's belief that 'Zeus is throwing thunderbolts' is a good indicator that there is thunder in his vicinity. So far, it seems Wilson's correlational point of view is providing a welcome elaboration of our disquotational outline rather than conflicting with it.

But Wilson has just begun. Suppose we suspend a rope between two nails. Eventually it settles down to a rest state that depends on gravity and the rope's stiffness. Wilson describes a method for computing that ultimate shape:

> Draw an arbitrary chain of broken lines between the two nails which we call G_1 (for guess #1). Compute how much energy is stored in G_1 ... Now wiggle some little portion of G_1 a wee bit, leading to a new shape estimate G_2. Compute G_2's stored energy. If it proves less than that of G_1, then G_2 probably represents a better guess as to the cord's true shape. Otherwise, wiggle G_1 in some other way. Proceeding thus, we can grind our way through a sequence of guesses that progressively carry us, in zig-zag fashion, closer to a good approximation to the rope's hanging shape. ... If our corrective instructions can be made *coercive*—that is, [if] we force the error to become smaller on every repetition— ... our broken line must zero in on a final answer ... which, if further conditions are met, will be ... correct. (Wilson [2006], p. 219)

Now let's ask our usual question of correct application, say about G_4. Unless the sequence of approximations reaches its fixed point very quickly, the shape described in G_4 cannot properly be applied to the rope, either at any intermediate moment as the rope wiggles toward equilibrium—the process doesn't track the rope's actual movements—or at the end. On the other hand, suppose that at step G_4 in the process, I suddenly wrote down the rope's actual intermediate position or its final shape. Correct as this might be as a description of the rope, it would be incorrect as a proper next guess in the algorithmic process.[11]

What we have here is a different mode of correlation for our linguistic behaviors:

which Sommerfeld's 'the exponential is such-and-such' should be interpreted as saying that the Bessel series comes to so-and-so, or the Japanese swordsmith's 'I must now plunge the metal ... ' should be translated as saying that he needs to quickly lower the outer temperature.

[11] Cf. the treatment of the algorithm for computing logarithms in Wilson [2006], pp. 172–173.

Two notions of 'correct answer' are evidently at play ... : a *distributed* one ('What is the correct sentence to write if the method is to achieve its final purpose?') and a *directly supported* one ('Which sentences qualify as true given the normal references of its component words?'). (Wilson [2006], pp. 173–174)

For present purposes, it seems the appeal to truth and reference in the final parenthesis is unnecessary; the relevant contrast is between terms whose directivities correlate 'directly' or locally with their supports—as in the cases we've considered up to now—and terms whose directivities correlate only via their role in more extended patches of discourse—as in the rope approximation case.

Using only the machinery of II.4, it seems the best description our disquotationalist can give of the case of the relaxing rope is to remark that a belief resulting at the end of the stage-by-stage application of the algorithm will be a good indicator of the final state of the rope. This leaves out the various facts—how the energy stored in the rope depends on its stiffness, how it tends toward the state of least energy, and so on, not to mention the 'coerciveness' of the approximation method—facts that play a crucial role in producing the resultant direct correlation. For Wilson, these extra items are part of the distributed support for this patch of language, part of what makes it work as well as it does, and a full description of the functioning of the word–world relations would have to include them. I see nothing that blocks our disquotationalist from helping herself to this Wilsonian insight, as a friendly and welcome amendment, from replacing her talk of 'indication relations' with the more general and flexible Wilsonian language of 'directivities' and 'supports', both direct and distributed. Let's consider the Second Philosopher's disquotationalism to be hereby so modified.

Wilson's most striking example of distributed correlation, perhaps his central inspiration,[12] is Oliver Heaviside's operational calculus. Heaviside was a physicist and engineer concerned to reduce distortion in long distance electrical transmission, especially for telegraph and later for telephone lines. The operational calculus resulted from his idiosyncratic, even bizarre approach to the relevant differential equations. To get a sense of how bizarre, recall a lesson from freshman calculus: that we shouldn't read too much into the Leibnizian notation 'dy/dx', that it's not a matter of

[12] See Wilson [2006], p. 519: 'It is largely from reading Heaviside that I have come to adopt the semantic opinions articulated here.'

one thing divided by another, really, still less of one product divided by another, but actually an infinite limit, properly understood in terms of the well-worn, ε's and δ's. Contrary to this wise counsel, Heaviside treated *d*, *x*, and *y* as if they were simply numbers, open to ordinary algebraic manipulation:

That is, beginning with the equation

$$dy/dt + y = t^2,$$

Heaviside will 'factor' it

$$[(d/dt) + 1]y = t^2,$$

then 'divide' it,

$$y = t^2/(d/dt + 1)$$

and finally 'expand' it [using the infinite series expansion for $1/(x + 1)$, with d/dt for x]. ... on the face of it, such procedures are about as sensible as dividing both sides of the movie star equation

Cary Grant = Archie Leach

by 'Cary' to derive a conclusion about our eighteenth president:

Grant = Archie(Leach/Cary).

(Wilson [2006], p. 519)

The great surprise is that by these and further 'algebratizing rules',

Heaviside invariably obtained correct answers ... Moreover, his algorithm generally found the right answer more quickly than orthodox methods (when the latter could be made to work at all). (Wilson [2006], p. 520)

Obviously Heaviside's machinations were onto something, but the worldly supports were completely unknown, indeed inaccessible to the mathematics of his day (1850–1925).

Such unorthodox and unfounded procedures naturally drew severe criticism, to which Heaviside responded with remarkable aplomb:

The rigorous logic of the matter is not plain! Well, what of that? Shall I refuse my dinner because I do not fully understand the processes of digestion? No, not if I am satisfied with the result ... First, get on, in any way possible, and let the logic be left for later work. (quoted by Wilson [2006], p. 521)

A man would never get anything done if he had to worry over all the niceties of logical mathematics under severe restrictions; say, for instance, that you are bound to go through a gate, but must on no account jump over it or get through the hedge, although that action would bring you at once to your goal. (quoted by Wilson [2006], p. 543)

Or my personal favorite:

Logic is eternal ... it can wait. (quoted by Wilson [2006], p. 28)

Wilson agrees that at such times 'an Ecclesiastesian season of semantic agnosticism is clearly mandated' (Wilson [2006], p. 544), and he highlights the way various episodes of semantic unclarity in the history of science have produced more insidious reactions, wherein '*philosophy* [is used] to patch over reasoning gaps that should be properly filled with more sophisticated *mathematics*' (Wilson [2006], p. 149). Let me sketch a bit of this case.

Around the turn of the nineteenth into the twentieth century, an alarming range of anomalies had arisen in classical physics: points at infinity and points with complex coordinates had been introduced with great benefit to fields from geometry to engineering, but without any clear picture of why this should work;[13] the basic entities of mechanics were ill understood (point particles? rigid bodies? elastic bodies?);[14] equations derived in one context were imported into others where the original grounding was no longer available;[15] and so on. Hertz, Helmholz, Duhem, Mach, and many others worried over such cases, sharing the vague feeling that a demand for strict semantic explication was perhaps too much to ask. This line of thought first led to the suggestion that a theoretical term is sufficiently specified by computational algorithms, then that it is implicitly defined by its role in an axiomatic theory, and finally to the vaguer notion that the content of such a term is grasped when we understand the informal theory in which it is embedded. If we add to this the thought that the goal of science is successful prediction, we arrive at instrumentalism, common

[13] See Wilson [2006], pp. 150–152, for examples from pure geometry (two circles always intersect at two points, but these may have imaginary coordinates) and engineering (designing a circuit to control the orientation of a telescope).

[14] Wilson [2006], pp. 360–362, illustrates (with the example of a vibrating bell) how explanations in classical mechanics were sometimes structured like a 'lousy encyclopedia': an account is given in terms of one fundamental notion with 'for more details, see ...'; the 'more details' are phrased in terms of another fundamental notion, with another 'for more details, see ...', ... and pursuing these references brings us back where we started. Wilson calls this 'ungrounded foundational looping'.

[15] Here the example is the Navier–Stokes equations, derived from Newton's $F = ma$ for rigid bodies, then transferred to viscous fluids, where the 'particles' supposedly acted upon by Newtonian forces are no longer ordinary masses, but moving spatial regions that gain and lose molecules while maintaining a constant total mass. See Wilson [2006], pp. 158–159, 175–176.

to Duhem and Mach. By this process, a mathematical gap has been filled with instrumentalist philosophy.[16]

Quinean holism represents a similar effort to combat excessive semantic demands[17]—like overcoming the Comparison Problem—by detaching scientific predicates from the world:

> Quine adapts to his own purposes the basic mechanism of predicates being supported semantically within a webbing of theory... The base idea is that, if we know how to manipulate syntax in response to natural conditions in a sufficiently rich way, we qualify as understanding that vocabulary fully... It is from their position within this gigantic snarl that specific predicates obtain their individualized personalities... only full-bore 'observation sentences' ('Lo, a rabbit!' is his favorite example) receive any worldly direct attachment and then through a process he vaguely calls 'conditioning to stimuli' ... [presumably] causally installed... [thus freeing] the component predicates within these observation sentences from any attachments of their own to attributes. (Wilson [2006], pp. 236, 237, 238–239)

The notion of distributed correctness is here expanded not just to embrace a local algorithmic patch or method—as with the hanging rope—but to include the entire 'web of belief' (see I.6). And the notion that such distributed correctness is underpinned by some distributed correlation with worldly supports, even if that support is for now unknown, drops out of Quine's picture entirely. As Wilson makes clear, such strange turns of doctrine are surprisingly common in a school of thought, American pragmatism, that presumably began from the down-to-earth idea that language is a tool for performing useful work.[18]

[16] Cf. Wilson [2006], p. 656, 658: 'Mach adopts a simple instrumentalist position... attempting to combat what is essentially a *mathematical misdiagnosis* ... with a *philosophical maxim* ... just because we presently lack the tools to resolve a problem, we shouldn't attempt to bridge the gap with slogans.' For more on the line of thought in this paragraph, see Wilson [2006], pp. 118–119, 125–127, 147–171, 555–557, 653–659. See also I.6, footnote 17, IV.4, footnote 48, IV.5, footnote 35.

[17] Wilson ([2006], pp. 236–240) characterizes Quine as reacting against the classical Russellian view described in the next paragraph, though something like the Comparison Problem seems to figure in his account of the motivations of many pragmatists (Wilson [2006], pp. 263–264).

[18] Wilson describes 'useful language' this way: 'such employments display recognizable *strands of practical advantage*—viz. the achievement of certain goals requires that certain sentences fall into proper place during their execution. The "work" accomplished in each case is certified by the desired condition achieved.... the sentences we string out in executing a strand of advantage each acquire a pronounced measure of top-down *distributed correctness* from their roles within the integrated routine, where a sentence may qualify as "correct" by these practicality-focused standards even if it reports a patent falsehood if evaluated by more conventional measures' (Wilson [2006], pp. 227–228).

If Quine's holistic semantics stand at one extreme, the other is occupied by a theory of Fregean concepts or Russellian universals, handy items that can be both grasped (by us) and exemplified (by things in the world). This 'classical glue', as Wilson calls it, gives a robust account of word–world connections, notable for its optimistic assumptions:

(i) that we can determinately compare different agents with respect to the degree to which they share 'conceptual contents'; (ii) that initially unclear 'concepts' can be successively refined by 'clear thinking' until their 'contents' emerge as impeccably clear and well-defined; (iii) that the truth-values of claims involving such clarified notions can be regarded as fixed irrespective of our limited abilities to check them. (Wilson [2006], p. 4)

Natural as this position might seem—'We feel instinctively convinced that we know what it's like for a stone to be **red** on the surface of Pluto'[19] (Wilson [2006], p. 33)—Wilson rejects it:

[We have a] very basic inclination to overestimate our human capacities for anticipating the unexplored, especially in linguistic matters. (Wilson [2006], p. 88)

This basic inclination, he argues, leads to grievous error, not only in philosophy, but in other pursuits as well.[20] The Wilsonian position we're in the process of sketching is designed to resist the temptations of both Quinean and Russellian extremes.

To get a better sense of this classical over-optimism, consider the case of the Druids, a tribe living on an isolated island in an old B-movie.[21] When, by some happenstance, an airplane lands in their midst, they immediately classify it as a 'great silver bird':

To these Druids, having never heard words like 'airplane' and having little contemplated the possibilities of machine flight heretofore, 'bird' seemed *exactly the right word* to capture the novel object that had just settled before them. (Wilson [2006], p. 34)

Wilson imagines an enlightened descendant of these original Druids retaining and defending this usage along the following lines:

[19] Wilson [2006], pp. 231–233, brings home how little we actually understand this by considering how we might go about designing a Pluto rover to search for rubies.

[20] See, e.g., Wilson's discussion of the ethicist who criticized the elderly Darwin for taking more pleasure in the study of earthworms than in music or poetry (Wilson [2006], chapter 2).

[21] This example dates back to Wilson [1982].

'Yes, I recognize ... that we do not want to place great silver birds ... into the same *biological class* as ... chickens. Nonetheless, my forebears have always employed "bird" with a more general meaning than do the Yankees and I respect their ancestral practices. For biological purposes, the technical term "aves" will do nicely. But why should we follow the Yankees otherwise in their strange classifications? After all, they are also inclined to dub flightless cassowaries as 'birds', a classification that Druids have always rejected as deviant (although we allow, of course, that these creatures belong to aves).' Wilson [2006], pp. 34–35)

So far, so good, but what if the plane had landed unobserved, and the Druids had instead discovered it in the forest with the flight crew living out of the fuselage?

The vehicle's arboreal mise en scene now suggests 'house' to these folks every bit as vividly as the airborne arrival had erstwhile prompted 'bird'. (Wilson [2006], p. 35)

This time, the Druids call the plane a 'great silver house', and the enlightened descendant reasons:

'Of course, silver houses aren't *birds*—did you ever see windows in a bird? ... our ancestors were right to characterize these flying devices as "houses" because they can be lived in. Our people have never intended "house" to be employed in the narrow, "silver house"-rejecting mode favored by the Yankees.' (Wilson [2006], p. 35)

The moral of the story is straightforward:

... neither set of alternative Druids has any psychological reason to suspect that they have not followed the pre-established conceptual contents of their words 'bird' and 'house', although the chief factor that explains their discordant classifications actually lies with the *history* of how they happen to approach the airplane. Both groups instinctively presume their societally established notion of ***bird*** has already determined within itself whether a bomber properly counts as a 'bird' or not.... The Druidic tendency to assign excessive credit to the realm of 'what we have been conceptually prepared to do' seems completely harmless, but it nicely illustrates a basic ... mechanism that allows us to misjudge the strength of our current conceptual grasp ... It is beyond human capacity to fully prepare ourselves to classify any damn thing that might come along, but we can easily *fool ourselves* into believing that we possess such secret capacities. (Wilson [2006], p. 36)

This is the underlying over-optimism that gives the classical picture its appeal, and which the classical picture encourages in return.[22]

Illuminating as the Druid case may be, Wilson immediately warns us against treating a thought experiment as evidence and gets down to discussing actual cases. So, for example, it seems water can take various solid forms that our ordinary notion of 'being ice' is no doubt unprepared to classify, though we might, Druid-like, imagine ourselves compelled in one direction or the other depending on how we first learn of them (see Wilson [2006], pp. 55–57). The stubbornness of these 'compulsions' is well illustrated in the case of 'weight', as in 'weighs 180 pounds'. We all know that a man's having this property is a matter of the earth's gravitational field, that if he were on the moon, with its weaker gravitational pull, he'd weigh 33 pounds. We also 'know' that an astronaut weighing 180 pounds on earth and 33 pounds on the moon is weightless in earth's orbit. These happy thoughts run into trouble only when we contemplate both the earth and the moon at once, and realize that 'weightless' should actually only apply to an object at the tipping point between earth's gravity and the moon's, a point much farther from earth than the spacecraft's orbit. We're faced with what Wilson calls 'multi-valuedness': we want to say that the man in the orbiter weighs 176 pounds (a bit less than his 180 pounds at sea level), but we also want to say that he's weightless.

Wilson's analysis of this case introduces his notion of a façade:

> In the days of old Hollywood, fantastic sets were constructed that resembled Babylon in all its ancient glory on screen, but, in sober reality, consisted of nothing but pasteboard cutouts arranged to appear, from the camera's chosen angle, like an integral metropolis. (Wilson [2006], p. 183)

Some scientific theories—parts of classical mechanics, for example[23]—have this structure:

> They represent patchworks of incongruent claims that might very well pass for unified theories, at least, in the dark with the light behind them. (Wilson [2006], p. 183)

[22] Despite some structure similarity to Kripke's Wittgensteinian worries about whether the rule 'plus 2' dictates that 1004, rather than 1002, follows 1000 (see Kripke [1982]), Wilson's interest is not in such 'artifically exaggerated doubt', but in cases 'where the underlying directivities seem genuinely unfixed' (Wilson [2006], pp. 38–40).

[23] Wilson's argument for this claim is the central theme of [2006], 4.vi–4.ix.

Our predicate 'weighs 180 pounds' displays an analogous patchwork quality. Any particular application tends to fall within a relatively local context of usage—our immediate earthly surroundings, some area of the moon's surface, the interior of the orbiter—all with more or less the same easy-to-use satellite directivities:

> In everyday life, we assign approximate 'weight' values to objects simply by roughly estimating the difficulty of performing sundry tasks that relate to [their] locomotion: How easily can the object be lifted or thrown? How much will it hurt when it falls on one's head? And so forth. (Wilson [2006], p. 328)

On the earth and the moon, these methods succeed in tracking the weight due to local gravitational forces, as we intended, but without our noticing, those same directivities are supported instead by mass when applied aboard the orbiter. This shift—Wilson calls it 'property-dragging'—leaps out when we move to a more inclusive patch including all three, and we find ourselves with conflicting judgments: one on the local orbiter patch ('weightless'), and another for the orbiter on the more inclusive patch ('weighs 176 pounds').

Of course this could be cleared up by studious maintenance of the distinction between weight and mass, but as Wilson points out, dragging 'weight' to mass in the orbiter is almost irresistible: 'Even proverbial "rocket scientists" cheerfully succumb to this tempting prolongation' (Wilson [2006], p. 330).[24] As long as most of our uses of the term are confined to one or another of the patches, no harm results from the 'multi-valuedness' of the overall façade. (If you only look at it from the right camera angle, it looks like the real thing.)

Wilson's most dramatic and revealing example of this phenomenon is the behavior of the term 'hardness' (Wilson [2006], 6.ix–6.x). Descartes took hardness to be a sensation, like color; Thomas Reid disagreed, saying it was an as-yet-unknown physical quality; Wilson argues that neither is right

> Because our usage of the predicate 'is hard' displays a fine-grained structure that we are unlikely to have noticed, for our everyday usage is built from local patches of evaluation strung together by natural links of prolongation [i.e.,

[24] He cites a textbook that carefully explains that the astronaut's condition in the orbiter is only 'apparent weightlessness' on one page, then goes on to discuss the effects of 'long periods of weightlessness' on the next (Wilson [2006], pp. 333–334)!

property dragging] ... in everyday contexts we adjudicate the 'hardnesses' of various materials ... through a wide variety of comparatively easy to apply tests—we might *squeeze* the material or *indent* it with a hammer; attempt to *scratch* it or *rap* upon it; and so on ... our choice of tests is likely to have been suggested by the material in question: we instinctively appraise a wood by rapping upon it, a rubber by squeezing, a metal by attempting to make a small imprint, a glass or ceramic by rapping lightly or scratching ... we are normally interested in comparing hardnesses mainly within natural groupings of stuffs of generally allied characteristics ... although interesting crossover cases also arise. (Wilson [2006], p. 336)

In response to the practical needs of various fields of manufacturing, an array of more precise tests has sprung up, many of them refinements of the informal directivities:

Brinell or Vickers indenters (vigorous squeezing and then releasing); superficial Rockwell testing (mild squeezing and partial releasing), durometer (squeezing without releasing), sclerometer (scratching), scleroscope (... rap[ping]), the Charpy impact test (hitting with a hammer) and so forth. (Wilson [2006], p. 337)

The test for each local patch acts more or less as a satellite directivity supported by some underlying physical property: for metals, the yield strength; for rubbers, Young's modulus of elasticity; and so on. Multi-valuedness will arise here and there, when a material can be evaluated in more than one grouping, but most of the time we stick to one patch or another and the predicate functions smoothly.

Two questions naturally arise, one scientific, the other linguistic. The scientific question is this: does a single physical feature—hardness—underlie these various patches? Reid thought so, as we've seen, and Wilson quotes writers from as late as the 1940s who maintained the hope of discovering the underlying condition and an 'absolute scale of hardness'. Consider, for comparison, the history of 'temperature', which was originally measured by

Sundry measuring instruments (early thermometers or thermoscopes that, in fact, half acted like barometers). ... Depending on the working fluid employed, thermometers in themselves supply inherently incongruent readings in a manner like that of our sundry hardness testers. As is well known, the nineteenth century flourishing of thermodynamic thinking allowed Lord Kelvin to articulate an 'absolute' approach to ***temperature*** that freed it from the shackles of instrumentation. (Wilson [2006], p. 348)

Alas, for hardness, in the years following the hopeful 1940s,

> dislocations, phase changes, thin layer lubrication and the army of other critical notions of modern materials science became recognized, crushing in their wake any hope that an absolute 'hardness'... might be produced. (Wilson [2006], p. 349)

Notice that the classical picture would encourage us to think that there must be a unified concept of 'hardness' and that we could discover it by conceptual analysis, neither of which is true.[25]

Which brings us to the linguistic question: why not replace the word 'hardness' with a directly supported predicate on each well-behaved patch: 'yield strength', 'Young's modulus', etc.? Despite his uncertainty that 'hardness' registers a single physical feature, one of Wilson's authors from the 1940s writes:

> Would it clarify our thinking if we eliminated the word 'hardness' from our scientific vocabulary? The author's point of view is that the term is so firmly entrenched in our everyday vocabulary, that we could not go far before we began to use it again. (quoted by Wilson [2006], p. 350)

Much like the 'weightless' astronaut, though with a greater store of practical application, the 'hard' material is simply too attractive to resist:

> Its associations with swift practicality guarantee that the term will never vanish utterly from the colloquial language of anyone who works with materials... A set of evaluative patches linked together... through instrumentally inspired continuations supplies a simple model of the semantic platform upon which... 'is hard' is deposited [i.e., its support]. We thereby obtain the foundations for a workable, practical usage, but with no single physical property supporting the predicate everywhere, hence no 'absolute hardness'. (Wilson [2006], p. 350)

Clearly our use of 'hardness' is registering objective physical features of the materials described, though in its patchwork way.[26] Wilson proposes the

[25] Cf. Wilson [2006], p. 351: 'façade or flat structure—that is largely an arrangement that word and world must work out between themselves... it is largely on the forges of practicality that a useful fit between our available linguistic tools and the physical facts gets gradually hammered out and our feeble original intentions or the semantic vows we frame in intervening years are unlikely to deter this pattern of final accommodation substantially from its appointed courses.'

[26] The distribution of patches and associated directivities of a façade may well depend on our interests (e.g., description, prediction, design, reproduction) and our limitations (e.g., what we can perceive, what our mathematics can do) while the information registered remains objective. Speaking of 'is red', Wilson ([2006], p. 467) writes: 'it manages to encode physical information quite nicely, albeit in

term 'quasi-quantity' for such 'uneven data registration' (Wilson [2006], p. 389).

Notice that our disquotationalist considered a simple example of façade-like support in her description of Priestley's use of 'dephlogisticated air': one collection of his utterances was supported by oxygen, another by oxygen-rich air, and another by no physical property at all. In a similar spirit (though in greater detail), Wilson discusses the complexities of the alchemist's use of the term 'mercury' (Wilson [2006], pp. 642–646). Both these analyses rely on specific attention to the subjects' goals and to the worldly supports that make their usage work, when it does, and fail, when it does. A suggestion of disagreement arises when Wilson ([2006], 10.vii) objects strongly to the idea that one aspect of understanding others in such cases involves translation or interpretation into our language—a move our disquotationalist regards as minor in comparison with the primary correlational information, employed only to settle the tangential question of what Priestley was referring to and whether or not what he was saying was true—but Wilson is actually reacting against those who would employ translation in ways entirely foreign to the Second Philosopher: as an excuse to ignore or dismiss the word–world correlations at play in the linguistic practice we hope to understand; as 'a unitary "mapping to English" scheme' (Wilson [2006], p. 639) that disallows translating the same word to different English equivalents in different local contexts of usage; as introducing holistic indeterminacy (in the radical Quinean sense).[27] In fact, our disquotationalist's modest notion that the translator's goals might

a shifty and multi-valued way. True, the ways in which its parcels of usage piece together very much have the signature of human capacity written all over them, but that fact alone doesn't mean that the data entered upon those sheets has become thereby corrupted.'

[27] Part of Wilson's aversion for translation-based accounts springs from the way 'many of Quine's strongest admirers have detected in his "mapping into the home language" musings confirmation of profound truths about the human condition that run deeper, perhaps, than any thesis that Quine himself intends ... I have particularly in mind the melancholy conceit that the characteristics we attribute to other people—indeed, to every worldly event that passes before our gaze—come irrevocably tinctured with the contributions of our own point of view—that our interpreting gaze forever locks our conclusions into an ego-centered orbit from which they can never escape' (Wilson [2006], pp. 649–650), a view with particularly disastrous consequences for such fields as music preservation, a hobby of Wilson's (see Wilson [2006], chapter 2 and *passim*). Obviously, this sort of thing plays no role in our Second Philosopher's thinking.

properly play a role receives a passing nod from Wilson himself.[28] So once again, I see no conflict between our disquotationalism and Wilson's analyses, and much for the disquotationalist to gain.

We arrive, at last, at the complete correlational picture of semantic development for predicates:[29]

> A façade structure commonly enlarges when certain strands of practical advantage ... utilized in some established patch ... are found to extend profitably beyond its natural boundary ... (Wilson [2006], p. 516)

... as the Druid's 'bird' comes to apply to the airplane, or our 'weight' shifts up to the orbiter, or the turn of the century dy/dx shifts into Heaviside's operational calculus. New directivities gradually accumulate, including controls that block exchange of clashing information or rules between old and new patches. At first, the supports for the new usage may well be unknown, its correctness may be entirely distributed; indeed, the very fact that a prolongation has occurred may pass unnoticed.

> Eventually, however, the disharmonies between old and new usage can become so pronounced that some more systematic resolution is required. (Wilson [2006], p. 516)

Thus, a period of semantic agnosticism gives way to a new semantic picture that supports both usages, old and new, perhaps with some corrections. And now the process begins again, the 'wandering significance' of the book's title. During stable patches, the classical picture will beckon (especially if the façade complications go unnoticed); during agnostic periods, holism, instrumentalism, formalism are tempting; but both these overlook the constant shifting back and forth, as language accommodates as best it can to its descriptive tasks.

[28] e.g., the consideration of 'manifest poetic difficulties' (Wilson [2006], p. 641) or the differing translations for an alchemical claim that emphasize 'chemical information' versus 'otherwordly connection' (Wilson [2006], p. 644).

[29] I should say, 'a partial picture, for some predicates', as Wilson repeatedly warns that this framework is 'rather rough and ready—any real life case is likely to exhibit special features beyond its reach' (Wilson [2006], p. 377) and that 'In dubbing this pattern "canonical", I merely intend that it roughly qualifies the shape of most of the linguistic developments we shall study, *not* that every predicate's progress fits its pattern' (Wilson [2006], p. 518).

All this suggests that Wilson and our disquotationalist will give the same account of the functioning of word–world correlations, or more accurately, that our disquotationalist can happily help herself to Wilson's illuminating analyses. In more general terms, they also agree in opposing the extremes on either end of the debate: on the one hand, the extreme deflationist, confused by the Comparison Problem, who rejects all word–world connections as 'metaphysical', and on the other, the correspondence theorist, misled by the classical picture, who expects a single 'mechanism of reference'.[30] If any disagreement remains, it may come in Wilson's idea that 'true' and 'refers' work on their own façades: 'It seems to me that... "refers" operates in a context sensitive manner very much like that of... "hardness"' (Wilson [2006], p. 430). If so, if, for example, 'refers' correlates directly with some physical property on each fixed patch of usage—much as 'hardness' correlates with yield strength for metals, Young's modulus of elasticity for rubbers, and so on—then perhaps there *is* more to say about reference than the disquotationalist's R-sentences. But the existence of such patches and correlated physical properties remains an open question.

Wilson's discussion suggests that his own sense of our disquotationalism's shortcomings has less to do with the machinery of word–world correlation than with natural language usage of the words 'true' and 'refers'. During periods of stable (if façade-like) usage, he writes:

> I might possibly allow that, if language usage could linger long in this cheery and static condition, most of the employments of 'true' we might witness therein would conform docilely to the deflationist pattern (I have not investigated this claim seriously, as it strikes me as unverifiably counterfactual in its asseverations). (Wilson [2006], p. 633)

But the policy of identifying 'true' more or less with the T-sentences 'is either unwise or irrelevant during epochs in which an old [semantic] picture need[s] to be detoxified and rebuilt' (Wilson [2006], p. 632). So it is during such periods of semantic agnosticism that 'true' purportedly takes on a non-disquotational role.

[30] Cf. Wilson ([2006], p. 558): 'There isn't a single "mechanism of reference" and there needn't be: a goodly swirl of distinct directive influences act in concert to park a terminology over some fairly determinate stretch of ocean floor.'

What Wilson has in mind here is the sort of thing we might want to say in times of semantic agnosticism, for example, in the period before Heaviside's calculus was given straightforward foundations. Here the disquotationalist's

'*dy*/*dt*' refers to *dy*/*dt*

is irrelevant—because it gives us no help in sorting out the supports for Heaviside's calculus—and unwise—because it might suggest a classical simplicity here, when in fact the support for the '*dy*/*dt*' on the right is itself unknown. Our disquotationalist doesn't dispute the charge of irrelevance, nor (obviously) does she lean toward classicism, but Wilson's claim is that a non-disquotational use of 'true' arises in such cases: examining a passage of mysterious reasoning with an eye toward eventual regularization, we might conclude that a particular claim *S* is absolutely central to the workings of the operational calculus, that it will almost certainly survive in any eventual reconstruction of the method, and we might express this by saying 'of all this, *S must* be true!'

> In using 'true' and 'false' thus, we certainly *seem* to be sorting out sentences according to their capacity to correlate with supportive fact. (Wilson [2006], p. 633)

This certainly seems right. 'There was much truth in what Priestley was saying' or 'Priestley was talking about (referring to) oxygen, though he didn't realize it' would seem to perform related functions.

I see no reason for anyone to deny that the words 'true' and 'refer' have these uses, but I also think that we are here uncovering subtle differences in the questions Wilson and our disquotationalist have been addressing. One question is: what notions of 'truth' and 'reference' do we need in our project of explaining how human language functions, of explaining the word–world relations involved? Our disquotationalist thinks the answer to this is that we don't need 'truth' or 'reference' at all, that the whole story is best told in terms of indication relations between bits of language and bits of the world, or better, in apparent agreement with Wilson, in terms of his directivities, supports, façades, and so on. Another question is: what notions of 'truth' and 'reference' do we need to say the things we want to say? Our disquotationalist thinks that we need only notions more or less exhausted by the T-sentences and the R-sentences, those employed for linguistic ascent, descent, and generalizing tasks. Asked about the uses of the previous paragraph, she suggests that these are really just loose paraphrases for the more straightforward and informative 'This move in the operational calculus looks to be one that has some sort of direct support' or 'Some of Priestley's claims are correlated with facts about oxygen', that we don't need these uses of 'true' or 'refers' to say what we want to say. Again, there is no clear disagreement with Wilson.

But there is a third question. We might ask—how are the words 'true' and 'refers' actually used in human speech?—as a special case of the original project of explaining the functioning of our language. In answer to this question, it's no doubt important to note that the words have the uses Wilson mentions in times of semantic stress.[31] Still, it's just as important to note that those uses don't correlate with any variety of inflationary truth; their support is actually to be found in the now-familiar machinery of directivities, supports, façades, and so on. Thus it seems Wilson's answers to the third question don't impugn the view that our best account of word–world correlations does not involve inflationary truth, or even the view that the only uses of 'true' essential to saying what we want to say amount to semantic ascent, descent, and various forms of generalization.[32] And this combination of views appears to deserve the label 'disquotationalism'.

In any case, I think we can safely conclude that Wilson's correlational point of view and the stance of our second-philosophical disquotationalist resemble each other far more than either resembles more familiar forms of inflationism or deflationism. Both take the serious work to be the investigation of word–world connections and both reject the call for a single 'mechanism of reference'. Our disquotationalist is perhaps more skeptical than Wilson's correlationalist on a prediction: that further study will uncover unified supports for word–world correlations in local patches. But, if the evidence goes against her, our disquotationalist will happily convert, because both these theorists carry out their inquiries in an entirely second-philosophical spirit.[33]

[31] Not to mention others: e.g., 'Trollope's novels are true to life', 'Joe is a true friend', and many more.

[32] See Wilson [2006], pp. 634–635, for Wilson's take on the second-philosophical disquotationalism, e.g.: 'Maddy ... seeks a rationale for *side-stepping* many of our key issues.'

[33] I'm grateful to Wilson for his patience with my questions during the drafting of this section.

PART III

A Second Philosophy of Logic

III.1
Naturalistic options

It's time, finally, to stop describing Second Philosophy and get down to producing some. I hope to say something useful about mathematics, and perhaps a bit, too, about natural science before this book ends, but it seems to me that in order to do either of these, we first need an understanding of the ground of logical truth. The logical truths I have in mind are the simplest, most uncontroversial examples:

If all oaks are trees and this is an oak, then this must be a tree,

If it's either red or green and it's not red, then it must be green,

or particular instances of modus ponens or modus tollens.[1] The questions I'm asking are the straightforward metaphysical and epistemological questions that arise at the beginning of any 'philosophy of *x*': what kind of truths are these? how do we come to know them? In the case of logic, there's also the special problem of determining the source of the logical 'must'.

Before tackling these questions head on, let me briefly review the familiar options that tend to recommend themselves to naturalists. I hope a sense of the discomforts of these positions will help motivate the search for an alternative.

[1] I concentrate on uncontroversial cases like these and leave aside the question of where the boundary falls between logic and non-logic. Though the decision seems to me largely terminological, my own inclination is to regard 'logic' as somehow 'obvious or potentially so' (in more or less Quine's phrase; see his [1954], pp. 111, 112), so as to make sense of the logicist's thought that reducing arithmetic to logic makes it epistemically transparent, or at least represents an epistemological gain. This suggests a line between first-order predicate logic and higher-order logics, but those who disagree should simply take what I offer here as an account of examples of the general sort listed above.

i. Psychologism

One early school of naturalistic thought is the so-called 'materialism' of mid-nineteenth-century Germany.[2] The movement began as a reaction against theology and speculative metaphysics (especially Hegel) and quickly developed to include the idea that:

> Philosophers can no longer be viewed in opposition to natural scientists, because any philosophy worthy of its name laps up the best sap of the tree of knowledge, and by the same token produces only the ripest fruit of that tree. (from Moleschott,[3] quoted in Gregory [1977], p. 146)

In plainer terms, philosophy uses the methods and results of science and what it produces is again science, broadly construed.

The leading materialists were physiologists—concerned primarily with issues like vitalism (see II.3), spontaneous generation, and the development of species—but they were also widely read popularizers of the burgeoning natural science of the time and opponents of all things supernatural. All knowledge arises from sensory experience, on their view, but this sensory experience is not understood as a stream of ideas (as in Locke, Berkeley, and Hume; see I.3, I.4), or of sensory 'data' (as in later empiricists like Quine; see I.6), but as a physiological process much closer to the Second Philosopher's story of light rays and retinas, neurons and cognition. These same materialists held that logic is a record of the laws of thought, discovered by observation of human inferential patterns, understood in the end physiologically, and hence ultimately in terms of chemistry and physics. This is a particularly straightforward version of psychologism.[4]

The best-known and most devastating criticisms of psychologism in all its forms were advanced by Gottlob Frege, the father of modern logic.[5] For Frege, 'logic is the science of the most general laws of truth' (Frege [1897], p. 128); it tells us, for example, that statements of certain forms are always true, that inferences of certain forms are truth preserving.

[2] For discussion of this school, see Sluga [1980], chapter 1, and especially Gregory [1977].

[3] The leading 'materialists' were Karl Vogt, Jacob Moleschott, Ludwig Büchner, and Heinrich Czolbe.

[4] Others versions were espoused by various neo-Kantians who replaced Kant's transcendental psychology with empirical psychology. For discussion, see Anderson [2005].

[5] Frege's main target was the neo-Kantians mentioned in the previous footnote.

I understand by logical laws not psychological laws of *holding as true*, but laws of *being true*. If it is true that I am writing this in my room on 13 July 1893, whilst the wind howls outside, then it remains true even if everyone should later hold it as false. If being true is thus independent of being recognized as true by anyone, then the laws of truth are not psychological laws, but boundary stones set in an eternal foundation, which our thought can overflow but not dislodge. And because of this they are authoritative for our thought if it wants to attain truth. (Frege [1893], p. xvi)

Truth is objective, so the laws of truth, that is, the laws of logic, are also objective, not psychological. It follows that logical laws are not merely descriptive, but normative:

Logic is concerned with the laws of truth, not with the laws of holding something to be true, not with the question of how men think, but with the question of how they must think if they are not to miss the truth. (Frege [1897], p. 149)

Since this normativity follows, for Frege, from the factual character of logic,[6] the difference between logical and other laws is only a matter of degree:

Any law that states what is can be conceived as prescribing that one should think in accordance with it, and is therefore in that sense a law of thought. This holds for geometric and physical laws no less than for logical laws. The latter then only deserve the name 'laws of thought' with more right if it should be meant by this that they are the most general laws, which prescribe universally how one should think if one is to think at all. (Frege [1893], p. xv)

Though Frege's understanding of logic as factual in more or less the same sense as physics is not widely shared,[7] his resulting opinion that logical laws are normative—and thus that psychologism is inadequate—enjoys a near-universal consensus to this day.[8]

[6] Anderson [2005], p. 295, and MacFarlane [2002], pp. 35–38, more or less share this straightforward reading of Frege on the normativity of logic, but these days no item of Frege interpretation is uncontroversial. I take no stand on these vexed issues: the position here ascribed to Frege serves to illustrate a particular approach to logical normativity; if it isn't really Frege, I'm happy to drop the ascription.

[7] As we'll see, the Second Philosopher is one of the few to follow Frege in this, though her logical facts aren't features of a 'third realm' of non-spatiotemporal, causal abstracta—e.g., thoughts and truth functions—as Frege takes them to be. Unlike some naturalists, the Second Philosopher doesn't rule out non-physical things from the start, but she will, in her usual commonsensical way, avoid them as unnecessary if she can do so without strain to other features of her theorizing.

[8] e.g., Musgrave ([1972], p. 606) writes, 'Nowadays only a few cranks officially subscribe to [psychologism] about the nature of logic. There is progress in philosophy after all!'

ii. Empiricism

Another acceptably naturalistic idea is that we learn logical truths by experience: we have good reason to believe that this is a tree when we've established that all oaks are trees and this is an oak because we've seen many cases of this form—where 'all As are Bs' and 'this is an A'—and every time it's turned out that 'this a B'. This sort of position is often called Millean, after John Stuart Mill, whose analogous view of simple sums was ridiculed by Frege as 'Mill's piles of pebbles or gingersnaps' (Frege [1884], §27), but in fact it seems Mill regarded deductive inferences as 'merely verbal'.[9]

Ironically, perhaps the clearest statement of simple inductivism comes from Bernard Bolzano, better known for foreshadowing the Fregean view that logic concerns the structure and interrelations of objective, non-spatiotemporal, acausal abstract entities.[10] For all his apparent Platonism,[11] when it comes to epistemology, Bolzano advocates 'the explanation that was given long before Kant':

> It has always been maintained that these sciences [logic, arithmetic, geometry, and pure physics] enjoy such a high degree of certainty only because they have the advantage that their most important doctrines can be easily and variously tested by experience, and have been so tested. ... The only reason why we are so certain that the rules *Barbara*, *celarent*, etc. [i.e., the various forms of syllogism] are valid is because they have been confirmed in thousands of arguments in which we have applied them. (Bolzano [1837], §315.4)

Alas, this straightforward empiricism faces an overwhelming obstacle: the very process of confirming a generalization by examination of its instances presupposes at least some logic,[12] as would the subsequent application of that generalization to the case of the oak tree.

In the Quinean web of belief,[13] logic is also empirically confirmed, but not in the straightforward sense of generalization from instances. Instead, it takes its place in the seamless whole of our best scientific theory:

[9] See Jackson [1941]. [10] See footnote 7.

[11] This term is used for any position that posits such abstracta. I say 'apparent Platonism' because, e.g., George [1972] has argued that Bolzano is no more a Platonist than Carnap.

[12] e.g., to recognize a reported non-black raven as disconfirming the hypothesis that all ravens are black, we need something like the Laws of Universal Instantiation and Non-contradiction.

[13] See I.6. The view of logic in the text might be called 'classic Quine'. For discussion of how Quine's thought evolved, see my [2005a], pp. 442–444.

A self-contained theory which we can check with experience includes, in point of fact, not only its various theoretical hypotheses of so-called natural science but also such portions of logic and mathematics as it makes use of. (Quine [1954], p. 121)

Given some recalcitrant datum, we could choose to revise even the laws of logic:

Revision even of the logical law of the excluded middle has been proposed as a means of simplifying quantum mechanics;[14] and what difference is there in principle between such a shift and the shift whereby Kepler superseded Ptolemy, or Einstein Newton, or Darwin Aristotle? (Quine [1951], p. 43)

However, the principle of minimum mutilation weighs heavily against any revision of logic, which would reverberate through the entire fabric of the web: 'The price is perhaps not quite prohibitive, but the returns had better be good' (Quine [1970b], [1986], p. 86). This, according to Quine, is why we tend to think of logic as necessarily true—because we're so reluctant to revise it.

On this view, the laws of logic occupy a central position in the web, relatively distant from the sensory inputs at the periphery, relatively insulated from the empirical buffeting that takes place there. This puts logic closer to mathematics and the most rarified of physical theorizing, which may seem odd, as basic logic seems particularly obvious, not the sort of thing justified by an elaborate theoretical confirmation.[15] But this concern is swamped by one more serious: Quine's story of how the web of belief itself evolves presupposes some principles of logic—the call for revision in light of recalcitrant experience, presupposes, for example, the Law of Non-contradiction—and it's hard to see how the principles of web maintenance could be revised without crippling the scientific enterprise as Quine's holist understands it. Thus, it seems neither the Quinean holist nor the simple inductivist can carry out his empirical testing of logical laws without presupposing at least some of those laws.

[14] Quine may have in mind an early, now defunct proposal of Reichenbach [1944] (see Putnam [1957], [1965], Gardner [1972] for discussion. The idea is now 'generally admitted to be a non-starter' (Gibbins [1987], p. 127). A more sophisticated, though still ultimately unsuccessful proposal appears in Putnam [1968]. See III.6 for more.

[15] For a similar concern about elementary mathematics, see my [1997], pp. 106–107.

iii. Conventionalism

Suppose for a moment[16] that Carnap intended to answer the very questions Quine posed: what makes logical truths true, and how do we come to know them? This imaginary Carnap might have answered that logic is true by virtue of our conventional choices, that we are free to adopt any logic we like without coming into conflict with any pre-existing facts, that we choose the logic we have for its various pragmatic virtues, that we both create and come to know logical truths through this act of decision. Quine considers this proposal carefully, as a potential solution to the problem of logical knowledge, and it retains its attractions for naturalists today, despite our clearer understanding that it isn't after all what Carnap himself had in mind. Still, against this imaginary target, Quine's critique seems to me conclusive.

The central idea dates back to Lewis Carroll's 'What the tortoise said to Achilles' (Carroll [1895]). Achilles is trying to persuade the tortoise that if

(A) Things that are equal to the same are equal to each other.

and

(B) The two sides of this Triangle are things that are equal to the same.

then

(Z) The two sides of this Triangle are equal to each other.

The tortoise accepts (A) and (B), but not the hypothetical, 'If A and B are true, then Z is true', and so denies that he is forced to accept (Z). Achilles responds by asking him to accept the hypothetical

(C) If A and B are true, then Z must be true.

The tortoise complies—he now accepts A, B, and C—but again doesn't see that he's required to accept Z, as he doesn't accept the hypothetical

(D) If A and B and C are true, then Z must be true.

Achilles politely requests that he do so, and the Tortoise amiably agrees, but once again denies that he's required to accept Z, because he doesn't accept a new hypothetical ... And so it goes.

Quine naturally puts this in more sober terms. We might begin selecting our conventional truths of logic one by one, 'but this picture wanes when

[16] Contrary to the conclusions of I.5.

we reflect that the number of such statements is infinite' (Quine [1936], p. 91). Instead, we need a finite list of general conventions that determine all the rest. But

> Each of these conventions is general, announcing the truth of every one of an infinity of statements conforming to a certain description; derivation of the truth of any specific statement from the general convention thus requires a logical inference ... (Quine [1936], p. 103)

> In a word, the difficulty is that if logic is to proceed *mediately* from conventions, logic is needed for inferring logic from the conventions. (Quine [1936], p. 104)

So we can't come to know logic by means of explicit conventions.[17]

David Lewis subsequently initiated the study of implicit conventions (Lewis [1969]), but these also seem to presuppose logic in the 'common knowledge' his agents share. More recently, Brian Skyrms proposed an evolutionary account of how an understanding of the logical particles might develop without explicit convention or common knowledge (Skyrms [1999]), but this account is no longer purely conventional: in addition to the conventional (arbitrary) signals used, knowledge of logical truth would also depend on the structure of the world, revealed in the form of evolutionary hard knocks.[18] Thus, a viable purely conventional account of logic seems unlikely.

iv. Analyticity

Another account of logical truth and knowledge that might have issued from our imaginary Carnap holds that logic is analytic, not in Kant's sense of conceptual containment, but in the contemporary sense of true-by-virtue-of-meaning:

> The principles of logic ... are true universally simply because ... we cannot abandon them without ... sinning against the rules which govern the use of language, and so making our utterances self-stultifying. (Ayer [1936], p. 77)

[17] See also Quine [1954], p. 115: 'The difficulty is a vicious regress, familiar from Lewis Carroll ... the logical truths, being infinite in number, must be given by general conventions rather than singly; and logic is needed then to begin with ... in order to apply the general conventions to individual cases.' In other words, we would have to know logic already in order to learn it from general conventions.

[18] Skyrms's suggestion for the evolution of 'proto-truth functional signals' (Skyrms [1999], p. 86) seems entirely consistent with the second-philosophical account of the ground of logical truth proposed below.

So logical truth is grounded in human language, and we know it because we know how to speak, for example, English. On this view, though it may be 'conceivable that we should have employed different linguistic conventions' (Ayer [1936], p. 84), this isn't central; the focus is on the natural language we have. Languages as spoken are far from systems of explicit conventions:

> There is a sense in which analytic propositions do give us new knowledge. They call attention to linguistic usages, of which we might otherwise not be conscious, and they reveal unexpected implications in our assertions and beliefs. (Ayer [1936], pp. 79–80)

Logic is implicit in our linguistic practices.

The idea here is that analytic claims, including logical truths, tell us nothing about the world; 'they are entirely devoid of factual content' (Ayer [1936], p. 79), because they reflect only the structure of language. It seems almost churlish to note that the structure of our human languages is itself a feature of the world, that natural languages in fact have the structures they do and not others. What's left unexplained by the proposed analyticity of logical truth is why languages are the way they are, or, as we might put it, why we use these meanings rather than some others. Once this question is raised, we can hardly help wondering if the evolutionary pressures noted by Skyrms haven't been at work here—which reopens the question of the world's contribution to the genesis of our logical beliefs.

I hope this brief survey creates or reinforces existing discomfort with the available options, and thus provides an incentive to search for an alternative. My own interest in finding another way springs partly from the shortcomings of views just rehearsed and partly from an embarrassingly hazy impression from elementary arithmetic: $2 + 2 = 4$ seems to me to report something about the world, and that something seems closely connected to logic (see IV.2.ii). But oddball motivations aside, my goal in what follows is to sketch a second-philosophical account of the ground of logical truth that differs sharply from those sketched above. I begin with a look at Kant's view of logic, with an eye to appropriating its most attractive points.

III.2
Kant on logic

Kant's critical philosophy seems to me to contain the raw materials for an account of logical truth considerably more subtle and intriguing than might appear at first glance.[1] My goal in this section is to sketch out this Kantian take on the ground of logic;[2] in III.3, I suggest that it can be naturalized, and the rest of Part III is a second-philosophical attempt to follow through on this suggestion. Those with no patience for or interest in Kant should skip immediately to III.3, but for the rest of us, I hope this brief Kantian sojourn might provide an illuminating lead-in to the Second Philosopher's proposal.

Any effort to understand Kant's stand on the nature of logic truth is inevitably hampered by the fact that Kant's own interest lies elsewhere. To see this, recall (from I.4) his division of judgments into analytic or synthetic, a priori or a posteriori. Any analytic claim—explicative or conceptual—must be a priori, because it takes no experience to examine our concepts. Ordinary empirical claims amplify our knowledge, go beyond mere explication of concepts, and depend on experience; these are synthetic a posteriori. The final combination—synthetic a priori—would consist of claims about the world that aren't merely conceptual but which are nevertheless known a priori, and it is the project of the *Critique* to explain how such knowledge is possible: in mathematics, in science, in philosophy (B14–24). Kant remarks that a study of the a priori in general 'is, so far as our aim is concerned, too broad in scope' because it includes 'both analytic as well as synthetic *a priori* cognition' (A12/B25). As the only slot

[1] This section descends from my [1999].

[2] I mean here what Kant calls 'general logic' (A50–55/B74–79)—which abstracts from all content of cognition and attends only to form—not 'transcendental logic' (A55–57/B79–82)—which concerns also the 'pure cognition ... by means of which we think objects completely *a priori*' (A57/B81).

plausible for logical knowledge is analytic a priori, we must be prepared to extrapolate, I hope credibly, from Kant's explicit doctrines.

i. Analytic a priori

Let's begin, then, by asking after the status of our sample claims 'if all oaks are trees, and this is an oak, then this must be a tree' and 'if it's either red or green, and it's not red, then it must be green': we want to know what kind of truths these are, how human beings come to know them, and what lies behind the 'must' involved. As we've seen, the obvious first step, from a Kantian perspective, is to classify them as analytic a priori,[3] so let's start with a closer look at this notion.

Commentators have much to debate concerning Kant's notion of analyticity, but for our purposes, I think we can work with the simple 'containment' analysis:

> In all judgments in which the relation of a subject to the predicate is thought ... this relation is possible in two different ways. Either the predicate B belongs to the subject A as something that is (covertly) contained in this concept A; or B lies entirely outside the concept A ... In the first case I call the judgment analytic, in the second synthetic. (A6/B10)

Unlike an intuition, which relates directly to an object, a concept, for Kant, relates to an object indirectly, 'by means of a mark [a feature or features], which can be common to several things' (A320/B377); in an analytic judgment, the predicate is one of the features already present in the concept of the subject (A6–8/B10–12). Kant's example is 'all bodies are extended'; 'extention' is one of the marks of the concept 'body'. 'Heavy' on the other hand, is not a feature of 'body', so 'All bodies are heavy' is synthetic.[4] In case

[3] Kant often classifies logical truth as a priori (e.g., A53–54/B77–78, A131/B170, [1792], p. 693, [1800], pp. 13–14). Oddly, as far as I can tell, the only place he comes near calling logic analytic is B16–17, referring to 'a=a', but 'a=a' is not part of syllogistic logic (the core of logic for Kant) and not a 'principle' (the concern of logic (A53/B77)). It's been suggested to me in discussion that Kant refrains from applying 'analytic' to logic because his logic consists of inferences, not judgments, but the same scruple should apply to 'a priori'. I certainly don't mean to suggest that Kant took logic to be synthetic; I just note the oddity.

[4] In his version of Kant's logic, Jäsche puts this symbolically: 'An example of an *analytic* proposition is, To everything x, to which the concept of body $(a + b)$ belongs, belongs also *extention* (b). An example of a *synthetic* proposition is, To everything x, to which the concept body $(a + b)$ belongs, belongs also attraction (c)' (Kant [1800], §36).

it isn't evident that one is a mark and the other isn't—I come back to this point in a moment—we might substitute Quine's more familiar example: 'Bachelors are unmarried'. If the marks of bachelorhood are being unmarried and being male, then this judgment comes out analytic, as we'd expect.

Now obviously this containment analysis is of no immediate use to us, because our sample logical truths are not of subject/predicate form; in fact, their complex forms would seem, pre-theoretically, to be the source of their logical truth. An equally obvious remedy would be to extend the containment analysis and take a judgment to be analytic if its truth depends only on the concepts involved. As a special case, we might add that an analytic truth is logical if its truth depends only on the logical concepts involved (that is, 'all', 'some', 'is', 'not', 'if/then', 'or', 'and', and so on). On this account 'bachelors are unmarried' and 'nothing is both crimson and not red' would be analytic, but not logical, and our samples would come out both analytic and logical, as they should.

To extract satisfying answers to our questions from an account along these lines, we need to know more about what is and isn't contained in our concepts. This is a complicated topic,[5] but for now, we should note that our access to this information depends for Kant on the type of concept involved. Those types are empirical and a priori, given and constructed, generating four distinct possibilities. Empirical constructed concepts are those we concoct out of elements taken from experience; for example 'bachelor' or 'keyboard'. Such concepts are easily defined: 'In such a case I can always define my concept; for I must know what I wanted to think, since I deliberately made it up' (A729/B757). Empirical constructed concepts, then, are the ones that generate our familiar analytic claims like 'all bachelors are unmarried'.

Empirical given concepts are those derived from experience but not constructed by us, including, for example, ordinary natural kind terms like 'water' and 'gold'. Such concepts, Kant believes, 'cannot be defined at all, but only **explicated**' (A727/B755) because, for example,

> In the concept of **gold** one person might think, besides its weight, color, and ductility, its property of not rusting, while another may know nothing about this. (A728/B756)

[5] See, e.g., Beck [1956a], [1956b].

But this lack of definition doesn't trouble Kant:

> What would be the point of defining such a concept?—since when, e.g., water and its properties are under discussion, one will not stop at what is intended by the word 'water' but rather advance to experiments, and the word, with the few marks that are attached to it, is to constitute only a **designation** and not a concept of the thing; thus the putative definition is nothing other than the determination of the word. (A728/B756)

In other words, we can define and redefine 'water' as we go along, in the interests of clarity, but we're really just fixing our use of the word, not the concept, and we should never let these temporary definitions impede our scientific inquiries ('water can't be H_2O—it's *defined* to be a simple substance!'). Thus it seems Kant would hardly be surprised or bothered by the quasi-Quinean observation that the classification of various judgments about water as analytic or synthetic might shift as science progresses.

A priori concepts, by way of contrast, depend on no experiential input; they are available a priori. According to Kant, some such concepts can be constructed in pure intuition:

> The pure form of sensible intuitions in general is to be encountered in the mind *a priori* ... This pure form of sensibility itself is called **pure intuition**. So if I separate from the representation of a body that which the understanding thinks about it ... as well as that which belongs to sensation [the matter of experience] ... something from this empirical intuition is still left ... These belong to pure intuition, which occurs *a priori*, even without an actual object of the senses or sensation, as a mere form of sensibility in the mind. (A20–1/B34–5)

Our pure intuition of space, for example, provides the arena for constructing geometric concepts, and again, as concepts 'I deliberately made up', these can be defined:

> Mathematics has definitions ... For the object that it thinks it also exhibits *a priori* in intuition, and this can surely contain neither more nor less than the concept, since through the explanation of the concept the object is originally given. (A729–730/B757–758)

A geometric concept can be thought of as a set of instructions for constructing the corresponding object in pure intuition, and as such, it can be defined simply by spelling out that recipe. Furthermore, since the construction process itself is constrained by the forms of intuition, many necessary properties of the resulting object are not contained merely in the definition (the set of instructions) alone, so most mathematical judgments come out synthetic.

The final variety, then, is a priori given concepts, those given to us, rather than constructed by us, and given to us a priori, 'through the nature of the understanding', rather than empirically, 'through experience' (A729/B757). Conspicuous among these a priori given concepts are the categories. Here Kant holds that

> No concept given *a priori* can be defined, e.g., substance, cause, right, equity, etc. ... since the concept ... as it is given, can contain many obscure representations, which we pass by in our analysis though we always use them in application, the exhaustiveness of the analysis of my concept is always doubtful ... Instead of the expression 'definition' I would rather use that of **exposition**, which is always cautious, and which the critic can accept as valid to a certain degree while yet retaining reservations. (A728–729/B756–757)

> Analytical definitions [that is, definitions produced by analysis, as opposed to definitions of constructed concepts] can err in many ways, either by bringing in marks that really do not lie in the concept or by lacking the exhaustiveness that constitutes what is essential in definitions. (A732/B760)

In the case of a priori given concepts, our original puzzle—how can we know what is and isn't contained in our concepts?—is exacerbated by the admission that our analysis of them 'can err in many ways' and is 'always doubtful'.[6]

Though a priori given concepts are determined by the structure of our understanding, they are not under our control (not constructed) and not transparent to us. They precede our efforts to analyze them—'these concepts, though perhaps only still confused, come first' (A730/B758)—and often frustrate those efforts. Their content is stable, unlike the definitional wanderings of 'water', but Kant tells us little about how we come to know what we do about it; what tells us, for example, that 'extended', but not

[6] Of the categories, Kant writes: 'in the presentation of the table of the categories, we spared ourselves the definitions of each of them, on the ground that our aim, which pertains solely to their synthetic use, does not make that necessary, and one must not make oneself responsible for unnecessary undertakings that one can spare oneself. This was no excuse, but a not inconsiderable rule of prudence ... But now it turns out that the ground of this precaution lies even deeper, namely, that we could not define them even if we wanted to' (A241/B299).

'heavy', is contained in the concept 'body'? Beck describes the situation this way:

> The rationalist tradition in which Kant wrote fixed many of the most important concepts by 'implicit' definition and common use or by nominal definitions that had become well established. Thus Kant could confidently decide that a given proposition is analytic without the necessity of referring to a 'rule book' of stipulative definitions. We, in a more conventionalistic period, are usually puzzled by some of his decisions, and can only feel that Kant and his contemporaries were committing what Whitehead called the 'fallacy of the perfect dictionary'—when the dictionary could not, in principle, exist for Kant at all. (Beck [1956b], p. 15)

In any case, it's clear that Kant does regard a priori given concepts as enjoying a fixed content, despite being indefinable, and that this fixed content is what determines which statements involving them are analytic and which are not.

With this taxonomy of concepts in place, we can now ask where the likes of 'all', 'some', 'not', 'if/then', 'and', and 'or' find their place within it. Logic, for Kant 'abstracts ... from all content of cognition, i.e. from any relation of it to the object, and considers only the logical form in the relation of cognitions to one another, i.e., the form of thinking in general' (A55/B79). So in pursuing logic, 'We abstract from all empirical conditions under which our understanding is exercised, e.g., from the influence of the senses, from the play of imagination, the laws of memory, the power of habit, inclination, etc.' (A52–53/B77). Thus logic could hardly involve empirical concepts of either sort. In addition, logic

> has nothing to do with this origin of cognition, but rather considers representations, whether they are originally given *a priori* in ourselves or only empirically, merely in respect of the laws according to which the understanding brings them into relation to one another when it thinks, and therefore it deals only with the form of understanding, which can be given to the representations wherever they may have originated. (A56/B80)

So logic cannot be concerned with the peculiarities of constructions in pure intuition, as in the study of constructed a priori concepts. By elimination, the only place left is the company of given a priori concepts.

The trouble with this conclusion, from the Kantian perspective, is that the famous Table of Categories (A80/B106) is supposed to be a complete listing of all pure concepts of the understanding, of all a priori given concepts relevant to theoretical knowledge,[7] and 'all', 'if/then' and company do not appear there. If the logical notions aren't concepts at all, then it's hard to see how our logical truths could be analytic. But before drawing any such dire consequences, we should take stock of the positive lessons of the discussion so far. If the logical notions were concepts, they would be given a priori, akin to the categories. The categories, as we know, are the pure concepts that the understanding must use in order to have any experience at all, to make any objective judgments. Logic, on the other hand, 'contains the absolutely necessary rules for thinking, without which no use of the understanding takes place' (A52/B76). So, as we suspected, the categories and logic *are* closely related: the concepts necessary for all judgment, and the forms necessary for all judgment.

This suggests that logic is true not by virtue of the content of concepts, but by virtue of the forms of judgment: 'Logic ... concerns itself merely with the form of thinking' (A131/B170). Here Kant is speaking not of thinking in general, but of discursive thinking; in parentheses after the remark just quoted, he adds 'of discursive cognition'. So the suggestion is actually that logic is true by virtue of the necessary structure of discursive cognition. Let's see what can be made of this idea.

ii. The discursive intellect

As we've seen (in I.4), a discursive intellect is one whose judgments relate to objects indirectly, by means of concepts.[8] Kant holds that any cognition requires that its object be directly presented to the mind, so a discursive intellect needs both concepts and singular representations or intuitions.[9] Before asking how logic must be built into the structure of any such intellect, let's take a moment to fill in the bare definition with a look at two contrasting cases.[10]

[7] As opposed to a priori given moral concepts like 'right', 'just', etc.

[8] See A68/B93: 'a cognition through concepts [is] discursive'.

[9] See Allison [2004], p. 13.

[10] For a similar treatment of the opposing options, see Allison [1983], pp. 65–66, 340–1, footnote 2, and [2004], pp. 13–14.

Notice first that an intuitive faculty can be either receptive, like ours, or spontaneous:[11]

A faculty of a **complete spontaneity** [as opposed to *receptivity*] **of intuition** would be a cognitive faculty distinct and completely independent from sensibility. (Kant [1790a], p. 406)

An intellect with such an intuition would not be passively affected; rather, it would produce its intuitions from itself:

As, say, a divine understanding, which would not represent given objects, but through whose representation the objects would themselves at the same time be given, or produced. (B145)

The object itself [would be] created by the representation (as when divine cognitions are conceived as the archetypes of all things). (Kant [1772], p. 71)

Such an intuition is 'original', that is, 'one through which the existence of the object of intuition is itself given' and 'that, so far as we can have insight, can only pertain to the original being' (B72). This formidable cognizer is what Kant calls an intuitive intellect: it would know objects directly, having created them; its knowledge would be unmediated by concepts.

At the other extreme lies another receptive[12] intuition, again one quite different from that of the discursive intellect:

If a representation is only a way in which the subject is affected by the object, then it is easy to see how the representation is in conformity with this object, namely, as an effect in accord with its cause. (Kant [1772], p. 71)

Here the receptive intuition is passively stamped with a copy of its object; no forms of intuition are in play. Such a cognizer might be called an empirical intellect, one that would gain knowledge of objects directly from sensory experience;[13] like the intuitive intellect, its knowledge would be unmediated by concepts. Of course, it's central to the *Critique* to argue that such an intellect is impossible, that knowledge of objects cannot be gained

[11] Spontaneous intuition is sometimes called intellectual, as opposed to sensible (see Allison [2004], p. 13).

[12] Or sensible. See previous footnote.

[13] In this pre-critical letter to Hertz (Kant [1772]), Kant actually calls this the *intellectus ectypi*, in contrast to the *intellectus archetypi* or intuitive intellect. He eventually denies the possibility of such an intellect (see below) and the term *intellectus ectypi* drifts to become a synonym for the discursive intellect (see Kant [1790a], p. 408). For discussion of this letter, see Beck [1965a], Allison [1973], pp. 57–59.

by pure sensation alone, but the letter from which this characterization is taken dates to the pre-critical Kant.[14]

So the receptive intuition of the discursive intellect doesn't create its objects—as would the spontaneous intuition of the intuitive intellect—nor does it reveal its objects as they are in themselves—as would the receptive intuition of the empirical intellect. Instead, it orders the raw matter of experience under a priori forms, tailoring it to accept the conceptual contribution of the spontaneous understanding. In the particular case of human cognition, the forms of intuition are space and time, but another discursive intellect could have others. (It must have some, to keep from being empirical.) The discursive understanding, on this picture, has several jobs to do—it synthesizes intuitions into the representation of a single object; it synthesizes various marks into a single concept—and it does these in the course of its fundamental act, the act of judging, when it synthesizes intuitions and concepts into objective cognition of the world.

But what does all this have to do with logic?[15] Kant's contention is that every synthesis is carried out according to a rule:

> We have ... explained the **understanding** in various ways—through the spontaneity of cognition (in contrast to the receptivity of the sensibility), through a faculty of thinking, or a faculty of concepts, or also of judgments—which explanations, if one looks at them properly, come down to the same thing. Now we can characterize it as the **faculty of rules**. (A126)

For example, the concept 'body' is a rule that unifies many spatiotemporal intuitions into a cognition of a single thing:

> All cognition requires a concept, however imperfect or obscure it may be; but as far as its form is concerned the latter is always something general, and something

[14] In this letter, Kant is groping for his critical turn. Having argued in his Inaugural Dissertation that space and time are a priori forms of intuition, he realized he hadn't answered the obvious question: how do we then know that objects really are spatiotemporal? If our intellect were intuitive, he argues, its representations would have to correspond to its objects, because the intuitive representing would so create the objects. Similarly, if our intellect were empirical, its representations would correspond to its objects, 'as an effect in accord with cause'. The Kantian problem arises only for an intellect that 'is not the cause of the object ... nor is the object the cause of the intellectual representations in the mind' (Kant [1772], p. 71), in other words, for the discursive intellect, i.e., for us. His ultimate answer, of course, is transcendental idealism: objects conform to our forms and categories because these help constitute the world we experience.

[15] As sources for what follows, see Reich [1932], Longuenesse [1993], and Brandt [1995]. See also Allison's discussion, informed by these works, in his [2004], chapter 6.

that serves as a rule. Thus the concept of body serves as the rule for our cognition of outer appearances by means of the unity of the manifold that is thought through it. (A106)

This conceptual synthesis occurs in the process of a judgmental synthesis, which again involves a rule:

Judgments, insofar as they are regarded merely as the condition for the unification of given representations in consciousness, are rules. (Kant [1783], §23)

To make a judgment, then, is to synthesize according to a certain rule, and the resulting judgment might be said to embody that rule.[16] Finally, the judgmental syntheses of the discursive intellect can be achieved in a fixed number of ways, by a fixed set of rules, and these are precisely the twelve logical forms of judgment listed in the famous Table of Judgments (A70/B95).

Widespread criticism of the Table of Judgments dates to the first appearance of the *Critique*.[17] Kant claims that it is 'Systematically generated from a common principle, namely the faculty for judging ...' unlike the classifications of Aristotle, which arise 'rhapsodically from a haphazard search' (A80–1/B106), but Hegel among many others accuses Kant of having done no better.[18] Others suggest that—despite Kant's stated admiration for ancient logicians (Bviii) and despite his claim to agreement with standard logics of his day (A70–71/B96)—his Table actually includes innovations introduced to adapt it to his transcendental purposes.[19] Building on work of Klaus Reich ([1932]), recent commentators—notably Beatrice Longuenesse ([1993]) and Reinhardt Brandt ([1995])—provide more sympathetic reconstructions of the derivation of the Table.[20] Fortunately, for

[16] e.g., as we'll see below, to judge that 'all oaks are trees' is to adopt a particular rule of the universal categorical form.

[17] Allison ([2004], p. 134) places it 'among the more widely rejected parts of the *Critique*'; Beck ([1992], pp. xii–xiii) dubs it 'one of the least esteemed parts of the *Critique*' and remarks that 'several eminent commentators have passed over it in polite silence ... and those who enjoy lording it over him are unanimous in their conviction' of its shortcomings. (Both are describing the Metaphysical Deduction as a whole, but the table itself is also singled out for ridicule.)

[18] According to Hegel, the Table is 'taken merely from observation and so only empirically treated' (as translated in Reich [1932], p. 2).

[19] The charge is that the Table of Judgments had been manipulated in light of the desired outcome for the Table of Categories (see the discussion of the metaphysical deduction in the next subsection). See Beck [1992], p. xv, for discussion and references; also Longuenesse [1993], p. 29, note 19, Brandt [1995], pp. 96–110, Allison [2004], pp. 140–141.

[20] See Allison [2004], chapter 6, for a synthesis.

our purposes, these debates can be passed over: because we're ultimately interested in the ground of contemporary logic, not the syllogistic of Kant's day, such use as we make of Kantian ideas will not be fine-tuned to the details of the Table. Let's simply grant that a discursive intellect, judging of objects mediately through concepts, must employ various logical forms more-or-less present in Kant's Table: subject/predicate, if/then, and so on.

To continue the Kantian story, logical inference rests on these rules present in judgment:

In every syllogism I think first a **rule** (the *major*) through the understanding. Second, I **subsume** a cognition under the condition of the rule (the *minor*) by means of the **power of judgment**. Finally, I determine my cognition through the predicate of the rule (the *conclusio*). (A304/B361)

For example, to form the universal categorical judgment, 'all oaks are trees', I apply the corresponding general rule (form of judgment), that is, I unite the concepts 'oak' and 'tree' so that the first is a condition for the second: the concept 'oak' synthesizes intuitions and establishes reference to the relevant objects and the judgment combines these under the concept 'tree'. The result—the judgment 'all oaks are trees'—is then a particular rule of universal categorical form—the rule that takes 'oak' as a condition for 'tree'—and it serves as our major premise. The particular categorical judgment 'this is an oak' brings a particular oak under the condition of that rule, providing our minor premise. Finally, the conclusion 'this is a tree' results when the rule of the major premise is applied to the particular cognition of the minor premise.

The picture, then, is this: the discursive intellect has a receptive intuition and a spontaneous understanding. That spontaneous understanding judges by synthesizing, which must take place according to one or another of the forms of judgment. The judgments so generated are particular rules—'all oaks are trees'—whose form is given by the corresponding form of judgment—universal categorical. Application of these rules yields logical truths like our samples. So the discursive intellect, by its very nature, is bound by the laws of logic:

Logic has nothing to do with the possibility of knowledge in regard to its content, but merely with its form in so far as it is a *discursive* knowledge. (Kant [1790b], p. 244)

Logical truth is grounded in the structure of the discursive intellect. The logical must is the must of rule-following.

Given that we humans are discursive intellects, we are thus bound to think logically. It's important to realize that this is a norm for how we ought to reason, not a description of how we do reason. If I judge that all oaks are trees, thus committing myself to the rule embodied in that judgment, and subsequently judge that this is an oak, it follows from these two beliefs of mine that this is a tree, but I may be too confused or distracted or inattentive to notice. In practice, we're led astray by a variety of factors, but in pure logic,

> We abstract away from all empirical conditions under which our understanding is exercised, e.g., from the influence of the senses, from the play of imagination, the laws of memory, the power of habit, inclination, etc., hence also from the sources of prejudice. (A52–53/B77)

The study of these factors is proper to applied logic, which

> is directed to the rules of the use of the understanding under the subjective empirical conditions that psychology teaches us. (A53/B77) It deals with attention, its hindrance and consequences, the cause of error, the condition of doubt, of reservation, of conviction, etc. (A54/B79)

Pure logic 'draws nothing from psychology' (A54/B78)—or better, it draws only from transcendental psychology, the pure analysis of discursive cognition.

iii. Analyticity revisited

Now recall that our road to the analyticity of logical truth was blocked toward the end of (i) by two realizations: if logical notions (or, not, if/then, all, some, etc.) are Kantian concepts and logical truths are conceptual truths about them, they must be a priori and given, in other words, they must be pure categories of the understanding; and they don't seem to appear on the Table of Categories (object-with-properties, ground-and-consequent, etc.). We've now seen that for Kant, logical truth arises from the structure of the discursive intellect, in particular, from the various forms discursive judgments can take. Under the circumstances, it's natural to ask if there is any connection between the logical forms of judgment and the pure categories of the understanding, and the answer, of course, is a resounding yes. In fact, the Table of Judgments provides the 'clue' or 'guiding thread' to the Table of Categories (A66–83/B91–116); the passage from the former to the latter is what's called the Metaphysical Deduction (B159). In Kant's scheme, the twelve logical forms of judgment stand in perfect one-to-one correspondence with the twelve pure concepts of the understanding.

Commentators predictably disagree on the source of this correspondence. A common suggestion is that the category is required in order to make an objective judgment of the correlative logical form.[21] For example, to judge that 'Socrates is mortal', a judgment of the logical form subject/predicate, one must conceive the subject, Socrates, as an object with properties, like mortality—which is to employ the category object-with-properties. Or, to take Kant's example of an if/then judgment (A73/B98): if one is to judge that 'if there is perfect justice, then obstinate evil will be punished', then one must conceive the punishment of obstinate evil as depending on perfect justice—which is to employ the category of ground-consequent.

Kant explains why there is such a close connection between form and corresponding category:

> The same function that gives unity to different representations **in a judgment** also gives unity to the mere synthesis of different representations **in an intuition** ... the same understanding ... indeed by means of the very same actions through which it brings the logical form of a judgment into concepts ... also brings a transcendental content into its representations by means of ... pure concepts of the understanding. (A79/B104–105)

When the understanding synthesizes 'Socrates' and 'mortality' into the judgment 'Socrates is mortal', it performs exactly the same function as when it synthesizes various intuitions into a representation of Socrates as an object with properties; when it synthesizes 'perfect justice' and 'punishment of obstinate evil' into an if/then judgment, it performs the same function as when it synthesizes intuitions falling under the former with intuitions falling under the latter as ground and consequent. The difference is that in one case that function is applied to synthesize concepts into a judgment, and in the other, to unite intuitions into a representation of an object:

[21] For discussion of the following two examples, see Allison [2004], pp. 148–150.

> **The categories**... are concepts of an object in general, by means of which its intuition is regarded as **determined** with regard to one of the **logical functions**. (B128) The **categories** are nothing other than these very functions for judging, insofar as the manifold of a given intuition is determined with regard to them. (B143)

This is why the Table of Judgments provides a 'clue' to the Table of Categories:

> In such a way there arise exactly as many pure concepts of the understanding, which apply to objects of intuition in general *a priori*, as there were logical functions of all possible judgments. (A79/B105)

In short, a particular logical form is the application of a particular function of the understanding in synthesizing judgments; the corresponding category is the application of that same function in synthesizing intuitions.[22]

There is obviously much to wonder at in this line of thought, but whatever we make of the Metaphysical Deduction, what matters for our purposes is simply the close association of a logical form and its corresponding category: subject/predicate and object-with-properties; if/then with ground-consequent; and so on. Though the forms aren't concepts, strictly speaking, they are rules for synthesis, indeed the same rules, in some sense, as the corresponding pure concepts, just applied in a different context. This near-concepthood of the forms leads Henry Allison to such locutions as 'the logical function or "judgmental concept" operative in the act of making judgments' (Allison [2004], p. 149). Previously, we adjusted the containment account of analyticity to allow for statements of other than subject/predicate form; it now seems at least as reasonable to extend the term to include judgments whose truth depends only on their logical form.[23] Granting this friendly amendment, our sample logical truths count as analytic after all.

What's important here is not so much the 'analytic' label, but a full understanding of the nature of the analyticity in question. In the taxonomy of concepts sketched in (i) above, each type generates a different brand of analyticity. We've seen that both varieties of constructed concepts allow

[22] See Allison [2004], pp. 152–156.

[23] Kyle Stanford first recommended this extension (see my [1999], footnote 32), and it seems even more appealing on the view sketched here of the close connection between the forms and the categories.

explicit definitions, and hence the sort of trivial definitional truths we tend to think of as analytic in the wake of Quine and the positivists: 'bachelors are unmarried', 'triangles have three sides'. In the case of empirical given concepts, like 'water', we're interested in the stuff itself, not the concept; we give tentative definitions and let science guide us on how to modify them in light of evidence. Here again, we have analytic truths, but they're shifting and insignificant, as well as trivially definitional. But the case of a priori given concepts—like the categories, and presumably, by extension, the logical forms of judgment—is dramatically different. Here we can't give definitions, we only grope toward 'expositions', and we can never be sure if our analyses are correct or complete.

Consider then one of our logical truths: if it's either red or green, and its not red, then it must be green. This is true, we're assuming, by virtue of the logical forms of judgment it involves—if/then, either/or, not, and—but though its truth is 'contained in the concepts', this is not the transparent containment with which 'unmarried' is contained in 'bachelor' (defined as 'unmarried male') or 'three-sided' is contained in 'triangle' (defined as 'three-sided planar figure'). Rather we discern it from our analysis of the relevant concepts (or 'judgmental concepts'), every element of which is a hard-won discovery about a content fixed prior to and independent of our inquiry. Though the relevant logical concepts are forms of our discursive intellect, they aren't subjective in the usual sense, and we can go wrong in our investigation of their features. In some ways, these non-trivial analytic truths more closely resemble synthetic judgments about empirical given concepts like 'water' than definitional truths about constructed concepts, empirical or a priori. Granted, synthetic truths rely on intuition[24] while these non-trivial analytic truths require only the tools of philosophical analysis, but both are substantial truths about things not of our manufacture, in sharp contrast to the empty definitional character of other analytic truths. So the first moral of this story is that logical truth, on this reading of Kant, is contentful and even elusive, despite being analytic a priori.

[24] Another route to the thought that logical truths should count as analytic would begin by defining 'synthetic' as depending on intuition, then considering anything else to be analytic. Given the overall project of the *Critique*, 'synthetic' would seem to be the more fundamental notion in any case. Kant in fact takes this approach in his reply to critics (Kant [1790b], p. 241): synthetic judgments are those '*that . . . are only possible under the condition that an intuition underlies the concept of their subject*, which, if the judgments are empirical, is empirical, and if they are synthetic judgments a priori, is a pure intuition a priori' (emphasis in the original).

To see the second moral, recall the teaching of transcendental idealism: because the world of experience is partly constituted by contributions from us, we can know a priori that the world we experience will conform to those contributions, but this is a priori knowledge of the world as experienced, not as it is in itself. In particular, since space and time are the forms of intuition, we can know a priori that everything we experience will be spatiotemporal, but

> If we remove our own subject or even only the subjective constitution of the senses in general, then all constitution, all relations of objects in space and time, indeed space and time themselves would disappear. (A42/B59)

So the synthetic a priori truths of mathematics are true only of the experience of a discursive intellect with our forms of intuition, not of the world considered in itself.

The contributions of the understanding also generate synthetic a priori truths about the world, but their status is more complex. We've seen that for Kant every discursive intellect, simply by virtue of its discursivity, must synthesize its judgments in conformity with the logical forms of judgment and their corresponding categories. It's tempting to conclude that the usual Kantian synthetic a priori claims of science—such as 'every effect has a cause'—are true of the experience of any discursive intellect,[25] but this would be wrong. To see why, notice that the pure concept involved here (ground-consequent) and its corresponding logical form (if/then) are both available to any discursive intellect whatsoever; they bear no relation to the spatiotemporal inputs we humans bring to them. To bridge the gap between the pure concepts of the understanding and sensible intuitions, those concepts must be 'schematized', that is, equipped with means of processing actual representations for classification:

> It is clear that there must be a third thing, which must stand in homogeneity with the category on the one hand and the appearance on the other, and makes possible the application of the former to the latter. This mediating representation ... is the **transcendental schema**. (A138/B177)

[25] Ordinarily, 'experience' in the Kantian context is generated when a logical form of judgment synthesizes sensible intuitions (ordered by space and time) and concepts, where the latter include not pure, but schematized categories (that is, pure categories prepared to classify spatiotemporal inputs). My usage here is broader, meant to include the analogous cognition of any discursive intellect, even one with non-spatiotemporal forms of intuition.

Discreetly averting our eyes from the question of how this is accomplished,[26] we note that schematizing the pure category ground-and-consequent yields the schematized category cause and effect.[27] So 'every event has a cause' depends not on the pure category, but on its schematized version, and hence, it depends essentially on our particular forms of intuition, space and time (as they partly determine the features of the required schema). This means that Kant's synthetic a priori principles of science arise from both the categories *and* our forms of intuition, and as such, are true only of the world as experienced by discursive intellects with spatiotemporal intuition.

So where does this leave logical truths? Though analytic, they aren't trivial truths arising from explicit definitions. Though a priori, they aren't synthetic, they don't depend on intuition, but they nevertheless do depend on the logical forms of judgment and the pure concepts of the understanding or unschematized categories. In other words, they depend on the structure of any discursive intellect regardless of its forms of intuition; they are true of the world as it is experienced[28] by any discursive knower. To paraphrase the passage quoted a moment ago:

> If we remove the discursive subject or even only the subjective constitution of the discursive understanding in general, then all constitution, all logical relations of objects, indeed logic itself would disappear.

This means that logic, for Kant, is transcendentally ideal, in the sense that it reflects features of the world as it is constituted by our cognitive machinery, rather than features of the world as it is in itself. But it is not transcendentally ideal in as strong a sense as mathematics and the law of cause and effect, because it depends only on the discursive features of our understanding, not on our particular forms of intuition. We might say logic is weakly transcendentally ideal.

Returning, finally, to the lessons of the benign reading (from I.4), what we say about the ground of logical truth depends on our level of inquiry. Just as the world is objectively spatiotemporal and causally ordered, speaking empirically, from the point of view of ordinary scientific inquiry, it is also

[26] See A137–147/B176–187. For discussion, see Allison [2004], chapter 8.

[27] Notice that Kant's example of a ground-consequent judgment—if there is perfect justice, then obstinate evil will be punished—is not causal. Also, the pure category object-with-properties schematizes to spatiotemporal-object-with-properties. In each case, there are three elements: the logical form of judgment, the pure category, and the schematized category.

[28] See footnote 25.

logically structured—it consists, for example, of objects with properties, standing in ground-consequent relations—and this is why logical laws are true. Speaking transcendentally, on the other hand, the logical structure of the world, much like the spatiotemporal and causal structure, is ideal, produced by the necessities of our discursive cognition.[29] Let's now see what the Second Philosopher might make of this intriguing position.

[29] The difference, again, is that the spatiotemporal and causal structures depend on our particular forms of intuition, in addition to the discursive structure of the understanding.

III.3

Undoing the Copernican revolution[1]

There's considerable appeal to the suggestion that logic depends on very general structural features of the world, and to the quite different idea that logic is embodied in our most primitive forms of conceptualization. Kant's combination of transcendental idealism and empirical realism accomplishes the neat trick of giving us both these at once: transcendentally, logic is dictated by the forms of judgment and pure concepts of the discursive intellect; empirically, logic describes the underlying structure of the world. But as we've seen (in I.4), the Second Philosopher makes no sense of transcendental analysis. The goal of the remainder of Part III is to forge a second-philosophical version of the Kantian position that preserves its merits while bypassing the transcendental. In place of Kant's two-level view, the Second Philosopher seeks one unified scientific account.

Obviously it won't do simply to jettison one of Kant's levels and treat the account proper to the other level as a straightforwardly scientific claim. This approach, applied with preference to the transcendental level of Kant's analysis, would give us an empirical psychology that goes something like this: any cognizer who uses our forms of judgment—subject/predicate, if/then—and the corresponding pure categories—objects-with-properties, ground-consequent—will be bound by the laws of logic. But this is just a form of psychologism: logic is grounded in the structure of human

[1] Kant compares the move to transcendental idealism to the Copernican revolution: 'Up to now it has been assumed that all our cognition must conform to the objects; but all attempts to find out something about them *a priori*... have, on this presupposition, come to nothing. Hence let us once try whether we do not get farther... by assuming that the objects must conform to our cognition... This would be just like the first thoughts of Copernicus, who, when he did not make good progress in the explanation of the celestial motions if he assumed that the entire celestial host revolves around the observer, tried to see if he might not have greater success if he made the observer revolve and left the stars at rest' (Bxvi).

cognition.[2] On the other hand, if we hew exclusively to Kant's empirical level, we get the robustly realistic idea that logic is true of the world because the world consists of objects-with-properties standing in ground-consequent dependencies, but this line of thought alone leaves us with no explanation of how we know this. Indeed, it was the hope of explaining how we could have knowledge in a priori disciplines that inspired Kant's critical philosophy in the first place! So it's clear that if the Second Philosopher hopes to retain the strength of the Kantian picture, she must find a way to combine the two levels into a single scientific theory.

As a first approximation, then, the Second Philosopher hopes to develop an account of logical truth with two components: (1) logic is true of the world because of its underlying structural features, and (2) human beings believe logical truths because their most primitive cognitive mechanisms allow them to detect and represent[3] the aforementioned features of the world. As soon as these two ideas are laid down, it's natural to hope that they can be further reinforced by a connection between them: (3) human begins are so configured cognitively because they live in a world that is so structured physically.

[2] As Anderson [2005] points out, some nineteenth-century neo-Kantians in fact took this route.

[3] The notion of mental representation or content raises questions analogous to those surrounding truth and reference, discussed at some length in Part II (see, e.g., Sterelny [1990], chapter 6). Recall from II.4 the correlational story of how Priestley's use of 'dephlogisticated air' functioned in his successful and unsuccessful interactions with the world, especially the fact that this explanation proceeded effectively without any ruling on whether one or another of his utterances did or didn't refer to oxygen. That judgment turned out to hinge on how he was subsequently interpreted, which might properly vary with the context or intentions of the interpreter. With simpler words like 'table' there presumably would be little room for such variation; any contemporary American interpreter in any context would quickly determine that his 'table' corresponds to our word 'table' and thus that he was referring to tables. Now suppose we ask instead about the role of a given brain state in Priestley's interactions with the world. We might be able to connect the state with a linguistic item and repeat the same analysis (cf. Field [1986a], pp. 77–78), but suppose instead that we ask about the brain state of a pre-linguistic infant (cf. III.5). Presumably, if we knew more about brain function than we do, we could tell a similar correlational story, a story that would explain effective and ineffective activations of that state without any ruling on whether or not it 'represents' this or that: e.g., a certain state might be causally linked to cats, which allows the infant to store information and react appropriately to cats, but it might also be activated when she's approached by an unusual cat-like wild creature, leading the child to react incautiously and dangerously. All this could be explained without settling whether the relevant brain state 'represents' both cats and these wild creatures (and the child mistakenly generalized from her experience with cats to expectations for the entire class) or 'represents' only cats (and the child mistakenly applied it to the wild creature). Perhaps, as in the case of reference, this determination depends on how we decide to map the child's brain states to our own, a decision open to a range of options in cases like this one, though also often as straightforward as the case of Priestley's 'table'. I won't try to fill in the details here, but talk of 'representation' in what follows should be understood in this deflationary/correlational spirit.

To begin the long process of filling in the details, let's turn first to identifying the relevant structural features that the world purportedly has and that our cognitive mechanisms purportedly reflect. The thought is to follow Kant here, adapting his forms of judgment and pure categories to play this role. So far, I've focused on the forms subject/predicate and if/then, and the corresponding pure categories object-with-properties and ground-consequent, though there is much more to Kant's tables than these. But, as contemporary students of the subject, we can happily help ourselves to post-Kantian developments as well, and most particularly, to Frege's astute improvements on traditional analyses of logical form. The idea, then, is to use Frege's work to amend and update the Kantian Tables.[4]

Frege's most fundamental insight was the realization that the subject/predicate form is superficial (see Frege [1879], §§3, 9). For Frege, the judgment expressed by 'Socrates taught Plato' breaks down into a function—x taught y—applied to two arguments—Socrates and Plato. This very judgment could be phrased as 'Socrates taught Plato', making 'Socrates' the grammatical subject, or as 'Plato was taught by Socrates', making 'Plato' the grammatical subject, but these variations are irrelevant to its logical relations. It can hardly be doubted that this innovation produces a better understanding of logical form,[5] so let's begin by replacing the Kantian form of judgment subject/predicate with the Fregean argument/function, and the corresponding pure category object-with-properties with objects-in-relations.

Something like the Kantian if/then is present in Frege, who once remarked that the conditional 'has a close affinity with the important relation of ground and consequent' (Frege [1880/1], p. 37). But Frege's function/argument analysis brings with it a second great advance in the understanding of generality, namely, his theory of quantification, which clarified traditional syllogistic statements and revealed the logical structure of iterated quantifications—'everyone has a parent'—for the first time. So we should also recognize as a form of judgment the generalization of an argument of a function, with the corresponding pure category, universality.[6]

[4] I don't suggest that Frege himself either was or wasn't a Kantian.

[5] Frege ([1879], p. 7) wrote, 'I believe that the replacement of the concepts *subject* and *predicate* by *argument* and *function*, respectively, will stand the test of time'. In this, he surely was right.

[6] The form is the universal generalization of an argument place in a function; the category is the universality of the property that an object has if it stands in the relation(s) specified by that function.

With these forms and categories, the resulting second-philosophical account holds that the world is structured in the ways they specify, that human cognition is also so structured, and that humans are this way because the world is this way. We'll examine the evidence for these claims in the next two sections, but first let's pause to ask what a world so structured would be like, and what logical truths could be said to hold there. In other words, let's consider for a moment, the logic structure of a KF (for 'Kant–Frege') world.[7]

Speaking in complete abstraction, then, a KF-world would consist of individual objects, *a*, *b*, *c*, ..., which enjoy various properties *P*, *Q*, ..., and stand in various relations with various numbers of arguments, *R*, *S*, ... It's not assumed that each object enjoys every property or stands in every relation to every other property, so this picture includes the possibility of an object failing to enjoy a given property or failing to stand in various relations. Also implicit is the idea that an object might stand in relation to more than one other object, or be related either this way or that, and so on. In other words, among the ways a KF-world can be, some situations stand as conjunctions, disjunctions, and negations of others.[8]

In addition, a property might hold universally in a given KF-world—for example, it may be that everything either fails to have *P* or bears *S* to *c*[9]—and existentials, as usual, come along with the universals.[10] Finally, some of the various states of our KF-world may be interconnected: for example, it might be that every *a* for which *Rac* is also an *a* for which *Pa*. This might be, so to speak, an accidental connection, but if *a*'s bearing *R* to *c* is the ground of its having *P*, then this is a ground-consequent dependency. In sum, then, our KF-world consists of a domain of objects that bear properties and stand in relations, perhaps some universal properties, plus compounds of these involving conjunctions, disjunctions

[7] The terminology acknowledges the genesis of this notion, though for a range of reasons, neither Kant nor Frege would embrace it.

[8] A reader who doubts that these compounds must be part of what it is to realize a category of objects-in-relations should take the above description as stipulating what the Second Philosopher takes her perhaps more generous category to include.

[9] Note that the formulation in this example captures the traditional 'All *P* are *Q*' (where *Q* is the property of bearing *R* to *c*). The more familiar version—(for every *x*)(if *Px*, then *Qx*)—is too strong when if/then is understood as involving a ground-consequent dependency.

[10] As usual, (there is an *x*)(... *x* ...) is just not-(for every *x*)not-(... *x* ...).

and negations, and some interconnections between these situations are robust ground-consequent dependencies.

Let's press a bit further into the KF-structure. A given object *a* might enjoy the property *P* and it might fail to enjoy that property, but nothing in this picture precludes the possibility of indeterminate cases.[11] So for any property *P*, the domain of a KF world will divide into those objects that have *P*, those that don't have *P*, and those for which *P* is indeterminate; and perhaps even these boundaries between these groupings are somewhat fuzzy.[12] The same goes for objects standing or not standing in the various relations of the KF-world. For compound states:[13] not-(...) obtains if (...) fails; fails if (...) obtains; and is otherwise indeterminate. ((...) and (__)) obtains if (...) and (__) do; fails if either (...) or (__) fails; and is otherwise indeterminate. ((...) or (__)) obtains if (...) or (__) does; fails if both (...) and (__) fail; and is otherwise indeterminate. A universal generalizes conjunction, so (every *x*)(... *x* ...) holds if for every *a* in the domain of our KF-world, (... *a* ...) obtains; fails if there is an *a* in the domain such that (... *a* ...) fails; and is otherwise indeterminate.[14] Dependencies are more troublesome, because their status isn't always settled by those of their parts. In fact, all that seems fixed is that if (...) and (if (...), then (__)) hold, then (__) must also—under these conditions, (__) cannot fail or be indeterminate.[15]

Given this notion of an abstract KF-world, the question of logical truth becomes: are there states—like (*Pa* or not-*Pa*) or not-(*Rbc* and

[11] At this point, the specification of what counts as an abstract KF-structure is constrained only by the hope of retaining what's useful from Kant, as modified in light of Frege, where 'usefulness' is determined by the persuasiveness of the cases required in III.4 and III.5. My reasons for thinking it wise to allow for indeterminacy unfold there.

[12] All I mean by 'property', here and elsewhere, is a classification of this form. I don't place any restriction on how that classification is generated (e.g., by a natural kind, a single feature, a family resemblance, etc.), and though I sometimes speak of things enjoying a property as 'similar', none of these theoretical ideas should be read into this locution.

[13] When I say, e.g., '(*Pa* and *Qa*) obtains', I only mean that *a* has two properties, *P* and *Q*. 'Obtains', 'fails', or for that matter, 'state' or 'situation' or 'fact', aren't intended to carry any metaphysical weight; the metaphysics consists of objects-in-relations, nothing more.

[14] These are modeled on the strong Kleene or Lukasiewicz connectives. (These two differ only on the conditional, which is not at issue here.) These connectives are considered standard when gaps are caused by (so to speak) metaphysically indeterminate cases, because, e.g., (...) can be indeterminate without necessarily undermining the determinacy of ((...) or (__)). By contrast, when gaps arise from failures of linguistic meaning, the weak Kleene tables are appropriate, because, e.g., if (...) is meaningless, so is ((...) or (__)).

[15] I won't attempt any deeper analysis of the ground/consequent relation, for reasons that should become clear below (see footnote 18).

not-(*Rbc*))—that must obtain in any KF-world[16] simply because of their form? If so, they would hold regardless of the particular details of the given KF-world, strictly by virtue of its form, and this would be a robust logical fact about that world. But, alas, apart (perhaps) from a few trivialities about identity—like $a = a$ and (for all x)($x = x$)[17]—there are no such purely formal facts. No compound state will hold simply on account of its form, because there are so many indeterminacies:[18] for example, if (...) is indeterminate, both ((...) or not-(...)) and not-((...) and not-(...)) are indeterminate.

But we might focus instead on valid logical connections rather than logical truths, that is, we might ask: are there situations such that, if one obtains in a KF-world, the other must also? For example, if (*Pa* and *Qa*) obtains in a given KF-world, must *Pa* also, regardless of what else is going on, simply as a matter of the respective forms of the two states? If so, this would be a robust logical fact about KF-worlds, and this time, the answer is a straightforward yes: in any KF-world, if a state of the form (*Pa* and *Qa*) obtains, then *Pa* and *Qa* do, too, so *Pa* must obtain—despite the fact that there is no simple logical fact of the form (if (*Pa* and *Qa*), then *Pa*).

The reason for this phenomenon is also straightforward: the assumption that the first state obtains is often enough to force out the sorts of indeterminacies that would compromise the obtaining of the second.[19] In this way, the logical structure of a KF-world underwrites validities corresponding to most classical inferences involving 'not', 'and', 'or', 'all' and 'exists': for example, not-((...) and (__)) implies[20] (not-(...) or not-(__)), and so on through the various DeMorgan laws; ((...) and ((__)or (//))) implies (((...) and (__)) or ((...) and (//))), plus the rest of the distributive laws; (not-not-(...)) implies (...); (for every x)(... x ...) implies (... a ...), and so on. But, obviously, ((...) or not-(...)) does not guarantee that ((__) or not-(__)) despite the validity of the corresponding

[16] Or rather, in any KF-world with objects a, b, c, property P, and relation R.

[17] See III.4, footnote 16.

[18] The fact that no classical tautology is always true in the presence of truth value gaps (see, e.g., Parsons [2000], p. 25, or Priest [2001], p. 121) transfers directly to our context for cases involving 'not', 'and', 'or', and 'every'. Our minimally constrained 'if/then' will also fail to generate such logical facts, but perhaps some deeper understandings of the ground/consequent relation would: e.g., (...) might be a ground for itself and (if (...), then (...)) thus hold come what may. I won't pursue this possibility because it seems to me pointless to try to be completely precise about the content of the rudimentary logic described in this section. See III.7.

[19] See Parsons [2000], p. 25.

[20] That is, if the first obtains, so does the second.

classical inference. And, finally, our ground-consequent if/then doesn't satisfy counterparts to the usual truth-functional equivalences: for example, (if (...), then (__)) doesn't imply nor is it implied by (not-(...) or (__)).[21] modus ponens is retained, but modus tollens is not.[22] Such familiar rules as reductio ad absurdum are also compromised in the presence of indeterminacies.[23] The result, then, is an array of validities corresponding to many classical inferences involving 'not', 'and', 'or', 'all', and 'exists', plus a few scattered bits involving conditionals. I won't try to make this any more precise, for reasons that will emerge later.[24] My purpose here is simply to suggest that some such rudimentary logic holds in any KF-world.

Perhaps the general idea of a KF-world and its corresponding logic can be illuminated by contrasting it with a different sort of abstract world. Recall Kant's idea of an intuitive intellect, an understanding that 'would not represent given objects, but through whose representations the objects would themselves at the same time be given, or produced' (B145). Consider what a world created by such an intellect would be like. Assuming the Creator[25] works in a simple time sequence, an object *a* has a property *P* at some point if the Creator so imagines it; a stands in relation *R* to *b* if the Creator so imagines it; and so on for all atomic cases. Now, given an object *c* and a property Q, under what conditions could it be said that *c* fails to enjoy Q? It won't be enough that the Creator hasn't so far imagined that *c* has Q—he might well go on to do so later—so *c* fails to have Q at some point only if the Creator considers the possibility of imagining *c* with Q at some future point and sees, in his imagination, that he would then have to imagine something he's dead set against imagining.[26] In that case, it is guaranteed he will never, in the future, imagine *c* with Q, so *c* fails to have Q.

Furthermore, to imagine ((...) and (__)), the Creator must imagine both (...) and (__); to imagine ((...) or (__)) requires him to imagine one or the other. To imagine (if (...), then (__)), he must consider what it would be like to imagine (...) in the future, and to see, in his imagination,

[21] One familiar connection does hold: ((...) and not-(__)) implies not-(if (...), then (__)).

[22] If both (if *Pa*, then *Qa*) and Pa, then *Qa* can't fail or be indeterminate; it must obtain. On the other hand, if both (if *Pa*, then *Qa*) and not-*Qa*, all that follows is that *Pa* must either fail or be indeterminate.

[23] See Parsons [2000], p. 25. Cf. footnote 28.

[24] See III.7 and IV.2, footnote 35.

[25] Kant characterizes the intuitive intellect as 'a divine understanding' (B145).

[26] One thing he's dead set against imagining is a contradiction.

that he would then have to imagine (__), as well.[27] To imagine (there is an x)(... x ...) is to imagine an object a such that (... a ...). To imagine (for all x)(... x ...), the Creator must see in his imagination that every thing, including anything he might newly imagine in the future, has or will have P.

What's been generated by this process—let's call it a 'Creator-world'—is a dynamic world, one that changes with time, but only in extremely orderly ways: once an object has been created, it is never destroyed; once (...) is imagined, it remains stable forever. For a situation to obtain in such a world is for it to obtain 'eternally', so to speak, that is, at every point in the time sequence; for one situation to imply another is for the second to obtain at every stage in every Creator-world at which the first obtains.

So, what is the logic of Creator-worlds? Indeterminacies are to be expected, as in a KF-world: for example, if the Creator hasn't imagined a to have P, but also hasn't seen that he will never do so in the future, then (*Pa* or not-*Pa*) does not obtain. More concretely, consider a very simple Creator-world with only two stages 1 and 2; suppose the Creator has imagined a at stage 1, but doesn't settle that a has P until stage 2. Then at stage 1, neither *Pa* nor not-*Pa* obtains, so (*Pa* or not-*Pa*) doesn't obtain, either.

But despite this similarity, the logic of Creator-worlds is not rudimentary logic: the same simple Creator-world shows that not-(not-*Pa*) doesn't imply *Pa*, as it does in KF-worlds. This opens the door for another disagreement, over not-not-(*Pa* or not-*Pa*), which fails along with (*Pa* or not-*Pa*) in KF-worlds. If the Creator considers what it would be like to imagine not-(*Pa* or not-*Pa*), he quickly realizes that he would then be forced to imagine (not-*Pa* and not-not-*Pa*),[28] which he cannot do. So not-not-(*Pa* or not-*Pa*) obtains no matter what in Creator-worlds.[29] Both Creator-worlds

[27] So, in general, the process of imagining not-(...) is the same as imagining (if (...), then A), where A is something impossible. See the special case of imagining the failure of *Qa* in the previous paragraph.

[28] He's considering what it would be like to imagine not-(*Pa* or not-*Pa*), and he then also considers *Pa*. If *Pa*, then (*Pa* or not-*Pa*), so he has attempted the impossible. This shows that if he imagines not-(*Pa* or not-*Pa*), he must also imagine not-*Pa*. Similarly, he must also imagine not-not-*Pa*, and thus, their conjunction.

[29] Why doesn't the Creator's argument for not-not-(*Pa* or not-*Pa*) work in rudimentary logic? Well, (not-*Pa* and not-not-*Pa*) is implied by not-(*Pa* or not-*Pa*) in rudimentary logic, but, as mentioned above, reductio ad absurdum doesn't work as usual: it follows that not-(*Pa* or not-*Pa*) doesn't obtain, but not that not-not-(*Pa* or not-*Pa*) must.

and KF-worlds support validities corresponding to the distributive laws, but they disagree on some DeMorgan laws.[30]

Perhaps the difference between KF-worlds and Creator-worlds is most starkly revealed by considering an existential case. In rudimentary logic, not-(for every *x*)*Px* implies (there is an *x*)not-*Px*. In contrast, the Creator's consideration of what it would be like to imagine (for every *x*)*Px* might show this to be something he will never do—so that not-(for all *x*)*Px* obtains at this point—without the Creator's having imagined a particular *a* with not-*P*—which would be needed for (there is an *x*)not-*Px*. More concretely, consider another Creator-world with only two stages: at the first stage, *a* has been imagined, but not with *P*; at the second stage, a new object, *b*, is imagined, and *a* is imagined to have *P*. In this world, not-(for all *x*)*Px* obtains at the first stage, because (for all *x*)*Px* never holds, but (there is an *x*)not-*Px* does not.

Of course, all this should sound familiar: 'the Creator' here is a caricature of the Creating Subject[31] of Brouwerian intuitionism and the course of his imaginings is modeled on Kripke's intuitionistic semantics, so the logic of a Creator-world is intuitionistic.[32] This shows that the rudimentary logic of a KF-world does depend on the particular structure of such worlds; if their structure were different, like that of a Creator-world, their logic would be different, too.

To sum up, then, our first approximation to a second-philosophical version of Kant's position is this: (1) rudimentary logic is true of the world[33] because it is a KF-world, (2) human beings believe the simple[34] truths of rudimentary logic because their most primitive cognitive mechanisms allow them to detect and represent the KF-structure of the world, and (3) the primitive cognitive mechanisms of human beings are this way because they live in a KF-world. Let's now see how well this proposal squares with reality.

[30] e.g., in a Creator-world, not-((...) and (__)) doesn't imply (not-(...) or not-(__)).

[31] This character is often called 'the creative subject', but van Dalen ([1999], p. 394) observes that Brouwer's own version, 'the creating subject', is more apt.

[32] See van Dalen [2001], pp. 237–239, or [2002], pp. 25–37, for discussion of the connection of the Creating Subject with Kripke models; the second reference gives completeness proofs.

[33] Let me note here some terminological looseness I'm going to allow myself: I use 'rudimentary logic' indifferently to cover the worldly relations between situations of various forms and the correlated relations between mental and linguistic representations, counting on the context to differentiate in cases where it matters; I use the claim, e.g., that disjunctive syllogism is 'true' then to cover the fact that, e.g., if a situation of the form '... or __' and a situation of the form 'not-...' both obtain, then '__' must obtain, as well as the more usual facts about representations and linguistic items.

[34] Complex validities of rudimentary logic might be beyond our powers of recognition.

III.4

The logical structure of the world

Is our world a KF-world? In what follows, I take up each of the defining features in turn: the existence of individual objects, their enjoyment of properties and relations, the presence of dependencies, and finally, the question of indeterminacy.

i. Objects

Common sense clearly endorses the idea that the world contains many medium-sized physical objects. Such things cohere, have boundaries, and move continuously as units; examples range from apples, chairs, and people, to boulders, books, and baseballs. When the Second Philosopher examines these beliefs more closely, she finds both confirmation and explication. Such objects are indeed distinct from their surroundings: they are composed of intricately arranged atoms dotted throughout largely empty space, and those atoms and arrangements differ starkly from the atoms and arrangements in the space nearby. Their cohesiveness comes from the bonds between their atoms; their solidity comes from the electromagnetic fields they generate; they move according to certain principles of motion, and so on. On further, more specialized investigation, she uncovers further such objects: planets, blood cells, spider mites. She concludes that ordinary physical objects are real structures in the world.[1]

[1] If this seems improbably straightforward, it may help to recall that the Second Philosopher isn't troubled by traditional threats from radical skepticism (see I.2, I.3) or transcendental idealism (see I.4). For skepticism about unobservables, see IV.1.

Notice that nothing the Second Philosopher discovers in this way will rule out the possibility that the world is also organized in ways that cross-cut its structuring into objects. Indeed, she finds it reasonable to suppose that the world, in its inherent complexity, has many features that we fail to notice, for lack of interest or ability. But this is not to embrace the entirely unsupported claim that the world somehow has every structure imaginable, or that it has no structure at all (see I.7). The Second Philosopher rests with her conclusion that the world has, at least, a distinctive structure of individual objects.

Here some will object on grounds of circularity—we use our beliefs about bounded, cohesive, spatiotemporally continuous objects to confirm that the world is populated with such objects—but this is just a special case of the circularity objection considered in Part I:[2] you're using science to justify science! We noted there that this objection often betrays the objector's underlying wish for a 'higher', extra-scientific justification of science, so that science can be preferred to pseudo-science (like astrology or creationism) on some neutral grounds. But the Second Philosopher finds no footing outside her own methods; she simply argues that astrology and creationism are wrong, on her own terms, unimpressed by the reply that they could run a parallel argument against her. She knows what's wrong with that argument, too!

A less sweeping concern about circularity is also possible, namely, the fear that such self-certification is automatic, just as *p* can always be inferred from *p*. But we've also seen that the Second Philosopher's justifications are not circular in this sense, that their success is not a foregone conclusion. So, for example, the scientific study of scientific method may well turn up biases and distortions, like the effect of unsubstantiated assumptions about sex roles on observations in primatology.[3] A scientific assessment of the

[2] Especially I.2 and I.7. See also IV.1.

[3] Hrdy writes: 'According to Darwin's brilliantly original hypothesis, males compete among themselves for sexual access to females ... elusiveness was as integral to the female sexual identity as ardor was to that of their male pursuers ... [but] what [are] we to make of brazenly assertive macaques and chimpanzees, and the not-so-coy solicitation of neighboring males by "harem-dwelling" langurs and monogamous titi monkeys?' (Hrdy [1999], pp. xiii, xiv). She traces the development of new views of females among primatologists and continues, 'How much did feminism have to do with this transformation? Feminism was part of the story, but not because women primatologists ... do science differently. Rather, women fieldworkers were predisposed to pay more attention when females behaved in "unexpected" ways. When, say, a female lemur or bonobo dominated a male, or a female langur left her group to solicit strange males, a woman fieldworker might be more likely to follow, watch, and

reliability of perceptual beliefs will make use of various perceptual beliefs, but along the way, it will use them to describe situations in which we are likely to be led to false beliefs by our usually reliable perceptual systems, and to explain why they are liable to break down in such cases. Our examination of the extent to which the world's structuring into objects can be pushed into the microscopic turns up an analogous breakdown, though it is—alas!—much less well understood than our perceptual failures.

To see this, recall the familiar twin slit experiment:[4] electrons are fired at a sensitive screen that registers hits; between the electron source and the screen is a barrier with two slits; when one slit is covered, the hits on the screen are densest directly opposite the open slit, with a steady falling off on both sides. What will happen when both slits are open? If the two slits are fairly close together, the area where most electrons would be expected to hit would be in the center of the screen, where the high densities from each slit haven't fallen off much and the overlap is great; there should then be a gradual decrease in both directions. And this is what we observe if we fire bullets at a wall instead of electrons at a screen. But, as sad experience tells us, this is not what we observe when we fire electrons; in that case, we get a more complicated pattern of densities.

Many suggestions have been considered. Perhaps electrons are not like bullets, after all, but more like waves; but if this were true, we should see a different pattern when one slit is closed. Perhaps electrons don't simply travel directly through one slit or the other, but take a more complicated path; but no such solution, consistent with the data, has been found. In the end, it seems we must give up the conviction that each electron travels through one slit or the other; we must give up the conviction that an electron has, at each moment, a position in space; we must give up the conviction that an electron leaving the source at one moment, arriving at the screen a brief time later, has traveled some continuous path from one point to the other.

wonder than to dismiss such behavior as a fluke ... any time wrong ideas are corrected, science wins. If biases were there in the first place because of sexism, and a feminist perspective helped to identify them, it is still science that comes out ahead when they are corrected' (Hrdy [1999], pp. xviii–xix). (I take this simple example from primatology to be uncontroversial, unlike related issues in evolutionary psychology.)

[4] In this paragraph and the next, I follow Feynman et al. [1965], chapter 1, and Hughes [1989], sections 8.3 and 8.4.

The unpleasant conclusion is that the micro-world is not structured into things of the familiar sort; though the world does contain numerous ordinary objects, it also contains phenomena that are not so structured. Despite our scientific predisposition to see the world in these terms, our pursuit of science itself has taught us that the world is not as we expect it to be, not in all its parts. This portion of our empirical hypothesis—that the world consists of coherent objects that move as units along continuous spatiotemporal paths—must be qualified. The world is structured into such objects at the macro-level,[5] but at the micro-level, all current evidence suggests that it is not.

Indeed, the statistics of quantum mechanics present another difficulty for our ordinary notion of an object, this time not for its spatiotemporal aspects, but for individual identity: if two particles have the same 'intrinsic' properties (e.g., mass, spin, charge), then there is no real difference between the state in which they are as they are and the state in which they are switched. One commentator writes:

> At this point it is clear that the appropriateness of the particle-concept itself becomes doubtful ... the very essence of the particle idea seems to be lost. (Dieks [1990], pp. 140–141).

Here not only the spatiotemporal features of objects are undermined, but the pure notion of an object as an individual thing.[6] Thus it seems the micro-world cannot be said to display an abstract KF-structure of individual objects.

ii. Properties and relations

Once again, common sense holds that objects have properties and stand in relations to one another—the apple weighs two ounces; this basketball

[5] Feynman writes: 'the peculiar quantum mechanical behavior of matter on a small scale doesn't usually make itself felt on a large scale except in the standard way that it produces Newton's laws—the laws of so-called classical mechanics. But there are certain situations [he's thinking of superconductivity] in which the peculiarities of quantum mechanics can come out in a special way on a large scale' ([1965], p. 21–1). I use 'macro', here and elsewhere, as shorthand for large scale phenomena that don't display quantum effects.

[6] In Kantian terms, not only the schematized category of object is undermined, but the pure category as well.

player is taller than that baseball player—and the Second Philosopher's further investigations ratify such claims. There also seems every reason to believe that the apple fails to be as big as the house, and that it has a shape in addition to a weight. Furthermore, a coin resting on the table will either show heads or tails; a functioning light switch is always either on or off. Among several apples and oranges, each object is either an apple or an orange. And so on.

All this sounds so obvious that we might be tempted toward notions of necessity or a priority or analyticity, but such confidence is premature. To see this, we turn, once again, to the micro-level, and consider a (simplified telling of) the famous Stern–Gerlach experiments.[7] If a specially prepared beam of electrons passes between the poles of a specially shaped magnet, it splits in two: for example, if the magnet is oriented vertically, one pole above the other, the electrons split into two beams; half are deflected upward and half deflected downward. If the magnet is rotated 90°, to a horizontal position, the beam against splits in two, half deflected to the right, half to the left.

Now this is already odd, as we might expect the magnetic axes of the electrons to be oriented at random, producing a range of outcomes rather than two distinct beams, but this much can be accommodated with some new theorizing: an electron has an intrinsic angular momentum, called 'spin', which gives rise to the observed behavior; the component of this spin in any single direction can take one of only two values. So, when the beam passes through the vertically aligned poles, those with 'spin up' in the vertical direction are deflected up and those with 'spin down' in the vertical direction are deflected down. Similarly for the horizontal set-up, 'spin left' and 'spin right'. We can check this, for example, by running the beam through the vertical system, then running the beam deflected up through a second vertical system: the spin up electrons will be deflected into the upper beam out of the first device; a stream of exclusively up electrons will enter the second device; all of them are again deflected up. The same goes for the horizontal alignment. On the other hand, if we run the up-beam from a vertical device though a horizontal device, the beam will split left and right. The spin up electrons are half spin left and half spin right.

[7] Here I follow Hughes [1989], pp. 1–8. See also Feynman et al. [1965], chapter 5.

So far so good. The trouble comes when yet another, third device is added to this sequence. The beam entering the first vertical device is split into spin up and spin down; the up beam is directed through the horizontal device, splitting the electrons further into spin left and spin right. The spin left beam exiting the second, horizontal device should consist exclusively of spin up, spin left electrons: if we run this beam through a third, vertical device, it should all deflect up; there should be no further splitting. But—alas!—this is not what happens. The beam that should consist exclusively of spin up, spin left electrons splits when sent through a vertical device into half spin up and half spin down.

We might try to account for this by theorizing that the horizontal device intervening between the two vertical devices has altered the electrons in the spin up beam, but for this to work we would have to explain why only half of them are altered, and this seems impossible. Instead, it seems an electron cannot have a vertical spin property and a horizontal spin property at the same time, much as in the familiar case of position and momentum. As R. I. G. Hughes concludes, 'Indeed, it is not clear in what sense these "particles" can be said to have properties at all' (Hughes [1989], p. 1). Despite the reliable behavior of the properties and relations of macro-objects, this structuring does not seem to extend into the micro-world. Once again, our claim that the world has KF-structure must be qualified.

iii. Dependencies

Common sense sees dependencies of various sorts between the properties and relations of some objects and the properties and relations of others: the vase falls because the cat pushes it off the table; a daughter must have parents; red squares are always red.[8] The Second Philosopher's more elaborate inquiries will second these and add a rich and ever-unfolding story of further interconnections. But, as we've come to expect, the good behavior of the macro-world breaks down in the micro-world, as one last excursion demonstrates.[9]

[8] We might describe these as causal, semantic, and logical dependencies, respectively.

[9] This is Bohm's version of the Einstein–Podolsky–Rosen thought experiment, as described by Hughes [1989], pp. 158–162.

The standard way of understanding the state of the electron exiting a vertical spin detector is this: it has the vertical spin indicated by the mode of its departure from the detector, up or down, but it has no horizontal spin at all; it is, rather, in a 'superposition' of horizontal spins, where that superposition dictates the probability of the possible outcomes of its subsequent passage through a horizontal spin detector (50% chance of left; 50% chance of right). Now it's possible to generate a pair of particles whose overall horizontal spin is zero: the pair is in a superposition of Particle 1 with horizontal spin left and Particle 2 with horizontal spin right, and the reverse, and the probability that either one of these will turn up as the result of measurement is 50%. It is also possible to separate these two particles widely in space, despite their properties' being so entangled. If we now run Particle 1 through a horizontal spin detector, we will get an outcome, left or right, and this will tell us the horizontal spin of Particle 2, namely, right or left. What's odd here is that before we tinkered with Particle 1, Particle 2 was supposed to have no horizontal spin property, so we can't be said to have discovered what it was. We find ourselves torn: it seems as if the features of Particle 2 depend on what we did with Particle 1, but the spatial separation makes this so unlike the case of the cat and the vase that our ordinary notion of dependency is violated.[10] Here again, the macro-structure of the world seems not to be reproduced at the micro-level. If there is a KF-dependency in this case, we have no idea of its mechanism.

iv. Indeterminacy

Given that the world, at least the macro-world, is largely structured into objects with properties standing in various relations, we now ask if this structure is sharply delineated. To put the question in a familiar contemporary idiom, given that the world has joints, are those joints fuzzy? Nothing in the account to come truly rests on this, but I confess it seems obvious to me that the answer is yes. There is undoubtedly an apple on the

[10] The correlation between the properties of the two particles clearly isn't a candidate for a semantic or a logical dependency, either. See footnote 8.

table, but exactly which small bits are and aren't part of it is indeterminate. The world includes living organisms and inanimate objects, but there are indeterminate borderline cases, both kinds of objects (some primitive items) and individual objects (living things at points in the process of dying) that aren't determinately living or non-living. There are clearly tadpoles (immature creatures) and frogs (mature creatures), but the border between these is blurred.

Now the topic of vagueness is much discussed in philosophy these days,[11] but a surprising and sizeable majority of these writers agree that vagueness is not a feature of the world, only of our representations or descriptions.[12] So, for example, Dummett, writes: 'the notion that things might actually *be* vague, as well as being vaguely described, is not properly intelligible' (Dummett [1975], p. 260). Now the Second Philosopher surely agrees that there is indeterminacy due to linguistic vagueness—indeed, as James Tappenden ([1994]) has emphasized, some legal terms, like 'all deliberate speed', are purposely designed to be vague—but she maintains that there is worldly indeterminacy, as well.[13] Given the breadth and depth of the conviction that this isn't so, we should ask what lies behind it.

One technical argument often considered in the literature on vague objects arises in a short paper of Gareth Evans ([1978]). His concern isn't directly with vague objects, but with indeterminate identity statements. So, for example, perhaps the identity of a, our vaguely bounded apple, and b, some precise batch of molecules which it approximates,[14] is neither true nor false. Evans argues that this is impossible: if it's indeterminate whether or not a is identical with b, then a has a property—being indeterminately identical with b—that b does not have (because b *is* identical with b), so a and b are distinct, after all. Many strong objections to this argument have been raised, attacking it at various points,[15] but even at best it only rules out indeterminate identities,[16] not vague objects. No obstacle is raised to

[11] See, e.g., two recent anthologies: Keefe and Smith [1997a], and Graff and Williamson [2002].

[12] Among the exceptions are J. A. Burgess [1990a], [1990b], Tye [1990], and Parsons [2000].

[13] Again, nothing in the remainder of the Second Philosopher's account of logical truth hangs on this; linguistic vagueness would be enough to underlie the considerations in III.7.

[14] In the jargon, b is a 'precisification' of a.

[15] e.g., see Keefe and Smith [1997a], pp. 51–56, for an overview.

[16] If identities are never indeterminate, situations like ($a = b$ or not-($a = b$)) will obtain by virtue of their form (see III.3, the text surrounding footnote 17).

the otherwise attractive idea that *a*, vague object, and *b*, an exact one, are simply distinct.[17] So let's put Evans's argument aside.

I suspect that a deeper concern, much less discussed, in fact underlies the majority opinion here. Dummett, even in the act of retracting the stark dismissal of worldly vagueness quoted a moment ago, still feels an 'extremely strong' pull toward the view that vagueness 'must be due to our own limitations':

It is natural to us to conceive of physical reality as, in itself, capable of description in absolutely precise mathematical terms, a description upon which any other would be supervenient even if imprecise. (Dummett [1981], p. 440)[18]

Rosanna Keefe and Peter Smith fill in this picture:

Suppose our world is constituted by fundamental particles and fundamental properties both of which are entirely determinate: for an object *a* and a property *P* in this catalogue of 'base level' items, it will either be a fact that *a* has *P*, or a fact that it does not. There will be no indeterminacy and no borderline cases at this base level, which is then naturally described as being completely precise. Suppose additionally that the totality of these base-level facts fixes everything else. We would still have reason, for everyday purposes, to pick out and talk about various large collections of atoms (e.g. clouds or mountains) whose boundaries are left fuzzy. But what is true, false, or left indeterminate about them would supervene on how things stand at the precise base level. (Keefe and Smith [1997b], p. 56)

J. A. Burgess gives a similar diagnosis of the underlying conviction:

the physical world is divisible, at some level of microscopicity ... into discrete (sharply bounded) objects ... [and] it is of these objects alone that vague macroscopic objects are ultimately composed, albeit of an indeterminate number of them. (J. A. Burgess [1990a], p. 279)[19]

From this point of view, our apple is vaguely bounded, but this fuzziness is superficial. The ultimate facts of the micro-world are exact.

[17] Parsons and Woodruff [1995], pp. 332–333, and Keefe and Smith [1997b], p. 51, leave room for this option; Tye [1990], p. 556, occupies it. The burden of Parsons [2000] is to show that indeterminate identities are possible.

[18] He has come to regard the inclination toward this view, however strong, as a 'prejudice' (Dummett [1981], p. 440).

[19] He quite reasonably adds (J. A. Burgess [1990a], p. 280): 'there would be no *a priori* reason to stipulate that the underlying precise objects were not themselves divisible into smaller precise objects, *ad infinitum*.'

The idea seems to be this.[20] We all agree that there being an apple on the table consists of there being various sorts of molecules, arranged in various ways, made up of various sorts of atoms, again arranged in various ways, above the table (whose existence consists in other arrays of molecules), with an electrostatic force preventing the apple's molecules from passing through the table's molecules under the force of gravity, and so on. According to the Second Philosopher, this apple is an instance of objective KF-structuring in the world. Once this is admitted, there seem no grounds on which to deny that this very apple has vague boundaries, that a large conglomeration hangs together there above the table, that it will move together, and so on, but that it is indeterminate whether some bits are part of this conglomeration or not, given that they will fall away under certain motions, and so on.

To undercut the Second Philosopher's view of the matter, the objective existence of the apple must be denied. An extreme version is the claim that, strictly speaking, the apple doesn't exist, only the molecules. On the assumption that the molecules and other micro-objects are entirely precise, this added claim delivers a world with no vagueness. A more conciliatory version has it that the apple exists, sure enough, but that it is merely an imaginary enhancement of the underlying reality, imposed by us, by our language, to help us grasp a situation whose real structure is too complex for our ordinary capabilities and purposes. On this view, the apple does have vague boundaries, but it's only a conceptual or linguistic creation; the objective world remains precise.

Now given the quantum mysteries we've so recently rehearsed, I trust the micro-structure of the world will appear an odd place to look in search of determinacy! The micro-world appears not to consist of particles with locations or properties at all, in the usual sense, let alone particles with sharp boundaries and properties of the sort imagined here. The truth is that quantum mechanics, for all its predictive success, tells us very little about what the structure of the micro-world is actually like. And insofar as it tells us anything at all, it strongly suggests that the picture evoked here—of sharply bounded particles, like tiny, perfect ball-bearings—is almost certainly not appropriate.

[20] The following discussion makes several points of contact with the final sections of J. A. Burgess [1990a].

But there's something more fundamental going wrong here than a misconception of micro-physics. The central suggestion—that ordinary objects of the macro-world are less than real, less than objective elements of the world, that this level of distinction is reserved for the micro-world—seems to me profoundly misguided. Whatever else is true, surely apples are objective items in the world; they play a crucial part of the reproductive cycle of apple trees. Similarly, the property of being a tadpole is a central feature in the maturation process of certain amphibians, and there really are homo sapiens alive today, despite its being indeterminate when the first one appeared. By the Second Philosopher's lights, all these are objective features of the world, not conceptual or linguistic impositions: surely the cat walking across the living room is an individuated physical object, regardless of how we think or talk about it. Whatever elaborate theories we may develop at the micro-level, they will have to reproduce ordinary, medium-sized physical objects at the macro-level, even if they lead us to think of them differently than we once did.

In sum, then, I see no grounds on which to deny the commonsense observation that there is vagueness in the world, a feature our world shares with KF-worlds. At the macro-level, it also shares the remaining structures—objects standing in relations, ground-consequent dependencies—but as best we can tell, these structures are not to be found in quantum mechanics. So the first component of our second-philosophical account of logic requires adjustment: our world is a KF-world in many, but not all, of its aspects; many of its phenomena are so structured, but some are not. Let (1) of our first approximation to a second-philosophical account of logic be replaced by (1′): rudimentary logic is true of the world insofar as it is a KF-world, and in many but not all respects, it is.

III.5

The logical structure of cognition

We now turn to the human side of the proposed second-philosophical account of logic: (2) human beings believe the simple truths of rudimentary logic because their most primitive cognitive mechanisms allow them to detect and represent the KF-structure of the world. Given, as we've just seen, that the world isn't entirely KF in its structuring, this should be revised to (2′): human beings believe the simple truths of rudimentary logic because their most primitive cognitive mechanisms allow them to detect and represent the KF-structures in the world. Then comes the question of the connection between parts (1′) and (2′) of the account, now modified to (3′): the primitive cognitive mechanisms of humans beings are this way because we live in a largely KF-world and interact almost exclusively with its KF-structures. I take up each of the KF-structures in turn—objects, properties and relations, and dependencies—then examine the grounds for (3′).

i. Objects

There can be little doubt that ordinary adults see the world in terms of individual objects, but there is more to the Second Philosopher's claim than this. Those same adults are probably also disposed to think that the sun rises in the east and sets in the west, but this belief is surely acquired or learned, not the product of their primitive conceptual mechanisms. So we need to ask how our adult comes to her way of viewing the world, in particular, how she comes to her notion of 'object'.

The starting point for the modern psychological study of the development of our tendency to conceptualize the world in terms of objects is Piaget's

The Construction of Reality in the Child.[1] The focus of this research is on infants' understanding of what happens when an object leaves the field of vision, as tested by so-called 'occlusion studies'. At issue is when the child begins to share the adult's notions that an object continues to exist even when our view of it is blocked, that an object retains its physical properties and continues to behave according to ordinary physical laws while out of sight. This conception is often called 'object permanence'.

The experiments of Piaget and his collaborators were largely based on testing the manual search behavior of infants. For example, does the infant, after watching as a desired object is covered by a cloth, search for it by lifting the cloth? After successfully recovering a desired object hidden under the blue cloth, does the infant watching the object hidden under the red cloth go on to search for it under the red cloth or return to the blue? When the infant sees the desired object placed in a container, the container then shifted behind a series of screens, and the container, at the end, turns out to be empty, does the infant search for the object behind one or another of the screens?

On the basis of experimental designs like these, Piaget concluded that the child's object concept develops through a series of stages, from the first stage (0–9 months), when an occluded object ceases to exist when it passes from view, through a second stage (9–12 months), when the occluded object continues to exist, but not as the occupant of an objective position in space, through a later stage (12–18 months), when the occluded object does occupy an objective position in space, but always at the position where it disappeared from view, until finally the last stage (18–24 months), when the object is understood to carry on its motions even when hidden from view. Thus the child's experience is seen to develop from its initial 'blooming, buzzing confusion' (James [1890], p. 488) to an adult conception of the world of external objects between birth and the age of about 2 years.[2]

These experimental findings of Piaget and his co-workers were largely confirmed in the years that followed, but other evidence based on visual tracking rather than search behavior, some of it from Piaget's own experiments, pointed toward an earlier emergence of object permanence. Unfortunately, the tracking results were methodologically suspect and

[1] First published in French in 1937, then in English translation in 1954.

[2] I discuss these and related theories of Piaget and his followers in chapter 2 of my [1990]. The summary here is based on the opening pages of Baillargeon [1993].

ambiguous in their interpretation, so progress on these questions awaited the development of a new experimental paradigm. One such emerged in the early 1980s, namely, habituation and preferential looking.[3] The underlying idea is simple:

> In this method, infants are shown the same event repeatedly and their looking times recorded. With each repetition their looking times decline, that is, infants 'habituate'. When infants reach a pre-set habituation criterion, they are shown two displays alternately, one consistent with adults' understanding of the event and the other inconsistent. If the infants have the same understanding of the habituation event as adults, they should look longer at the inconsistent display as opposed to the consistent one. (Xu [1997], p. 372)

More generally, once the child is accustomed to seeing one event, its subsequent looking times will tell us which events seem the same as the habituation series and which seem different. The novel stimulus will attract a longer gaze.[4]

One early use of this new paradigm appears in a study of object permanence by Baillargeon, Spelke, and Wasserman ([1985]): infants around 5 months of age were habituated to the motion of a screen on a horizontal hinge that moved forward and backward, toward them and then away from them, through a 180° arc, like a drawbridge.[5] With the screen lying flat in its forward position, a small box was placed behind it. The screen was then raised, hiding the box, as if to turn once again through its 180° backward arc. The possible event had the screen stopping when it touched the top of the box; the impossible event had the screen moving freely until it lay flat again, in its backward position, away from the child. The experimenters write:

> Our reasoning was as follows. If infants understood that (1) the box continued to exist, in its same location, after it was occluded by the screen, and (2) the screen could not move through the space occupied by the box, then they should perceive the impossible event to be novel, surprising, or both. On the basis of the

[3] See Baillargeon [1993], pp. 267–272, Xu [1997], pp. 371–372, for discussion of these developments and references. Cohen [1988], pp. 211–214, discusses the origins of the habituation paradigm in earlier memory research.

[4] For methodological discussions of habituation and preferential looking, see Borstein [1985] and Spelke [1985], respectively. This approach is especially useful because 'these preferences emerge so early relative to other behavioral systems' (Spelke [1985], p. 324).

[5] If the set-up is unclear, imagine a long hinge placed left to right, lengthwise, on the table. Now attach a screen oriented vertically. This screen can now hinge forward, toward us, to lie flat on the table. It can also hinge backward, away from us, again to lie flat on the table.

commonly-held assumption that infants react to novel or surprising events with prolonged attention, we predicted that infants would look longer at the impossible than at the possible event. On the other hand, if infants did not understand that the box continued to exist after it was occluded by the screen, then they should attend to the movement of the screen without concerning themselves with the presence of the box in its path. Since the screen movement was the same in the impossible and the habituation events (in both events the screen moved through a 180-degree arc), we predicted that the infants would look longer at the possible event, which depicted a novel, shorter screen movement. (Baillargeon et al. [1985], pp. 195–196)

In fact, the infants looked longer at the impossible event, evidence for their understanding of something like (1) and (2).[6] Further exploitation of the drawbridge set-up suggested that the infants expected the box to retain its properties while occluded, for example, its height.[7]

Given Piaget's conclusion that infants only reach this degree of understanding considerably later on, we have to wonder if the infants' failures on his tests had more to do with the extraneous difficulties of searching than with their grasp of object permanence. A number of theories have been proposed, the most viable of which involves glitches in means/ends reasoning. So, for example, there is resistance, continuing into adulthood, to beginning the pursuit of one's goal by doing something that seems to conflict with it (e.g., by grasping the cloth cover when you want to grasp the object underneath) and there is a tendency, also continuing into adulthood, to use a previously successful strategy even in cases where it isn't appropriate (e.g., searching under the blue cloth instead of the red cloth).[8] It now seems Piaget's time line of conceptual development was artificially skewed toward later ages because his experimental tasks required more than mere understanding of object permanence.

Still, even on the most generous interpretation of the experiments considered so far, we remain some distance from all that would seem to be involved in a full adult concept of 'object'. Perhaps the subjects understood

[6] Baillargeon and her co-workers anticipate the worry that infants look longer at the impossible event because the 180° movement is simply more interesting than the shorter movement and conducted an additional experiment to rule this out (Baillargeon et al. [1985], pp. 196–197). In my descriptions here, I only touch on a tiny portion of the many experiments conducted, and I can't do justice to the many, carefully designed preliminary and subsequent experiments done to rule out alternative explanations of these findings. The interested reader is commended to the many fascinating papers referenced here.

[7] See Baillargeon [1993] for a summary.

[8] See Baillargeon [1993] for a survey of this debate. Gopnik and Meltzoff [1997], pp. 86–92, and Meltzoff and Moore [1999], pp. 71–72, are skeptical of such attempts to explain Piaget's results.

there was some obstacle to the screen's motion, but did they view that obstacle as a stable individual object, as a unit? We might break this into two questions: has the infant separated the box off from the rest of the visual scene, as starting here and stopping there ('individuation')? Does the infant regard the box before occlusion as the very same thing as the box afterwards ('identity')?

Available theories of individuation and identity divide roughly into general theories, according to which some single set of rules or principles governs these judgments, and sortal[9] theories, according to which perceiving and recognizing objects depends on our knowledge of particular kinds of objects. So, for example, on a sortal view, we perceive an area of the visual field as a dog and recognize it as the same dog at a later date by virtue of our understanding of the kind 'dog', which includes an understanding of the ways a thing can change—in size, shape, etc.—while remaining the same dog.

A general account was proposed by the Gestalt psychologists:

> Perceivers inherently tend to organize the surrounding layout into the simplest, most regular units. This tendency can be expressed as a set of principles such as *similarity* (surfaces lie on a single object if they share a common color and texture), *good continuation* (surfaces lie on a single object if their edges lie on the same line or smooth curve), *good form* (surfaces lie on a single object if their edges can be joined to form a region with a symmetrical shape), and *common fate* (surfaces lie on a single object if they move together). ... Gestalt principles of organization were also thought to underlie perception of object identity over successive encounters. When an object appears successively in different locations, perceivers were said to perceive a single, persisting body by grouping its appearances into the simplest patterns of motion and change. (Spelke et al. [1995a], pp. 300–301)

As the past tense in this passage suggests, this general theory has been superseded, but theories of both sorts have their difficulties.[10] To see how informed opinion now stands, let's return to the infant experiments.

Beginning with the question of object unity, here it seems the Gestaltist's 'common fate' is the decisive factor.[11] In Kellman and Spelke [1983], infants

[9] A sortal corresponds to a count noun (like 'dog'), as opposed to a mass noun (like 'water'). For such an account, see Wiggins [1980].

[10] For further discussion, see Spelke et al. [1995a], pp. 300–305.

[11] See Kellman [1993], p. 129. For discussion of the following experiments, see this paper by Kellman, and Spelke [1985].

were habituated to a rod moving back and forth behind a panel, with both ends showing, above and below the panel.[12] They were then shown the same scene without the panel, either with a single rod or with two rod pieces with a gap where the panel had been. The result was that

> The infants... looked longer at the rod with the gap. The experiment provides evidence that the infants perceived the ends of the original rod to be connected behind the [panel]: They perceived the complete shape of this partly hidden object. (Spelke [1985], p. 330)[13]

Oddly enough, these young infants (4 months) did not seem impressed by the color, texture, or shape of stationary objects: for example, when habituated to a partly occluded triangular figure, they showed no preference between a complete triangle and a triangle with a gap where the occlusion had been. On the other hand, even objects of very irregular color, shape, and texture were perceived as units if they were in common motion: for example, when habituated to a figure with a rod on top and an irregularly shaped, textured, and colored blob on the bottom moving behind a panel, the infants still looked longer at the display with a gap where the panel had been. So similarity, good continuation, and good form seem not to be used to determine object boundaries by infants this young.[14]

Related results turn up in investigations of infants' perception of object boundaries in scenes involving adjacent objects. Methods vary from reaching trials to a range of habituation/preferential looking experiments;[15] Spelke and Newport summarize:

> All these studies reveal that young infants perceive the boundaries between two objects if the objects are separated by a gap in three-dimensional space or if they are adjacent to one another but undergo separate motions, as when one object slides across the top of another... In contrast, infants sometimes fail to perceive the boundary between two objects that are adjacent and stationary, even

[12] If the set-up is unclear, imagine a metal rod oriented vertically. Place a screen in front of the rod so that the upper end of the bar is visible above the screen and the lower end of the bar is visible below the screen. Now move the bar back and forth, left to right and right to left, behind the screen.

[13] As a reminder of the many safeguards I'm not mentioning (see footnote 6), let me note that before performing the experiment just described, Kellman and Spelke checked to be sure that the broken rod could be discriminated from the unbroken one, that neither the complete rod nor the broken one was more interesting in and of itself, and that attention is directed to the visible parts of the occluded object.

[14] This isn't because the infants don't perceive these other factors. See Spelke et al. [1995a], pp. 310–311, for summary and references.

[15] For review, see Spelke et al. [1995a], pp. 305–314.

if the objects differ in color, texture, and shape ... (Spelke and Newport [1998], p. 292)

In addition to lack of spatial separation, it is again common fate (moving together), not the other Gestalt principles, that appears primary in determining object unity.

A particularly striking experiment of Xu and Carey[16] (from 1994) asked whether objects of different, familiar kinds would be distinguished without the help of 'common fate':

Ten-month-old infants were presented with two toys—for example, a yellow rubber duck and a red metal truck—arranged so that one toy stood on top of the other. (Spelke et al. [1995a], p. 310)

Some of the infants were familiarized[17] with the sight of the duck sliding back and forth along the top of the truck; the rest saw only the stationary display. In the test events, a hand grasped the duck and raised it: in one set-up, the truck stayed in place; in the other, it came along with the duck.

Infants who had been familiarized with the duck and truck undergoing relative motion looked longer at the second event, suggesting that they had perceived the two toys as separate objects and expected them to move independently. In contrast, infants who had been familiarized with the duck and truck without motion looked longer at the first event, suggesting that they had perceived the two toys as a single object. (Spelke et al. [1995a], p. 310)

Notice that the infants here are as old as 10 months; the previous tests involved infants as young as 4 months.

So far, then, it seems infants judge object unity by spatial contiguity and common fate, not by Gestalt similarity or smoothness, and not by familiar kind discriminations. Now consider object identity: when are this and that regarded as appearances of the very same thing? In another experiment from the mid-1990s (see Spelke et al. [1995b]), 4-month-old infants were familiarized with one of two scenarios: in the first, an object moved at a steady rate from left to right, passing behind two spatially separated panels not much wider than the object, so that it appeared to

[16] For references, see Xu [1997], p. 379, Spelke et al. [1995a], p. 310.

[17] 'Familiarization' involves showing the event a fixed number of times, without the requirement that looking time fall below a pre-set threshold, as in habituation, so this is a slightly different preferential-looking paradigm.

move continuously; in the second, the same object moved from left to right until it passed behind the left panel, then emerged from the right panel, with no object visible in the gap between them. The infants were then tested with two displays without the panels, one with one fully visible object, the other with two. The result was that the infants exposed to the first scenario looked longer at the two-object display than those exposed to the second:

> Infants presented with continuous motion appeared to perceive a single object that moved in and out of view [while] those presented with discontinuous motion appeared to perceive two distinct objects. (Spelke et al. [1995a], p. 316)

In related experiments, infants seemed not to be influenced by whether or not occluded motions could be understood to be smooth (constant in speed and direction), again contrary to Gestalt thinking.

In addition, just as in the case of object individuation, it seems that information about color, shape, texture, etc., unlike information about motion, has little effect on infants' determinations of object identity.[18] To test whether or not membership in easily recognized kinds might be used for judging object identity, Xu and Carey ([1996]) familiarized 10-month-old infants with a single wide screen, behind which sometimes one, sometimes two objects were discovered. They were then familiarized with the following events: a toy duck emerges from the left side of the screen then returns behind the screen; a ball emerges from the right side of the screen, then returns behind the screen. In the test events afterwards, the screen was removed to show either the duck and the ball (the possible event) or the duck alone (the impossible event). The researchers write:

> Given that the babies had a strong preference for two-object displays, success at this task is not looking longer at the unexpected outcome of one object, but rather overcoming the baseline preference. The major result of this experiment is the failure of 10-month-old infants ... to do so. ... In this experiment, babies failed to demonstrate that they could use the differences between a yellow rubber toy duck emerging from one side of the screen and a white Styrofoam ball emerging from the other side of the screen to infer that there must be at least two objects behind the screen. (Xu and Carey [1996], p. 129).

[18] See Spelke et al. [1995c], p. 166, Spelke et al. [1995a], p. 314, Spelke and Newport [1998], p. 293.

Incidentally, when the same experiment was run on 1-year-olds, they did overcome the baseline preference for two objects. The kind difference was enough to lead these older infants to expect two objects behind the screen.

One final experiment of Huntley-Fenner and Carey (see Huntley-Fenner et al. [2002]) suggests that young infants are truly representing discrete individual objects. In an important earlier experiment of Wynn ([1992]), infants were shown a single object; they watched as a screen lowered from above to block it from view, and as a second object was then placed behind the screen. As previously described experiments would lead us to expect, the infants looked longer at a subsequent display of one object than of two (Wynn [1992]).[19] But, and here's the point of Huntley-Fenner and Carey, if a similar experiment is repeated using piles of sand rather than objects—even if the two sand piles are hidden behind two screens with a gap between them—no consistent preference results. This suggests that the infants in the first experiment were not simply tracking the amount of stuff involved, but discrete individual objects.[20]

All this and much more points to the conclusion that infants as young as 4 months are able to individuate and identify objects using spatiotemporal criteria: spatial contiguity, common fate, and continuous motion. No such object can be in two places at once; two such objects cannot occupy the same location; all such objects travel on continuous paths (cf. Xu [1997], p. 370). They are 'complete, connected, solid bodies that persist over occlusion and maintain their identity through time' (Spelke [2000], p. 1233). Bloom calls these 'Spelke-objects':

> What is a Spelke-object? Such entities follow principles, the most central one being the *principle of cohesion*. To be an object is to be a connected and bounded region of matter that maintains its connectedness and boundaries when it is in motion. With objects of the right size, this suggests a crude test of objecthood: grab some portion of stuff and *pull*; all the stuff that comes with you belongs to the same object; the stuff that remains behind does not ... By this [criterion], heads are not typically objects; if you tug on a person's head, the rest of their body follows. When a head is severed, however, it is an object. A man on horseback is two objects, not one, because the man can move and be moved independently from the horse and vice-versa. (Bloom [2000], p. 94)

[19] This experiment also uses a slightly different preferential-looking paradigm, sometimes called 'violation-of-expectation transformation studies' (Feigenson et al. [2002], p. 34).

[20] For discussion, see Xu [1997], pp. 374–375, Spelke [2000], p. 1234.

He goes on to discuss other principles—continuity, solidity, and contact—but argues that cohesion is fundamental. Though these various characterizations may not coincide perfectly in emphasis or detail, I think a coherent and familiar notion has emerged.

If the notion of a Spelke object seems to provide a general criterion of object unity and identity to replace the Gestalt principles, we also have some further information about the sortals central to the alternative view. Preparatory investigations for the experiments described above and other work to be discussed below have verified, for example, that young infants can classify ducks separately from balls, but the upshot of the Xu and Carey experiments is that they do not use these kinds for determining identity, even up to 10 months of age. So it seems likely that these children have at best the notion of a property—duckness or ballness—and no sense that an object is unlikely to change suddenly from one with duckness to one with ballness.[21]

On the other hand, the 1-year-old babies did use the difference between ducks and balls to determine that this duck cannot be the same object as that ball. They might do this on the basis of a full sortal concept, or they might simply be more aware of the unlikelihood of an object changing its properties so dramatically. There is intriguing evidence in the direction of the full sortal hypothesis in the fact that the very period—between 10 and 12 months—during which infants begin to make the kind-based distinctions, they are also learning their first words, and those first words are nouns for sortals.[22] In any case, it must take considerable learning, of language and more, to realize that a crushed car is no longer a car, but a seedling grown into a tree is still the same plant. So, while the development of sortal concepts is a complex, at least partly linguistic achievement, it seems clear that individuation and identification of Spelke objects precedes language use.[23]

There is no doubt that sortals play a major role in the object concept of mature adults:

[21] Cf. Xu [1997], pp. 380–383, for discussion.

[22] There is some suggestion that the difference in performance between individual infants on the Xu and Carey tests correlates with the number of such nouns those children had learned. See Xu and Carey [1996], pp. 145–147, Xu [1997], 378–379.

[23] Cf. Bloom [2000], p. 94: 'it is not an accident that a notion of object developed from studies of how babies see the world so elegantly captures the word-learning biases we find in children and adults ... humans are naturally predisposed to see the world as composed of Spelke-objects—and this explains the object bias present in early word learning'. See also Spelke and Newport—'processes for perceiving object boundaries and object identity appear to guide language learning, rather than the reverse' ([1998], p. 294)—and the references cited there.

Why do we perceive a car and a trailer, instead of a front bumper and a bumperless car-trailer? ... no general principles readily explain why we are inclined to judge that a car persists when its transmission is replaced, but would be less inclined to judge that a dog persists if its central nervous system were replaced. ... Object perception may depend on processes of object recognition, which depend in turn on the perceiver's vocabulary of internal representations, or models, of the kinds of objects that furnish our surroundings ... Once an object is recognized, perceivers can apply their knowledge of the properties and the behavior of that kind of object ... Because we know that cars have bumpers but not trailers as proper parts, we perceive a car but not a bumperless car-trailer ... Because we know that dogs but not cars have behavioral and mental capacities supported by certain internal structures, we consider certain transformations of dogs to be more radical than other, superficially similar transformations of cars. (Spelke et al. [1995a], pp. 301–303)

But it seems this can't be the whole story of adult object perception: we can perceive a 'thing'—say 'at the corner of our eye'—before we have a sense of what it is, and we can change our opinion of what a thing is while holding the thing itself constant—'it's a bird, it's a plane, it's Superman!'[24]

In fact, there's evidence from studies of adult visual processing that we use two distinct systems for object individuation: the sortal system, and another more primitive system:

When an event suddenly occurs in the visual field, human perceptual systems appear to make a very rapid decision: Has a new object appeared, or has a previously visible object changed state or position? (Spelke et al. [1995a], p. 323)

Studies of this phenomenon suggest a

mid-level vision system (mid-level because it falls between low level sensory processing and high level placement into kind categories) that establishes object

[24] Xu [1997], pp. 383–387, discusses cases of this kind in the course of her argument that the notion of a Spelke object is used by adults to individuate and identify (though see footnote 26). The Superman example also appears in Kahneman et al. [1992]; after more than forty pages of experiments and discussion, they conclude as follows: 'Onlookers in the movie can exclaim "it's a bird; it's a plane; it's Superman!" without any change of referent for the pronoun. If the appropriate constraints of spatiotemporal continuity are observed, objects retain their perceptual integrity and unity. Since neither spatial location, sensory properties, nor even the most appropriate label need remain constant, we are forced to attribute any object-specific perceptual phenomena to some form of object-specific representation, addressed by its present location and by its continuous history of travel and change through space over time' (Kahneman et al. [1992], p. 217).

file representations, and that indexes attended objects and tracks them through time. ... This ... system ... privileges spatiotemporal information in the service of individuation and numerical identity. Individual objects are coherent, spatially separate and separately movable, spatiotemporally continuous entities. Features such as color, shape, and texture may be bound in the representations of already individuated objects; they play a secondary role in decisions about numerical identity, when spatiotemporal evidence is neutral. (Carey and Xu [2001], p. 181. See also Kahneman et al. [1992].)

To get a feel for how this works, suppose you are shown a panel with a black rabbit in the upper left-hand corner, a white bird in the lower left-hand corner, and a chair in the center. If you are later shown a second panel, with a black rabbit in the lower right-hand corner, a white bird in the upper right-hand corner, and a chair in the center, you will probably report that the rabbit and the bird have traveled diagonally across the panel. This is the action of the sortal-based system. If, however, the chair is removed, and you simply stare at the center while the first and second panels alternate:

Rather than seeing a bird and a rabbit each moving diagonally, you see two individuals each changing back and forth between a white bird-shaped object and a black rabbit-shaped object as they move side to side. (Carey and Xu [2001], p. 183)

This is the mid-level system at work, 'minimiz[ing] the total amount of movement' (Carey and Xu [2001], p. 183).

Given the data, researchers now hypothesize that the mechanisms underlying the infant's concept of a Spelke object and the adult's mid-level object tracker are the same.[25] (So, for example, the infants in the Wynn [1992] experiment described above have represented the two objects behind the screen by opening two separate object files.) The emerging view is that infants begin with something like the mid-level system, individuating and identifying Spelke objects, then gradually develop the sortal-based system, starting about the time that language emerges (around 1 year of age).

This picture helps answer the most vexing question about sortal theories, namely, how could a child ever come to perceive objects in the first place?

[25] See Spelke et al. [1995a], pp. 323–324, Spelke [2000], p. 1235, and especially Carey and Xu [2001]. See also the introduction to Huntley-Fenner et al. [2002] for a survey and other references.

If processes of object recognition underlie perception of the boundaries and identity of objects, then one can perceive objects only if one has a vocabulary of object models. But how do children develop this vocabulary? Because it is hardly likely that humans possess innate knowledge of the visual appearance of cars and trailers, children must have the means to learn about these objects as they encounter them. ... Because every visual scene presents a novel arrangement of objects, a perceiver who possesses neither a rich store of object models nor a set of general principles for organizing arrays into units seems doomed to experience a meaningless succession of novel arrays. (Spelke et al. [1995a], p. 303)

The leg-up is provided by the mid-level system, which continues to function, side by side with sortals, in adults.[26]

In sum, then, research of the sort reported here clearly suggests that the object concept does not depend on many of the experiences once thought to be required, such as manipulating objects,[27] or acquiring sortal concepts, or learning the quantification and identity syntax of natural language.[28] Instead,

A basic process for perceiving spatiotemporally connected and continuous objects arises early in development, without significant tutoring. (Spelke and Newport [1998], p. 297)

Cross-cultural language studies suggest that

This process is likely to be universal across human cultures, leading all people to perceive, act on, and talk about the same spatiotemporal bodies. (Spelke and Newport [1998], p. 297)

Divergences are to be expected, and are found, in the implementation of the second system of objective individuation, the sortal-based system that

[26] There is some debate over whether 'Spelke object' should be understood as an alternative to sortals or as the first and most basic of sortals. Xu [1997] defends the latter position (see also Xu and Carey [1996]); see Ayers [1997] and Wiggins [1997] for discussion. Spelke (e.g., in Spelke et al. [1995a]) tends to contrast 'general' and 'kind-based' processes of object perception, without addressing the issue of sortals. I'm not sure what hangs on this, but if 'Spelke object' is taken as a sortal, it won't be of the same conceptual level as the more familiar sortals (assuming it's backed by the mid-level system), it won't correspond to a simple word, learned early on, and so forth. In other words, lumping it with the familiar sortals would seem to ignore many important differences.

[27] As Piaget believed.

[28] Quine most often takes this view, but in a more guarded moment, he writes, 'To what extent the child may be said to have grasped identity of physical objects (and not just similarity of stimulation) ahead of divided reference, one can scarcely say without becoming clearer on criteria' (Quine [1960], p. 95). The experimental designs described above can be understood as developing those clearer criteria.

begins to emerge in 1-year-olds as they acquire language.[29] The object properties and classifications involved seem sensitive both to experience and to cultural variations.

In short, then, it seems humans are so configured, biologically, that they come to perceive a world of Spelke objects, without instruction, given ordinary maturation in a normal environment. Or, in the Second Philosopher's terms, the ability to perceive Spelke objects is part of a human being's most primitive cognitive equipment.

ii. Properties, relations, and dependencies

Some preparatory experiments for the studies described above involved showing that young infants are able to distinguish, for example, ducks from balls, though they don't use this information for determining object identity. Our concern now is with properties: are 'duckhood' and 'ballhood' understood as features that various Spelke objects can have in common, that a given Spelke object can enjoy or fail to enjoy? In work addressed directly to this issue, Cohen and his co-workers began with a habituation study:[30] one group of 7-month-old infants was repeatedly shown a picture of a particular stuffed animal; a second group was shown a series of pictures of different stuffed animals. After all had habituated to these stimuli, they were shown two pictures: yet another stuffed animal and a rattle. The first group, the one habituated to a single stuffed animal, looked equally at both:

> Both were novel, and the infants could discriminate the test stuffed animal from the one they had seen during habituation. (Cohen and Younger [1983], p. 201)

The second group, habituated to a variety of stuffed animals, found the rattle more interesting: 'For them the test stuffed animal was familiar enough to be treated as something they had seen before' (Cohen and Younger [1983], p. 201). This study initiated the accumulation of evidence that infants do classify objects.

Much work has been done since then with younger infants on more and more difficult tasks. So, for example, in Quinn et al. [1993], the question is

[29] See Spelke and Newport [1998], p. 299.

[30] For a survey of this experiment and others, see Cohen and Younger [1983].

whether or not infants of 3 to 4 months can form separate classifications for perceptually similar objects in such a way that each class explicitly excludes members of the other. In a series of experiments, these subjects were able to form classes of dogs and of cats, both of which excluded superficially similar members of the other. As the researchers note, this is 'an impressive accomplishment in light of the complexity of the various exemplars' (Quinn et al. [1993], p. 473). It seems even young infants manage to collect items into classes according to their features.[31]

Granting, then, that young infants can and do classify similar objects together, notice that we have understood this activity to involve recognizing both that some objects belong to the class and that some do not. Unless some metaphysical weight is attached to the term 'property',[32] we seem entitled to say that these infants conceptualize the world in terms of Spelke objects that can enjoy a property or lack it. Do they also represent combinations of properties?

Work of Cohen and Younger on habituation to correlated features suggests that slightly older infants are also capable of classifying by conjunctions of familiar traits.[33] This work involves drawings of imaginary animals that vary along several axes already familiar to the subjects: giraffe, cow, or elephant body type; feathered, fluffy, or horse-like tails; webbed, club, or hoofed feet; antlers, round ears, or human ears; two, four, or six legs. Infants were habituated to examples with correlations between these features; for example, giraffes with feathered tails and webbed feet, varying between antlers and round ears and between two and four legs, and cows with fluffy tails and club feet, also varying between antlers and round ears, two and four legs. They were then tested on three types of examples: one with the same correlations as in the habituation sequence; another with the same features (giraffe or cow, feathered or fluffy tails, webbed or club feet, antlers or round ears, two or four legs) but without the correlations; and a

[31] See Quinn [2003], pp. 51–53, for a summary of recent work in this area.

As in the case of object perception, here too there is a Piagetian argument that class formation is not understood until a much later age, namely about 7 years. The evidence presented actually has to do with confusions about classes and their subclasses, for which plausible alternative explanations have been suggested, some having to do with problems of language comprehension rather than class formation. For discussion and references, see Markman [1983], Braine and Rumain [1983], pp. 298–302, Macnamara [1986], pp. 163–167.

[32] See III.3, footnote 12.

[33] Younger and Cohen [1983] and [1986]. See also Cohen and Younger [1983], pp. 212–216, for description, discussion, and references.

third entirely novel (e.g., an elephant with a horse tail, hoofs, human ears, and six legs).

Cohen and Younger reason:

If the infants perceived the correlation, the uncorrelated test animal would appear sufficiently novel to the infants to elicit an increase in looking, whereas the correlated test animal would not. In contrast, if the infants were only remembering something about the specific attributes but not the relationship among them, the uncorrelated animal should look as familiar to the infants as the correlated animal. (Cohen and Younger [1983], p. 214)

As it happened, for 10-month-old infants, the looking time for the uncorrelated animal matched that for the novel animal, but the 4- and 7-month-olds looked longer at the novel animal and found both the correlated and the uncorrelated animals to be roughly equally familiar. This suggests a maturation period between 7 and 10 months, but notice this sensitivity to combinations of features still pre-dates the apparently language-linked individuation using sortals.

These studies leave open the question whether the infants are responding to conjunctions of properties—giraffe-like body and feathered tails and webbed feet—or to unanalyzed, overall similarities. Younger points out that in some of the Younger and Cohen [1986] studies

by at least two methods of computing similarity, the 'uncorrelated' test stimulus (i.e., the stimulus that violated the previously experienced pattern of correlation) was more similar overall to the set of habituation stimuli than was the 'correlated' test stimulus. (Younger [2003], p. 83)

One of the similarity measures mentioned here is simple enjoyment of the features most frequently displayed in the habituation sequence; the other a more complex 'exemplar-based classification method' (Younger [2003], p. 99). Further studies showed that though most infants favor the analytic property-by-property approach, some follow a more holistic pattern; such variations are termed 'stylistic'.[34]

Another series of experiments seems more directly relevant to the question of disjunctions of properties. Some of these involve the same imaginary animals considered above (see Younger [1985]); more recent studies along these lines use simpler stimuli (see Younger and Fearing [1999]). So, for example:

[34] See Younger [2003], pp. 86–88, for discussion.

'Infants were presented with photographs of cats and horses during the habituation phase ... or with photographs of male and female faces' (Younger [2003], p. 95). Would they form a single overarching category—'four-legged animal', 'human face'—or would they habituate to a disjunctive category—'cat or horse', 'male or female face'? The answer, once again, depends on the age of the infant. Four- and seven-month-olds followed the more general classification, but ten-month-olds were different: for them, in contrast to the younger infants, a dog appeared novel in the first experiment, a face ambiguous as to gender appeared novel in the second. So there is some evidence supporting the idea of early disjunctive representations.

As for relations between objects, our notion of classification already involves some of that; as Cohen remarks: 'The essence of a concept or category is that one treats as equivalent items that are clearly discriminably different' (Cohen [1988], p. 218). Furthermore, in several of the experiments on object perception, infants discerned relations between objects: for example, in the drawbridge scenario, the relations between the bridge and the box behind it were critical. The point is particularly well illustrated by a related experiment of Macomber and Keil.[35] Four-month-old infants were habituated to a ball dropping from above and passing behind a screen; the screen was then lifted to show the ball resting on the floor. A table was then placed behind the screen. The ball was dropped again, passing again behind the screen, after which the infants were presented with two test scenarios: in one, the screen was lifted to reveal the ball resting on the table; in the other, the screen was lifted to reveal the ball resting on the floor beneath the table. Despite the fact that the second scenario shows the ball in its familiar position, on the floor, infants looked longer at this version, presumably because the ball should not have passed through the table. This substantiates Baillargeon's conclusions—that young infants represent hidden objects and expect them to behave as solid Spelke objects—but for our purposes, the point is that the spatial relations between the ball and the table are clearly perceived.

Quinn has worked directly on the question of infant representation of spatial relations.[36] His experiments began by habituating 4-month-old

[35] For discussion and references, see Spelke [1988], p. 177.

[36] Beginning with Quinn [1994]. See Quinn [2003] for discussion of this and subsequent experiments, and for further references.

infants to a horizontal bar with a single dot above it in various locations; he then showed them a bar with a dot above it in a novel position and a bar with a dot below it. The infants

> Displayed a preference for the novel spatial category—a result consistent with the idea that they had formed categorical representations for the above and below relations between the dot and the horizontal bar. (Quinn [2003], p. 56)

This result was confirmed in a number of different variations. Similar experiments to test for perception of the relation of betweenness also found a preference for novel test stimuli, but only in slightly older, 7-month-old subjects. In further research, the objects in the habituation series were varied while the spatial relations remained the same, testing for a more abstract 'aboveness' or 'betweenness' relation. Positive results were obtained in 7-month-olds for 'above' and in 10-month-olds for 'between'.

Thus experimental evidence supports the claim that infants conceptualize the world in terms of Spelke objects bearing certain properties and standing in various relations, in such a way that one such object can have or lack a property, can stand or not stand in a relation to another, can enjoy a combination of properties or relations, or one or another of several properties or relations. I know of no such research on representations involving quantification: for example, might an infant habituated to scenes of all ducks, all balls, all red objects, have a preferential reaction to an image in which every object but one is square, indicating that it appears novel? Obviously, speculation is pointless in these matters, so I must leave this part of the Second Philosopher's empirical hypotheses unsupported.

Finally, the subjects in many of the experiments we've reviewed seem to perceive dependencies between the properties and relations enjoyed by some objects and the properties and relations enjoyed by others—for example, the movement of the drawbridge depends on the location of the box behind it, the location of a table determines whether or not it is possible for a ball to reach the floor—but the most explicit research on this topic has involved causal dependencies. So, for example, Leslie has argued that infants perceive causal relations, using experiments involving films of billiard balls:[37] in the first, a white ball enters from the left, comes into contact with a black ball in the center of the scene, after which the white

[37] Leslie and Keeble [1987]. For summary and discussion, see Leslie [1988].

ball remains in place and the black ball exits to the right ('direct launch'); in the second, exactly the same events happen, but there is a time delay between the moment the white ball arrives and the moment the black ball departs ('delayed reaction'). (To adults, the first will appear as a causal interaction and the second will not.) Leslie's idea was to habituate infants to one or the other of these scenarios, then show them its time reversal:

> If infants perceive causal direction only in direct launching and not in delayed reaction, they will be differentially sensitive to their reversal. They ought to respond to causal *and* spatiotemporal reversal in the case of direct launching, but only to spatio-temporal reversal in the case of delayed reaction. Reversal of direct launching should therefore produce greater recovery of interest. (Leslie [1988], p. 190)

And this was observed.

Subsequent studies by Oakes and Cohen use a more familiar design: groups of 10-month-old infants were habituated to either a causal event (direct launch) or a non-causal event (either delayed reaction or launching without a collision)[38]; they were then shown causal and non-causal events as test stimuli.

> The results ... clearly supported the conclusion that infants at 10 months of age ... perceived the causality of the launching sequences. ... When habituated to the causal event, the 10-month-olds dishabituated[[39]] to both novel noncausal ones, but when habituated to a noncausal event, they dishabituated only to the novel causal event and not to the novel noncausal one. This pattern of results occurred even though there was greater physical dissimilarity between the two noncausal events than between either noncausal event and the causal one.[[40]] (Oakes and Cohen [1990], p. 205)

This provides evidence that the subjects were classifying the events in terms of the presence or absence of causal dependencies.[41]

[38] That is, an object moves in from the left, stops at some distance from the second object, and the second object immediately moves off to the right. Oakes and Cohen's experiments also used familiar toys rather than billiard balls, a more complex visual presentation.

[39] That is, they preferred or looked longer at these outcomes.

[40] That is, 'the two noncausal events differed ... on two dimensions, space (... separation vs. no separation) and time (... delay vs. no delay). In contrast, the noncausal events differed from the causal one only on a single dimension, either space or time. Thus, even though the 10-month-olds had to be sensitive to these dimensional differences to determine whether or not the event was causal, they used that information to organize the events on the basis of causality, not on the basis of dimensional differences per se' (Oakes and Cohen [1990], p. 205).

[41] Note the suggestion that Hume was wrong to suppose that our impression that A causes B is due to much experience of Bs following As.

In sum, then, recent research on infant cognition suggests that humans are able to detect and represent some of the world's KF-structures from a very young age, in the course of normal development, without training or extensive sensory-motor experience. These structures in turn support the validities of rudimentary logic, which is why, the Second Philosopher claims, at least the simplest of these will seem obvious to us: suppose, for example, that we represent a situation as involving an object that is either red or green, and further represent it as not red; if we then entertain the possibility of its being green, the very structure of our representations inclines us to think that surely must be. (For the record, this isn't intended as a theory of human reasoning[42]—it's just an attempt to account for some simple inferences that we find compelling.) This leaves the question: what is the connection between the world's KF-structures and our capacity to detect and represent them?

iii. From the world to cognition

Finally, what evidence is there for the Second Philosopher's (3′), the claim that humans are so configured because the macro-world is so structured? Developmental psychologists tend to begin from the observation that the notion of an object is obviously dependable:

> Perceiving objects makes sense, because a single object usually forms a more stable configuration than does a scene as a whole. If we follow the scene in the figure [a Cézanne still life] over time, the bowl and the table may part company, but the objects themselves are likely to persist. (Spelke et al. [1995a], p. 297)

Furthermore, the spatiotemporal information central to the notion of a Spelke object is generally more reliable than the information infants

[42] There is considerable debate in the psychological literature over the role of logic in human reasoning: one school of thought posits a mental logic, a system of general purpose inference rules, often much like natural deduction rules, that are used in human reasoning; another—the mental model view—imagines the reasoner constructing a model and manipulating it to draw conclusions; a third sees human reasoning as domain specific, carried out by the use of particular domain-sensitive schemas; still others highlight heuristics and biases to explain how people reason. (See Evans et al. [1993] for a survey.) While the Second Philosopher's belief in some rudimentary logical structuring near the basis of all cognition bears some family resemblance to the first two schools, it does not preclude very different mechanisms playing central roles in developed theories of how people in fact tend to infer and reason.

eschew. So, for example, property- and sortal-based judgments can be misleading:

Suppose that we see a person with long black hair, wearing a red jacket and a black skirt. She then walks out of the room. A few minutes later a woman with a pony tail and glasses comes in, wearing a blue jacket and a blue skirt. We first note the differences in appearance ... Is she the same person or not the same person? ... detecting perceptually discriminable properties does not warrant the inference of numerically distinct individuals. (Xu [1997], p. 382)

And the types of spatiotemporal information infants prefer—cohesion, solidity, continuity of motion—are more reliable than those they tend to ignore—similarity, good continuation, good form—

Although not all objects are regular in shape and substance, move on smooth paths, or maintain constant shapes, colors, and textures throughout their existence, all objects move cohesively and continuously. (Spelke et al. [1995c], p. 176)

Objects are more likely to move on paths that are connected than they are to move at constant speeds ... and they are more likely to maintain their connectedness over motion than they are to maintain a rigid shape. Infants' perception appears to accord with the most reliable constraints on objects. (Spelke et al. [1995a], p. 320)

Kellman particularly emphasizes what he calls the 'ecological validity'—that is, 'the accuracy, in ordinary circumstances, of the relation between perceptual information and facts about the environment' (Kellman [1993], p. 122)—of principles like 'common fate':

When visible areas share identical motions in space (or rigid motions in general), it is highly likely that they are connected. ... The parts of connected entities will almost invariably move in connected ways. ... Connected motion will rarely occur for separate entities. Even separate objects falling under the influence of gravity, or flocks of birds headed in the same direction will ordinarily be detectably inconsistent with rigid unity. ... The ecological root of [that is, the worldly structure corresponding to] a common fate principle is that object motions ordinarily result from the application of forces. The likelihood of forces being applied by chance to separate objects so that their motion paths are rigidly related must be vanishingly small.[43] (Kellman [1993], p. 127)

[43] Cf. Kellman's wry observation: 'Those who have attempted to arrange common motion of visible objects by hidden mechanical means, e.g., in perception laboratories, can attest that it is a painstaking task' (Kellman [1993], p. 127).

By contrast, similarity works only when objects are relatively homogeneous, good continuation when they have smooth boundaries, and good form when they have regular shapes—all much less common than coordinated motion of an object's parts.

Granting their effectiveness, how do we come by these cognitive mechanisms, our first access to KF-structures in the world? Observers have often suggested that the ability to detect and represent Spelke objects is learned through our early interactions with the world:

> Helmholtz [[1867]] proposed that children learn to perceive objects by handling and moving around them, observing changing perspectives that active movements reveal. ... Piaget [[1937]] proposed that object perception results from the child's progressive coordination among activities such as reaching, grasping, sucking, manipulating, and visual following. Quine [[1960]] proposed that object perception results from the acquisition of language, particularly from linguistic devices for distinguishing one object from another and for distinguishing bounded from unbounded stuff... Wiggins [[1980]] and others [the sortalists] have proposed that object perception results from the acquisition of systematic knowledge of object kinds, such as *chair*, *pot*, *tree*, and *dog*. (Spelke and Newport [1998], p. 291)

As we've seen, the line of research sketched here undercuts all these views. The infants in the relevant experiments, those who perceive Spelke objects, are 4 months old or younger, too young 'to reach for, manipulate, and locomote around objects' as Helmholtz and Piaget require.[44] Furthermore, the abilities in question are present across cultures and in non-human animals and pre-linguistic infants, so they aren't acquired along with our native language.[45] And finally, these subjects are not yet making use of sortal information.

[44] Spelke and Newport [1998], p. 291. See also Spelke et al. [1992], p. 627: 'At 3 and 4 months of age, infants are not able to talk about objects, produce and understand object-directed gestures, locomote around objects, reach for and manipulate objects, or even see objects with high resolution.'

[45] Quine seems aware of some of this work; in his [1990], p. 24, he writes: 'True, an infant is observed to expect a steadily moving object to reappear after it passes behind a screen; but this all happens within the specious present, and reflects rather the expectation of continuity of a present feature than the reification of an intermittently absent object.' He gives a similar reading of various abilities of dogs. What he thinks the infant cannot do is anything comparable to his own ability to 'try to decide whether the penny now in my pocket is the one that was there last week, or just another one like it' (p. 25). But it seems infants are doing something very like this, e.g., in Spelke et al. [1995b], described above: whether or not the object that appears on the left is perceived as the very same object as the one that disappeared on the right is what determines which of the one- and two-object displays is novel.

What happens in younger infants, in newborns, is less clear.[46] Some evidence indicates they are less inclined to complete an occluded figure, at least before 2 months of age, but this deficit might trace to early limitations on visual abilities, not absence of the object concept. It could be that very young infants undergo learning from experience,[47] but it might also be that their sensory organs simply undergo ordinary maturation: perhaps there is an underlying competence in object perception whose performance is being blocked by other factors. Distinguishing performance from competence is difficult, but inventive researchers have found ways. For example, Spelke and Newport describe ground-breaking work of Thelen on how infants learn to walk.[48] Most animals learn their species-specific form of locomotion much earlier than humans; for that matter, humans begin with a form—crawling—quite similar to those of many other animals; only later do they develop the distinctively human mode of walking upright. It seems unlikely that infants come equipped with an underlying competence for walking upright that takes time to manifest itself, but this is the thesis that Thelen defends with her ingenious experiments.[49]

Efforts to test the hypothesis of underlying competence in the present case have so far been less successful:

> ... limits on a newborn infant's visual sensitivity may be so great that no ordinary visual experiment will reveal some of the perceptual competences that are present and waiting to be exercised. If that is the case, then the study of the origins of object perception must await the emergence of an investigator with Thelen's genius—someone who can devise situations that circumvent the sensory limitations preventing newborn infants' inherent perceptual capacities from functioning in natural contexts. (Spelke and Newport [1998], p. 295)

[46] The following two paragraphs follow Spelke and Newport [1998], pp. 294–297.

[47] Spelke et al. [1992], pp. 627–628, give reasons for doubting various accounts of this purported learning process.

[48] See Spelke and Newport [1998], pp. 278–282.

[49] To illustrate just one aspect of this work, consider this problem: why do infants stop kicking (a highly organized pattern of motion similar in structure to locomotion routines in humans and other animals) around 2 months of age only to begin again later? Answer: because their growing legs are temporarily too heavy for their developing leg muscles. How can this hypothesis be tested? 'In a simple and elegant way, Thelen [et al.] tested the hypothesis that infants who no longer step in an upright posture maintained the underlying competence to do so: They plunged the infants' legs under water, reducing the force needed to lift them. Under these circumstances, stepping re-emerged!' (Spelke and Newport [1998], p. 280).

It seems the question—what combination of genetic endowment, maturation, and experience is responsible for the infant's ability to detect and represent objects around 4 months of age?—must be considered open, at least for now. And similar if not more vivid uncertainties no doubt arise for our ability to perceive other aspects of KF-structuring—like compound properties and dependencies—that develop in later infancy, from 7 to 10 months. Experience of the various sorts usually posited as responsible for these achievements would appear to be irrelevant, but the possibility that very early post-natal stimuli play a role cannot be ruled out.

For that matter, though the accumulation of evidence does count heavily against the claim that these cognitive abilities are acquired with language, the possibility remains that they are of a piece with the underlying capacity to acquire language, part of what's often called 'the language learning device'. This idea founders on evidence from non-human animals.[50] So, for example, newborn chicks provide a fertile ground for study of object perception, given their capacity to imprint: chicks raised in isolation imprint on whatever objects they're presented with; the nature of a chick's representation of the object can then be uncovered by noting its subsequent reactions to test objects. Experiments show chicks treating a partly hidden triangle as if it were complete, searching for imprinted objects behind a screen, and so on, suggesting that they, like human infants, 'represent [the object's] continued existence over occlusion'.[51]

More precise comparisons of human infants with non-human animals require a new methodology:

> We faced a serious methodological challenge. Could we use looking time methods with nonhuman primates? In the wild? In the laboratory? Would monkeys be curious about the magic tricks we show infants? Would they sit still and look at the outcomes? (Hauser and Carey [1998], p. 62)

[50] This recalls the Stoic logician Chrysippus, of whom Mates reports: 'He claimed to have noticed that when a dog is chasing an animal and comes to a point where the path he is following divides into three, if he sniffs first at the two paths which the animal did not take, he will dash off down the other way without pausing to test it' (Mates [1972], pp. 215–216). The history of this apparent application of rudimentary logic is sketched in Blackburn [1994], p. 17.

[51] See Spelke and Newport [1998], p. 292, for this quotation and for further description, discussion, and references.

Hauser and Carey[52] managed to design experiments suitable for both wild and laboratory primates—both closely human-related (rhesus monkeys) and more distantly human-related (cotton-top tamarins)—and these subjects performed much like pre-linguistic infants on such tasks as Wynn's one-object-and-another-object:

> These experiments show that ... rhesus monkeys ... and cotton-topped tamarins represent [Spelke objects] with spatiotemporal criteria for individuation and numerical identity. (Hauser and Carey [1998], p. 67)

Finally, on the matter of detecting and representing properties, Cohen writes:

> Before one becomes too enamoured of the cognitive sophistication of the infant, one should examine the animal literature on categorization ... (Cohen [1988], p. 226)

He goes on to cite work with pigeons[53] and Japanese quail, and concludes:

> We seem to be drawn to the conclusion that categorization, like habituation and memory, is a basic adaptive mechanism available to a wide variety of species. (Cohen [1988], p. 226)

Wynn describes the situation this way:

> The fact that a wide range of humans and other vertebrate species possess similar ... abilities suggests that these abilities may have developed far back in our evolutionary history, at a point prior to the branching off of these different species. (Wynn [1998b], p. 120)

She goes on to point out that this isn't the only explanation possible; it could be that similar abilities 'may have developed independently in different species, as an easily instantiated biological solution to a common problem space' (Wynn [1998b], p. 120).[54] The task of finding evidence that discriminates between these two hypotheses remains open.

The evolutionary hypothesis is certainly appealing: the advantages these abilities confer were presumably as strong in our ancestral environment as

[52] See Hauser and Carey [1998], pp. 61–67, for summary and references.

[53] See Pearce [1988] and the work of Herrnstein cited there.

[54] Wynn is actually discussing the analog device discussed in IV.2.ii, but presumably the point carries over to the present context. Thanks to Kyle Stanford for his advice on the following discussion of the evolutionary hypothesis.

they are today;[55] it seems unlikely that we acquire them via more general learning mechanisms;[56] and so on. But fortunately the specifics here aren't crucial to our purposes. Whether their presence in us is due to evolutionary pressures on our ancestors or to our own early perceptual experiences or to some combination of the two, it remains the case, as (3′) requires, that human beings have primitive cognitive mechanisms capable of detecting and representing KF-structures because they live in a largely KF-world and interact almost exclusively with its KF-aspects. The evidence clearly suggests that we—like the chicks, pigeons, and monkeys—are responding directly to some of the world's most elementary features.

[55] As Sober points out, more refined aspects of human cognition such as exercise of the scientific method aren't obviously survival enhancing even now—'discovering beautiful theorems is central to science, but peripheral to reproductive success' (Sober [1981], p. 98)—and would have been even less so in ancestral environments. Skepticism among philosophers about evolutionary origins for human cognitive achievements tends to concern, e.g., higher, uniquely human capacities (e.g., science or language)—at issue here are more humble abilities that we share with a wide range of non-human animals—or the correctness of our classifications (e.g., Stich [1990], pp. 61–62)—at issue here is the claim that we conceptualize objects as having properties, not the accuracy of any particular property attributions.

[56] See Andrews et al. [2002] for discussion of the place for general learning mechanisms in our assessment of evolutionary hypotheses. Andrews and his co-authors review the strengths and weaknesses of various types of evidence often cited for the claim that a given trait is an evolutionary adaptation; the sort of case in question here comes closest to those discussed in Andrews et al. [2002], §3.1.6.1, which they classify among the potentially more reliable approaches.

III.6
The Status of rudimentary logic

We've surveyed the evidence and concluded that a modified version of the second-philosophical account of rudimentary logic proposed at the end of III.3 is viable: (1′) rudimentary logic is true of the world insofar as it's a KF-world, and in many but not all respects it is, (2′) human beings believe the simple truths of rudimentary logic because their most primitive cognitive mechanisms allow them to detect and represent KF-structures in the world, and (3′) the primitive cognitive mechanisms of human beings are this way because we live in a largely KF-world and interact almost exclusively with its KF-structures.[1] Given this picture, what can we say about the status of the validities of rudimentary logic?

To begin with metaphysics, we must come to terms with the fact that even this rudimentary logic cannot be assumed to apply effectively in every situation; it will only deliver validity when the relevant KF-structure is present. We might think of KF-structuring as a template that fits the world in many of its aspects—from the planets of the solar system to the cards in a bridge game—but not in all—for example, not in the recalcitrant quantum world. Applying this logic successfully, then, requires identifying, most often implicitly, a particular context of objects, properties, relations, and dependencies. For a simple example, we might speak of cards, their ranks and suits, for a complex example, of animals, their species and habitats;

[1] Those with balky intuitions might meditate on the following. Suppose that instead of our normal, continuous experience, we experience only instantaneous snapshots of the world. Suppose, for simplicity that inputs only come in red or green, round or square: when we think about color while (doing what feels like) opening our eyes, we get a color reading; when we think about shape and do the same, we get a shape reading. Now suppose we perform a series of experiments and get results analogous to those rehearsed for spin measurements (in III.4): if we do two shape readings in a row, the shape stays constant, but if we do a color reading in between, half the once-square items come out square and half come out round—and so on, through the various quantum oddities. It seems unlikely that we would have developed our KF-based cognitive machinery in a non-KF world like this, where it would have led us astray. What we might have developed instead is an open question.

as long as the KF-structuring is present, logic will serve us well. But it will lead us astray when we attempt to speak of subatomic particles, their properties and trajectories, because the required KF-structuring isn't there.

To spell this out, consider a simple disjunctive syllogism: the last card in her hand is either a club or a heart; it can't be a heart (because Joe played the last one); so she's holding a club. The cards and bridge players here are objective individuals; each card has a rank and suit; each card either has been played or remains in one and only one player's hand; and so on. The world of the bridge game is a simple KF-world, and in KF-worlds, disjunctive syllogism is valid: if (*Ca* or *Ha*) and not-*Ha*, then either *Ca* or *Ha*, and *Ha* fails, so *Ca* must obtain.

What I've given here is a soundness proof for rudimentary representations: if an infant (or a dog, for that matter[2]) detects situations of the forms (*Ca* or *Ha*) and not-*Ha*, then he can reliably conclude that *Ca*.[3] Commentators[4] are often troubled by arguments of this form—more sophisticated versions, in defense of logical rules expressed in natural or formal languages[5]—on grounds of circularity: my attempt to justify a disjunctive syllogism made use of that very principle; how can such an argument establish anything? But notice, the seemingly transparent soundness proof can be carried out only because the necessary objects, properties, relations, dependencies, are available in the context, that is, because the necessary KF-structure is there to be exploited.[6] That we tend not to notice this only

[2] See III.5, footnote 50.

[3] Again, I intend here no claim about how reasoning is actually carried out (see III.5, footnote 42). My point is that such an inference would, as a matter of fact, be valid.

[4] See Dummett [1973b] and Haack [1976], [1982] for discussion.

[5] Notice, for the record, that the Tarskian machinery of truth-in-an-interpretation used in such proofs is available to the disquotationalist of II.4: as noted in II.2, the inductive clauses follow from the disquotational account of truth and the base clauses are disquotational in form (this was Field's complaint). So, e.g., a compositional language with predicates '*C*' and '*H*' and name '*a*' can be interpreted in the domain of cards in the bridge game by associating '*C*' with $\{x|x$ is a club in her hand$\}$, '*H*' with $\{x|x$ is a heart in her hand$\}$, and '*a*' with the last card in her hand.

[6] By his own route, Wilson comes to a complementary position. In the terminology of II.6, notice that inferring '… and__' from '…' and '__' may lead us astray if '…' and '__' come from different patches of a word's façade structure (e.g., 'the astronaut is weightless' and 'the astronaut weighs 176 pounds' from different patches of the 'weight' façade), not to mention the dangers posed by familiar logical manipulations in periods of semantic agnosticism (think of Heaviside's calculus). Wilson writes: 'our soundness proof sketches the *minimal structure* that we must locate within a broader scale picture to be able to announce, "Whew! At least we won't have to worry about failures of logical reasoning here" ' (Wilson [2006], p. 628). I say 'complementary' because Wilson's concern is with logical failures brought on by faulty word–world relations, not by anomalies (i.e., non-KF-structuring) in the world itself. (For worries closer to Wilson's, see III.7, the text preceding footnote 6.)

shows how thoroughly we're disposed to think in terms of KF-structures; the presupposition that they are present is nearly invisible.

So the validity of rudimentary logic is contingent on very general structural features of the situation in which it's applied, in particular, on its KF-structure. What happens, then, in cases where this structure is lacking, for example, in the case of electron spins discussed in III.4? There we attempt to impose a KF-structure, like that of the bridge game: we take the electrons to be objects, their vertical and horizontal spins to be properties. Treating these in the usual ways, we assume every electron has a vertical spin, either up or down, and a horizontal spin, either left or right. Proceeding as usual, we have ((vertical spin up or vertical spin down) and (horizontal spin left or horizontal spin right)).[7] By our rudimentary logic, this implies ((vertical spin up and horizontal spin left) or (vertical spin up and horizontal spin right) or (vertical spin down and horizontal spin left) or (vertical spin down and horizontal spin right)); this is just an application of the distributive laws. But this conclusion is precisely what we don't find: the electron doesn't seem to enjoy any combination of a particular vertical spin and a particular horizontal spin. The KF-structure has broken down here, and with it, the distributive laws.

Still, we tend to overlook the contingency of rudimentary logic; indeed, KF-structuring (and more) is built into our very notion of a 'possible world'! This impression of necessity isn't surprising, given that KF-structuring is so fundamental to our cognition—so fundamental that we find it difficult even to imagine a world with significantly different structure. Creator-worlds are conceivable,[8] though some may regard them as artificial, not to be taken seriously.[9] In any case, it seems the psychological realities described in (2′) and (3′) provide adequate explanation for the common view that logic is necessary, true in every possible world.[10]

[7] For readability, I abbreviate 'the electron has vertical spin up' as 'vertical spin up'.

[8] Perhaps part of the reason that Creator-worlds are psychologically accessible to us is because their logic differs from rudimentary logic largely over matters involving negations. Even a strict KF-thinker can imagine alternatives there, as we're sometimes inclined to consider a borderline bald man to be not-(not-bald). The first 'not' is different from the one taken as primitive in rudimentary logic: it applies when the negated situation is either false or indeterminate.

[9] See footnote 11.

[10] Beyond this psychological 'must', the normativity forthcoming on this factualist account of logic differs only in degree, not kind, from what's present in other fields: you must admit these things if you want to stay on the right side of the truth. Frege's third realm factualism (see III.1.i) seems to yield similar sentiments: 'Any law that states what is can be conceived as prescribing that one should think

Another sense of quasi-necessity can be traced in the Kantian roots of the notion of a KF-world. Recall that the forms of judgment and pure categories are independent of the forms of intuition, that is, a discursive intellect could have different forms of intuition, but must still use the same logical forms and categories. Translated to the second-philosophical context, with the Fregean updates thrown in, this becomes the claim that KF-structures in our world are independent of its spatiotemporality. So logical validity is contingent, but it isn't contingent on any particular facts, and it isn't contingent even on spatiotemporal structure. Rudimentary logic will hold even in a world that violates the laws of our spacetime physics, as long as it's a KF-world.[11]

Turning to epistemology, the logical validities of rudimentary logic are built into the structuring of our innate, or nearly innate, representational capacities; our understanding of the world is subject to them as soon as we detect and represent anything at all. As a result, so the Second Philosopher claims, the simplest such validities will seem obvious or self-evident to us. At whatever point we should be counted as having beliefs,[12] we will believe them; indeed, it seems appropriate to say that we come to these beliefs a priori.

Whether or not this counts as a priori knowledge will depend on issues in the theory of knowledge proper.[13] An externalist[14] might argue that the evolutionary processes (and early experience and maturation) that most likely lead to our current cognitive make-up are reliable, and hence, that the true beliefs they produce count as knowledge. An internalist, on the other hand, would insist that we have access to an explicit justification. Perhaps a simple soundness proof of the sort rehearsed above would be enough, or perhaps a full investigation of the ground of logical truth, like the one undertaken here, would be required. Having outlined the relevant facts of the case, I leave to the epistemologists any further judgments on where to employ or withhold the honorific 'knowledge'.

in accordance with it, and is therefore in that sense a law of thought. This holds for geometrical and physical laws no less than for logical laws' (Frege [1893], p. xv; see also Frege [1897], pp. 145–146).

[11] Perhaps those who dismiss Creator-worlds are reacting to their dynamic structure, relying on some intuition that logic should be independent of temporal considerations.

[12] Some might say we believe them as soon as we detect and represent, that is, pre-linguistically; others might think beliefs require language.

[13] See IV.4, footnote 60, for a different sort of candidate for a priori knowledge.

[14] See II.2, footnote 42.

Granting, then, that rudimentary logic is contingent, and in one sense a priori and another sense a posteriori, we should consider the questions so much discussed in the philosophy of logic: is it empirical, is it subject to revision? Clearly it is not empirical in a psychologistic sense: though we determine empirically that humans tend to think in rudimentary logical terms, this isn't what makes its patterns valid—that job is done by the KF-structures in the world—our thinking in these ways is a mere consequence. And it also isn't empirical in the sense of simple inductivism: we don't come to believe in disjunctive syllogism because we've applied it successfully in a series of cases; we so believe because of the structure of our cognitive machinery. The empirical investigation required to justify our true logical beliefs involves verifying the KF-structure of the world, not simple inductive testing of instances of the syllogism. Nor does this second-philosophical position coincide with classical Quinean empiricism. Our embrace of rudimentary logic is of a piece with our most primitive perceptual abilities; it doesn't derive from the role of logic in some central theoretical block, alongside physics, mathematics, and so on, that's confirmed ever-so-indirectly by sense experience. Quine's monolithic web, as a whole, describes the world, as a whole; the Second Philosopher's rudimentary logic reflects a particular aspect of the world, namely, its KF-structuring, and is confirmed insofar as we confirm that structuring. Where Quine explains our tenacious adherence to logic as a pragmatic choice, driven by the principle of minimum mutilation, the Second Philosopher sees a symptom of just how deeply rudimentary logic is embedded in our most primitive ways of representing the world.

But once again: is it revisable, could empirical evidence dictate a revision of logic? As we've seen (in III.1), Quine takes this to be possible in principle, but highly unlikely in practice:

> Whatever the technical merits of the case, I would cite ... the maxim of minimum mutilation as a deterring consideration. ... let us not underestimate the price of a deviant logic. There is a serious loss of simplicity ... And there is a loss still more serious, on the score of familiarity. Consider ... the handicap of having to think in a deviant logic. The price is perhaps not quite prohibitive, but the returns had better be good. (Quine [1970b], [1986], p. 86)

Putnam's position is more complex. He holds that some logical beliefs could not be disconfirmed by observation and experiment alone, but only 'by thinking of a whole body of alternative theory as well' (Putnam [1995], p. 272). His idea is that we can imagine using familiar empirical methods, like observation and experiment, to falsify a claim like 'the moon's core is made of such-and-such', but in a case like the law of non-contradiction—not-((...) and not-(...))—'we do not today know how to falsify or disconfirm [it], and we do not know if anything could (or would) disconfirm [it]' (Putnam [1995], p. 272). In other words, though the law of non-contradiction is not necessary, though it could be false, we couldn't become convinced of this by ordinary empirical considerations; we would need some further instruction on how to conceptualize its falsity.[15]

Now there is much in the spirit of this position that the Second Philosopher finds congenial.[16] Rudimentary logic is contingent, not necessary—it would not be applicable if the world were different in fundamental ways[17]—but we are hard put to imagine finding out that our world has no KF-structuring, after all.[18] The Second Philosopher doesn't go as far as Kant—who holds that we simply cannot think of the world without using the forms of judgment and categories[19]—but she admits that all evidence suggests it would be hard for us to do so. Our continued failure to come to an understanding of the micro-world suggests how difficult it can be to think in non-KF-terms.

But there's more to Putnam's position; he also holds that some logical beliefs *can* be overthrown by ordinary empirical methods. Given that KF-structuring seems to break down in cases like electron spin, it's not surprising that the empirically driven revision Putnam advocates is a new

[15] Putnam [1978a], pp. 111–112, admits to being 'torn' on such questions, so my straightforward characterization of his views here doesn't include or even square with much of what he says in the relevant papers (Putnam [1962], [1978a], and [1995]). But this simple idea is one thread in his thinking, the one relevant here.

[16] The Second Philosopher can't use Putnam's example, the law of non-contradiction, as is, because this is not a fact of rudimentary logic. Still, ((...) and not-(...)) is never true, and not-((...) and not-(...)) is never false, so maybe this is enough. If not, she can switch to an example involving validities, like the distributive laws or DeMorgan equivalences.

[17] I've been saying that under certain circumstances, rudimentary logic would not be applicable; Putnam speaks of disconfirmation and falsification. I come back to this point in a moment, but I don't think it affects the current discussion.

[18] Again, we can imagine some parts of rudimentary logic failing in a Creator-world, but it's hard to imagine our world turning out to be one of those (cf. III.7, footnote 35).

[19] He needs the strong claim to conclude that we can know facts about the world of experience a priori (see I.4).

logic for quantum mechanics.[20] As we've seen, this requires the failure of at least some distributive laws, but Putnam regards these as different in kind from the law of non-contradiction—only the latter are what he calls 'necessary (relative to a body of knowledge)'.[21] Now we've seen that the Second Philosopher can also distinguish some parts of rudimentary logic as 'less necessary' than others—for example, the intuitionistic notion of a Creator-world has shown her how not-(not-(...)) might fail to imply (...)—but the distributive laws, common to rudimentary and intuitionistic logic, are not among these. This certainly appears to be a case in which we would need some 'alternative theory', say a picture of the world and its structure comparable to KF-worlds or Creator-worlds, to help us understand how the distributive laws could go wrong.[22] Perhaps oddest of all, rather than presenting us with a non-distributive and hence non-KF picture of the world, Putnam hopes to reinstate the idea that the electron has both a definite vertical spin and a definite horizontal spin, to reinstate, in other words, one of the KF-structures that seems in fact to be absent!

This sounds impossible, and in fact, Putnam's grand effort does not succeed. The promise is that the introduction of quantum logic will resolve the quantum anomalies and reinstate KF-structuring, and the fact is that it does neither.[23] There is a large literature on this subject which I don't pretend to have surveyed, but a few points now seem uncontroversial. We've seen how application of a distributive law to ((the electron has vertical spin up or down) and (the electron has horizontal spin left or right)) yields a string of disjuncts—like (the electron has vertical spin up and horizontal spin right)—none of which obtains. Putnam's quantum

[20] See Putnam [1968]. (This is not Reichenbach's simple three-valued approach, mentioned in III.1, footnote 14.) In Putnam [1995], even as he defends the necessity (relative to a body of knowledge, see below) of the law of non-contradiction, he remarks—'I do not claim ... that *no* revisions in classical logic are conceivable'—and reiterates his 'sympathy for ... quantum logic' (Putnam [1995], p. 281, footnote 10).

[21] Where this means: relative to our current state of knowledge, we would need some new theory to understand how they could be false.

[22] Putnam gives a list of principles that purport 'to be the basic properties' of the connectives and don't include the distributive laws (Putnam [1968], pp. 189–190). On the one hand, it's hard to see what makes these special; as Gibbins writes in his discussion of this point, 'why should just these properties ... be the essential properties which define the connectives? Putnam's claim seems to be simply arbitrary' (Gibbins [1987], p. 156). On the other hand, even if some case could be made that dropping a distributive law doesn't 'change the meaning' of the connectives, we're still left without a way of understanding how such a law could fail.

[23] Hughes [1989], p. 212, writes: 'the package we have bought seems markedly less attractive than the product which was advertised.'

logic lacks the relevant distributive law, which blocks this problem, but it includes others, and Peter Gibbins ([1987], pp. 147–151) shows how a closely related quantum anomaly arises in the new context. In addition, the sense in which particles are said to have determinate properties comes to the observation that (the electron has a vertical spin and a horizontal spin), despite there being no particular vertical spin and horizontal spin that occur jointly. Gibbins concludes that 'quantum logical realism is therefore a realism in name only' (Gibbins [1987], p. 152). There is much more to the story of quantum logic, but for purposes of this discussion, let me quote the conclusion drawn by Maria Luisa dalla Chiara and Roberto Guintini in their recent survey of the field:

> It seems to us that quantum logics [there are now many] are not to be regarded as a kind of 'clue', capable of solving the main physical and epistemological difficulties of [quantum theory]. This was perhaps an illusion of some pioneering workers in quantum logic. (dalla Chiara and Guintini [2002], p. 225)

The footnote to this passage cites Putnam.

Before leaving the topic of quantum logic, at least for now,[24] it's worth asking how we manage to develop our quantum mechanical theories if rudimentary logic fails and we have no viable quantum logic. The answer is that we have a workable theory not of particles in relations with dependencies, but of state vectors residing in a mathematical Hilbert space. This formalism issues forth predictions of astounding accuracy. And, happily, it displays an abstract KF-structure—mathematical objects with properties, standing in relations, with logical dependencies—so rudimentary logic can be safely applied. But this move by no means solves the underlying riddle of quantum mechanics. As Hughes puts it:

> The theory uses the mathematical models provided by Hilbert spaces, but it's not clear what categorial elements we can hope to find represented within them, nor, when we find them, to what extent the quiddities of these representations will impel us to modify the categorial framework within which these elements are organized. (Hughes [1989], p. 176)

The term 'categorial framework', due to Stephan Körner,[25] descends from Kant by a somewhat different route than our Second Philosopher's theorizing, but her notion of a KF-world would seem to qualify. As this

[24] It comes up again briefly in III.7. [25] See Körner [1969].

structure isn't present, we need an alternative to describe the structure of the quantum world, and this is what we don't have and can't seem to imagine. If such a thing were to track the mathematical formalism closely enough, rudimentary logic might still apply, but nobody knows.

To return, at last, to the question at hand: is rudimentary logic empirical, can it be revised? Though there doesn't seem to be any persuasive empirical case for revising rudimentary logic at this point, the Second Philosopher considers this a possibility in principle. She recognizes, however, that such a revision—especially one that reaches into principles that rudimentary shares with intuitionistic logic—would require considerable conceptual flexibility, perhaps more than our feeble human brains will be able to achieve.

Finally, in this catalogue of the status of rudimentary logic, I should touch on the traditional question of analyticity: is rudimentary logic true by virtue of meaning? If it is, then the rudimentary logician, the intuitionist, and the quantum logician don't disagree; they simply assign different meanings to the logical particles.[26] Having no functional theory of meaning, suspecting in fact that there is no viable theory that would do the work philosophers ask of it, I find this question hard to pin down, but at the very least, I think we can now reformulate our concern from III.1.iv in more poignant terms.

Suppose, for the sake of argument, that we manage to assign our meanings so that rudimentary logic comes out true. No amount of success in this endeavor will change the fact that the micro-world does not display the characteristics of a KF-world. In other words, regardless of how we specify our meanings, the question of whether or not the resulting logic applies in a given context remains open. Even if logical truth *were* a matter of meanings, we would still have to ask, why adopt *these* meanings as opposed to *those*? And our answer would still be, because these meanings are applicable in any KF-world, and much of our world displays this structure.

To come at the issue from another direction, consider what we should say about the truth value of rudimentary logic in contexts where it is inappropriate, like our micro-world or any Creator-world. As it happens, we seem to have two entrenched ways of reacting in such cases: sometimes, as with phlogiston theory, we tend to say that the theory has been falsified; other times, as with Euclidean geometry, we prefer to say that the theory

[26] This isn't Putnam's view of the matter. See footnote 22.

isn't false, it just doesn't happen to apply as expected to the physical world. In cases of the second sort, we might even go so far as to say, for example, that the parallel postulate is true in Euclidean geometry because this is part of, or follows from, the meaning of 'straight line' in that theory, or that it's true in any Euclidean space. If we were to follow the second course in the case of rudimentary logic, we'd say that it's not applicable in certain cases, but it's still true to the meanings of the logical particles, or still true in any KF-world.

In fact, I doubt that the differences between cases we describe as falsifications and those we describe as failures of application are as weighty as they might seem.[27] One important factor in determining which depiction to use seems to be our judgment of the likelihood that the theory in question will continue to be useful, in other contexts, as an approximation in the original context, or even as an interesting object of pure mathematical study. On these grounds, Euclidean geometry is a clear winner, and thus it isn't counted amongst the straightforwardly falsified;[28] in contrast, the theory of phlogiston failed in the one instance for which it was designed, and the chances that it would work elsewhere, or have any interesting mathematical features, are nil. Rudimentary logic is more like Euclidean geometry than phlogiston theory on this measure, so we prefer 'it doesn't apply in these cases' to 'it's false in these cases'.

If this is right, then the urge to classify logic as true-by-virtue-of-meaning or true in any KF-structure is a way of protecting it from the seedy company of phlogiston theory. But both moves seem to me to be overreactions. We have these theories—of phlogiston, of geometry, of logic—and we recognize that they don't apply in situation x or y. In some cases—Euclidean geometry and rudimentary logic—the theories are of continued interest for one reason or another. I see nothing wrong in preserving such theories on these grounds alone; I see no need, in other words, to manufacture an imaginary sense in which they continue to be true or imaginary worlds in which they continue to apply.

[27] Notice that it won't do to say 'we count phlogiston theory and not geometry as falsified because geometry is a formal (or pure or ...) theory and phlogiston theory is out to describe the world'. The move to reclassify geometry as formal is exactly the move we make to save it from being falsified, a move we chose to forgo in the case of phlogiston theory. The question at issue is why we rescue the one and not the other. (Resnik's happy term for such a move is 'Euclidean rescue'; see Resnik [1997], p. 130.)

[28] I suspect this is why it sounds so jarring when Putnam (in his [1968]) speaks of Euclidean geometry or classical logic as 'falsified'. See footnote 15.

In sum, then, the Second Philosopher's rudimentary logic is contingent on the KF-structuring of the world; it is in some senses a priori, and in other senses a posteriori; it is empirical, in the sense that it could be overturned on empirical grounds, but some extra theorizing beyond observation and experiment would be needed to help us see how it could go wrong; and finally, there seems little to recommend the view that it is analytic, but even if it were, this would not settle the extra-linguistic question of whether or not it applies in a particular context.

III.7

From rudimentary to classical logic

We've sketched the Second Philosopher's case for a rudimentary logic validated by the structural features of many aspects of the world, and for this reason, embedded in our systems of mental and linguistic representation.[1] Still, for all its virtues, rudimentary logic is a rather crude and unwieldy creature. We've already seen that it generates no logical truths,[2] that the ground-consequent conditional has few interesting properties, and that familiar rules like modus tolens and reductio ad absurdum don't correspond to rudimentary validities. And this is just the beginning.

One venerable difficulty arises from the vagueness of so many properties. The problem is familiar. Suppose *a* is a tadpole at t_0 and a frog at t_n. Suppose also that the interval between t_i and t_{i+1} is small enough to fall within the 'tolerance' of being a tadpole, that is, it is not enough time to change an object from tadpole to frog.[3] Then it seems '*a* is a tadpole at t_0' and the simple series 'not-(*a* is a tadpole at t_i and not-(*a* is a tadpole at t_{i+1}))'[4] together imply that '*a* is a tadpole at t_n', which fails. In fact, this is a validity of rudimentary logic, but it isn't sound; some of the premises in the middle series are indeterminate, because both components are indeterminate. But, given that the interval from t_i to t_{i+1} is within the tolerance of 'tadpole', there is no distinct *i* at which this happens; the borderline between premises that hold and indeterminate premises is itself indeterminate. This makes the

[1] See III.3, footnote 33.

[2] Except identities, like $a = a$, if these count as logic (see III.3, the text surrounding footnote 17, and III.4, footnote 16).

[3] The term is due to Wright [1975].

[4] Characterizing the premises this way avoids irrelevant worries about the ground-consequent if/then. Some versions of the puzzle use a single quantified premise here—(for all t_i)(so-and-so)—but this brings in further irrelevant concerns about quantification over times, mathematical induction, and so on.

notion of validity murkier than we'd like: it may be indeterminate whether or not a premise is true or indeterminate.

Furthermore, rudimentary logic in its representational and linguistic versions introduces further potential pitfalls: for example, a purported representation or naming expression can fail to correlate with anything at all, in which case a complex representation containing it is neither correct nor incorrect, introducing an additional source of indeterminacy.[5] Furthermore, for simple soundness proofs (of the sort given in III.6) to apply to linguistic representations, the predicates must succeed in classifying objects (that is, some objects fall inside the classification, some fall outside, and some can be indeterminate, where the boundaries between these groupings can be indeterminate, as well). The trouble is that resources of natural language allow us to form paradoxical predicates—like '*x* is the shortest person who can't be described in ten words'—that fail to do this job.[6]

In light of these various shortcomings, it's not surprising that logicians have sought stronger, more manageable systems of logic. The considerable distance between rudimentary logic and modern, first-order predicate logic ('classical logic') can be bridged by a number of restrictions and idealizations, all more or less traceable to Frege. Let's look at each of these in turn.

First, Frege is well aware of the possibility of failures of reference and resulting indeterminacies:

> languages have the fault of containing expressions which fail to designate an object ... although their grammatical form seems to qualify them for that purpose. (Frege [1892], p. 163)

His hope is to remove this 'imperfection of language', to devise 'a logically perfect language' in which

> every expression grammatically well constructed as a proper name out of signs already introduced shall in fact designate an object, and ... no new sign shall be introduced as a proper name without being secured a [referent]. (Frege [1892], p. 163)

[5] Just as a reminder, this sort of talk is intended in the spirit of II.4, II.6, and footnote 3 of III.3: strictly speaking, to say that such an utterance isn't correct, or isn't true, is to say that its most literal translation/interpretation isn't true; in the looser sense touched on toward the end of II.6, it could mean roughly that there's no correlational story of the utterance's successful use that involves correlation of the name in it with a thing; to say that a mental representation is incorrect would mean something analogous for the relevant brain state.

[6] Cf. III.6, footnote 6. My worries here, like Wilson's, concern word–world relations.

The *Begriffsschrift* (Frege [1879]) is his attempt to devise such a language, which he uses to found arithmetic in his *Grundgesetze der Arithmetik* (Frege [1893], [1903]).

Sadly, as we now know, Frege's efforts were unsuccessful.[7] Eventually, he came to see himself as having fallen victim to the very error he'd cautioned against:

> One feature of language that threatens to undermine the reliability of thinking is its tendency to form proper names to which no objects correspond ... A particularly notable example ... is the formation of a proper name after the pattern of 'the extension of the concept a' ... Because of the definite article, this expression appears to designate an object; but there is no object for which this phrase could be a linguistically appropriate designation. From this has arisen the paradoxes of set theory which have dealt the death blow to set theory itself. I myself was under this illusion when, in attempting to provide a logical foundation for numbers, I tried to construe numbers as sets. (Frege [1924/5a], p. 269)

In the end, Frege came to agree with Kant that 'the logical source of knowledge ... on its own cannot yield us any objects' (Frege [1924/5c], pp. 278–9), and to speculate that arithmetic should be founded on geometry.[8]

Despite this sad history, the Second Philosopher's approach roughly parallels Frege's here:[9] there's nothing wrong with the underlying rudimentary logic—Pa does imply (there is an x)Px—but certain linguistic forms are deceptive; that is, a linguistic expression that seems to represent a situation of the form Pa—tempting us to apply the corresponding validity—may not in fact do so. Just as we must be careful not to apply our logic in contexts without the requisite KF-structuring, we must also take care that our naming expressions do succeed in naming. The case of paradoxical predicates is similar: unless our predicates manage to classify the objects

[7] The trouble was Russell's paradox, of course. See my [1997], pp. 3–8, for a brief historical summary.

[8] See Frege [1924/5b], p. 277: 'The more I have thought the matter over, the more convinced I have become that arithmetic and geometry have developed on the same basis—a geometrical one in fact—so that mathematics in its entirety is really geometry.' See also his [1924/5c].

[9] This doesn't include agreement with Frege's view that fictional discourse is outside the range of logic: 'In myth and fiction thoughts occur that are neither true nor false. Logic has nothing to do with these' (Frege [1906], p. 198). One of the conceits of realistic fiction is that it is describing a world sufficiently like ours that ordinary logic applies. The issue in the text is the application of logic to the actual world (though see IV.4).

of the KF-structuring, our logic will not apply. Another way of thinking of this approach is that we restrict our logical attention to cases in which the language is functioning properly: the names naming, the predicates classifying.

Frege takes a similar position on vagueness, excluding vague predicates,[10] too, from logic:

If we represent concepts in extension by areas on a plane ... To a concept without sharp boundary there would correspond an area that had not a sharp boundary-line all round, but in places just vaguely faded away into the background. This would not really be an area at all; and likewise a concept that is not sharply defined is wrongly termed a concept. Such quasi-conceptual constructions cannot be recognized as concepts by logic. (Frege [1903], p. 259)

His reason is simple:

as regards concepts we have a requirement of sharp delimitation; if this were not satisfied it would be impossible to set forth logical laws about them. (Frege [1891], p. 141)

'Impossible' seems too strong—see below—but it can hardly be denied that a logic without indeterminate cases would be considerably less complicated, enjoying Quine's 'sweet simplicity' (Quine [1981a], p. 32). So the Fregean proposal, parallel to the case of empty names, is that we restrict our logical attention to predicates that aren't vague.

The trouble with this policy, from the Second Philosopher's point of view, is that vagueness is not a minor aberration of language, like the occasional empty name; vagueness is ubiquitous. Many commentators argue that linguistic vagueness is pervasive,[11] but as we've seen (in III.4), the Second Philosopher goes further, holding that the objects of our world are vaguely bounded in spacetime and that many objective properties of those

[10] From here on, I focus mainly on vague predicates. The fuzziness of the spatiotemporal boundaries of objects can be treated as vagueness in the predicate 'x is part of y'. Indeed, Parsons sees indeterminacy as a feature of states of affairs, and does 'not attempt the additional step of blaming it on either the objects or the properties or relations making up those states of affairs' (Parsons [2000], p. 29).

[11] See, e.g., Russell [1923] or Wright [1975]. Quine thinks some vagueness is inevitable, and that 'good purposes are often served by not tampering with vagueness' (Quine [1960], p. 127). See also Dummett [1975], p. 109: 'we feel that certain concepts are ineradicably vague. Not, of course, that we could not sharpen them if we wished to; but, rather, that, by sharpening them, we should destroy their whole point.' Tappenden [1994] gives examples of predicates deliberately designed to be vague that would otherwise not serve their purpose.

objects are vague in extension. Perhaps Frege can afford to ban vagueness, given that his interest centers on mathematical language,[12] but to extend this decree, to make it universal, would be to render logic irrelevant to most of our dealings with the world.

A less dramatic suggestion, from Carnap this time,[13] is that we replace vague predicates with precise ones before we apply our logic. Those who hold that language is essentially vague will consider this to be impossible, as no precise terms can be found to do the job; the Second Philosopher, with her picture of a truly vague world, will no doubt find herself in sympathy with this point of view.[14] But even if such precisification, as it's now called, were possible, it's not clear that it would be a good idea, as the sharp predicates would not track the underlying properties at work in the world. The result would be a kind of distortion: the item at t_i, on one side of the new boundary, would be classified as more like the tadpole at t_0 than it is like the object at t_{i+1}, on the other side of the new boundary, which seems quite wrong. After all, the difference between t_0 and t_n is what's biologically significant; the difference between t_i and t_{i+1} is not.

A better approach is to regard the assumption that our predicates have sharp boundaries as an explicit idealization, like the assumption that the earth is flat (when calculating the trajectory of a cannon ball), or that water is an ideal fluid (when describing how it drains out of a sink), or that there is no friction (when a steel ball rolls down a polished steel track). None of these assumptions is literally true, as we well know, but they all simplify situations that would otherwise be too complex for effective treatment, and they do so without introducing relevant distortions. Likewise, though we know full well that 'tadpole' doesn't have a sharp boundary, we often treat it logically as if it did, for the sake of 'sweet simplicity', when the situation we're describing and the problems we're attending to make this a reasonable and effective idealization.

The advantage of this approach over Carnap's is that we don't lose sight of the fact that we are idealizing; we don't lose the true underlying structures, and we remain alert to potential dangers. If we're analyzing the logical structure of the reasoning in one of the psychological experiments described in III.5, we can safely pretend that '4-month-old infant' is a

[12] See IV.4. [13] See Carnap [1950b], chapter 1.

[14] Notice that the usefulness of vague language would be enough reason to preserve it, even without vagueness in the world (cf. III.4, footnote 13).

perfectly precise property; to do otherwise would be to muddy the waters to no relevant purpose. On the other hand, in the context of a moral or legal argument, this same fiction—that there is a precise moment of birth—might seriously distort the debate. Here, just as in the rest of science, we must be sensitive to the benefits and the dangers of idealization, and satisfy ourselves that the idealization in question is appropriate to the case at hand.

Once referentless names have been ruled out and vagueness idealized away, all potential truth value gaps have been eliminated and our logical theory is much improved: all classical tautologies and inferences in 'and', 'or', 'not', 'for all', 'exists', and 'equals' are now available. But we are still one step away from the fully truth-functional treatment of all logical connectives and their interconnections, which must be counted among Frege's greatest contributions. To complete the journey from rudimentary to classical logic requires one additional idealization, the truth-functional or material conditional: (if (...), then (__)) understood as (not-(...) or (__)).

Here again, we follow Frege, who explicitly notes the paradoxes of material implication;[15] he points out, for example, that there need be no 'causal connection' between 'the sun is shining' and '$3 \times 7 = 21$' for 'if the sun is shining, then $3 \times 7 = 21$' to be true, regardless of the time of day (Frege [1879], p. 14). We shouldn't pretend that the material conditional captures our rudimentary notion of ground-consequent dependency, but we can and should recognize it as a serviceable approximation in many cases. For the purposes of logic, of course, it is a boon, bringing the last of the basic connectives into the truth-functional fold and supporting the full range of classical tautologies, including the interdefinability of the connectives. In fact, once this second and final idealization is in place, logical validity and logical truth go hand in hand: the argument from φ to ψ is valid if and only if 'if φ, then ψ' is a tautology.

The upshot is that in order to move from the robust but unwieldy rudimentary logic to the power and flexibility of modern, first-order predicate logic, we must agree to steer clear of empty names and defective predicates and to adopt two highly non-trivial idealizations. These last two take us beyond a logic that's literally true of many of the world's

[15] These include '__ implies (if..., then __)' and 'not-... implies (if..., then __)'. It follows that a contradiction implies anything (called 'explosion').

phenomena, but they do so for the sake of a vastly more effective instrument. The justification must be, as always, that they make it possible to achieve results that would otherwise be impossible or impractical, and that they do so without introducing any relevant distortions. So, if classical logic is to apply to the world in a given context, several conditions must be met: there must be underlying KF-structures present; the language must be functioning properly, the names naming, the predicates classifying; the idealizations of bivalent predicates and the truth-functional conditional must be appropriate, that is, both effective and non-distorting. In such cases, our familiar logic can be trusted.

We have, then, a characterization of classical logic as grounded in a rudimentary logic that's both true of the world and embedded in our most primitive modes of cognition and representation. Given this strong endorsement, the Second Philosopher owes some account of the many deviant logics that depart from this standard. This is obviously a huge genus, with numerous species and sub-species, so all I attempt here is a brief survey of how the field looks from the point of view of this second-philosophical account of classical logic.

The most straightforward of the deviant logics are those that simply reject one or another of the required restrictions or idealizations. So, for example, the free logician claims that we cannot responsibly ignore contexts with referentless names. Notice that there is no disagreement here about the scope of classical logic—the classical and the free logician agree that classical logic is simply not designed for such cases—the disagreement is over whether or not important logical relations are being ignored by the restriction to non-empty names. Thus, Karel Lambert argues that classical logic:

> cannot be applied to the inferential ruminations of astronomers prior to the discovery ... that there is no object that is Vulcan ... it cannot discriminate between inferences ... whose validity does not require that their constituent singular terms have existential import [and] those ... whose validity does. (Lambert [2001], p. 263)

There is room, no doubt, for the classical logician to respond to such claims, for example, to argue that the inferences of these scientists should be analyzed from their point of view, on the (false) assumption that Vulcan exists, or to propose treating apparent singular terms as predicates

('Vulcanizes') until existence has been established. If a persuasive case can be made that such expedients obscure key features, there is room for the position that classical logic should be understood as a limiting case of a broader free logic, applicable in most of the conditions we encounter, analogous to Euclidean geometry or Newtonian mechanics. I won't try to resolve this dispute; my aim is simply to isolate its locus: are there or are there not important logical phenomena that can't be treated effectively under the assumption that all names refer?

Another group of deviant logicians rejects the idealization of bivalent predicates, insisting that vagueness be represented in our logic. It's important to realize that the classical logician (as understood by the Second Philosopher) fully agrees that there are vague predicates, perhaps everywhere, perhaps inevitably. The debate isn't over the truth or falsity of the idealization—as an idealization, it is admittedly false—but over the effectiveness and appropriateness of the idealization in a range of applications. Our classical logician grants that her system is artificial for vague predicates, and more importantly, that it introduces misleading distortions in some contexts (like the 'instant of birth' case mentioned earlier). The only room for disagreement is over the viability of the cure.

The literature on logics of vagueness is immense, but most systems begin with an underlying structure like that of rudimentary logic:[16] a predicate holds of some objects, fails of others, and leaves some indeterminate. In some systems, all the indeterminate objects are lumped together, resulting in three possibilities—true, false, and indeterminate; in others, there are degrees of indeterminacy between 0 (false) and 1 (true).[17] Rudimentary logic is non-committal enough to include only the validities common to these various systems, but it has the added feature of insisting that the borderlines between true and indeterminate, indeterminate and false, are themselves vague. Without this higher-order indeterminacy, vagueness seems not to be taken seriously—there would still be a sharp boundary

[16] The exception is epistemic theories (see the brief discussion below). Supervaluational theories also begin from this picture, though they use a more complex procedure to assign truth values: given the sets of objects with P, without P, and indeterminately P, we say that 'a has P' if a has P on every precification, that is, on every admissible way of divvying up the indeterminate objects between those with P and those without. Then 'for all x, x has P or x doesn't have P' comes out true, despite there being a's for which neither 'a has P' or 'not-(a has P)' are true. See Fine [1975].

[17] See Urquhart [2001] or Malinowski [2001] for surveys of multi-valued logics. See Keefe and Smith [1997b], pp. 35–49, for a look at their applications in the logic of vagueness.

between *a*'s being a tadpole and its being indeterminate whether or not *a* is a tadpole[18]—but given that our most effective way of pinning down the structure of our logical systems is to employ a precise mathematical meta-language, a coherent account has been devilishly hard to achieve.[19] This is one reason why, in III.3, I left the description of rudimentary logic ... well ... vague!

One conspicuous reaction to this situation comes from Timothy Williamson:

> The use of non-classical logic or semantics has been advocated for vague languages. New and increasingly complex systems continue to be invented. What none has so far given is a satisfying account of higher-order vagueness. In more or less subtle ways, the meta-language is treated as though it were precise. ... Such proposals underestimate the depth of the problem. (Williamson [1994], p. 3)

He considers the possibility that 'the nature of vagueness might be to defy perspicuous description', but this 'counsel of despair', he thinks, 'overestimate[s] the depth of the problem' (Williamson [1994], p. 3). His solution is simple: vagueness is an epistemic phenomenon; that is, there *is* a precise moment at which the tadpole becomes a frog, but we don't (can't) know what it is. Even Williamson recognizes that 'the epistemic view of vagueness is incredible' (Williamson [1994], p. 3). Nevertheless, he writes

> The most obvious argument for the epistemic view of vagueness ... [is that it] involves no revision of classical logic and semantics; its rivals do involve such revisions. Classical logic and semantics are vastly superior to the alternatives in simplicity, power, past success, and integration into theories in other domains. In these circumstances it would be sensible to adopt the epistemic view ... (Williamson [1992], p. 279)

[18] Notice that even on supervaluational theories, there is still a sharp point at which '*a* is a tadpole at t_i' goes from being true to being indeterminate, and a similar sharp drop-off between true and indeterminate in the series of premises in the sorites paradox. Dummett, who proposed the supervaluational idea only to reject it, holds that its merits are 'gained at the cost of not really taking vague predicates seriously, as if they were vague only because we had not troubled to make them precise' (Dummett [1975], p. 108).

[19] See, e.g., Sainsbury [1990] for an argument that it can't be done. J. A. Burgess [1990b] argues that higher-order vagueness, at least for secondary quality predicates, dies out by about order five; he can then appeal to Fine's supervaluational treatment of higher-order vagueness (Fine [1975], pp. 140–150). Tye ([1990], [1994]) makes an attempt at a vague meta-language, with his theory of vague (not fuzzy) sets.

To save ourselves from deviant logics of vagueness, we are to swallow the epistemic view.

Much as the Second Philosopher sympathizes with the goal here—preserving classical logic—she resists the epistemic theory on the straightforward grounds that it is obviously false. Perhaps a perspicuous logic of vagueness will emerge; perhaps such a logic will be effective, more effective than classical logic, in those contexts where the idealization of bivalence seems downright misleading. Until then, she is inclined to hew to our powerful, idealized logic, but to apply it with care and sensitivity.[20] J. A. Burgess speaks of 'the lamentable tendency ... to pretend that language is precise when dealing with problems not explicitly to do with vagueness' (J. A. Burgess [1990b], p. 434), but to the Second Philosopher, this seems no more distressing than pretending friction is absent when it can be safely ignored. In any case, once again, this is the locus of the debate: are there deviant logics that are more effective in (some? all?) contexts where the idealization of bivalence is a distortion?[21]

The other idealization of classical logic, the truth-functional conditional, also has its detractors, who propose a bewildering array of alternatives: probabilistic analyses, modal analyses, intuitionist conditionals, relevance entailment, and so on.[22] These are designed for a wide range of different purposes, but the last, the conditional of relevance logic,[23] may serve as an example of an alternative that aims—as the Second Philosopher might expect—for a conditional closer to the underlying notion of a

[20] Sorensen [2002] credits H. G. Wells as the first to suggest that in reaction to vagueness, 'we must moderate the *application* of logic'. He shares this delightful quotation from Wells: 'Every species is vague, every term goes cloudy at its edges, and so in my way of thinking, relentless logic is only another name for stupidity—for a sort of intellectual pigheadedness. If you push a philosophical or metaphysical enquiry through a series of valid syllogisms—never committing any generally recognized fallacy—you nevertheless leave behind you at each step a certain rubbing and marginal loss of objective truth and you get deflections that are difficult to trace, at each phase in the process. Every species waggles about in its definition, every tool is a little loose in its handle, every scale has its individual.'

[21] Before leaving the topic of vagueness, let me add one passing note on truth: one of the general objections to deflationism is that it doesn't function well with vague predicates (if 'Joe is bald' is neither true nor false, then Joe is not-bald and Joe is not-not-bald, a contradiction; see Field [1994b], [1998a], Horwich [1990/1998], §§26–28). I suggest that it's best to think of the disquotational truth predicate of II.4 as part of the idealized package of classical logic: it is a logical device, after all; the claim is that it functions best when vague predicates have been idealized away.

[22] Edgington [2001] gives a helpful survey and references. See also Priest [2001].

[23] There's some disagreement over terminology, between 'relevance logic' and 'relevant logic'. I use the former, in agreement with Dunn—'I dislike the persuasive definition aspect of 'releva*nt*' (Dunn [1986], p. 124)—consistent with Mares's geography: 'these systems ... are called "relevance logics" in North America and "relevant logics" in Britain and Australia' (Mares [2006], p. 1).

ground-consequent dependency. Thus, relevance logicians reject the truth-functional conditional on the grounds that the antecedent can be entirely irrelevant to the consequent:[24]

It is a natural thought that for a conditional to be true there must be some connection between its antecedent and consequent. It was precisely this idea that led to the development of relevant logic. (Priest [2001], p. 173)

Of course, once again, the classical logician (as understood by the Second Philosopher) fully recognizes the paradoxes of material implication; she simply holds that truth-functionality is worth the sacrifice, that the advantages outweigh the disadvantages. Relevance logicians see the debate in the same terms:

Logic is the *science* of argument, and like any other science, Logic has a right to simplifying assumptions.... But it also has an obligation to enrich that formalism, the better to separate good arguments from bad. (Mares and Meyer [2001], p. 281)

So the question, once again, is whether relevance logic offers a better way to do this.

In practice, this has turned out to be a difficult job, as Graham Priest admits in the continuation of the above quotation: 'A sensible notion of connection is not so easy to spell out, however' (Priest [2001], p. 173). One obstacle is obvious:

... there should be a connection between the content of the antecedent and the content of the consequent. This connection might seem difficult to enforce. For *content* is a semantic notion. The notion of a logic, on the other hand, is usually taken to be ... syntactic. (Mares and Meyer [2001], p. 283)

A step in that direction for propositional logic is the requirement that the antecedent and the consequent must have at least one propositional

[24] Rejection and replacement of the truth-functional conditional isn't the only adjustment of classical logic that the relevance logician recommends. Where the classical logician distinguishes 'if φ, then ψ' (true unless φ is true and ψ false) from 'φ implies ψ' (any situation in which φ is true is one in which ψ is also true), ' "hard core" relevance logicians often seem to luxuriate' in identifying the two, holding that there is one 'generic conditional-implication' (Dunn and Restall [2002], p. 3). One might say that there is a strong relevance conditional, and that the Deduction Theorem transfers this strength to the implication relation (see Dunn and Restall [2002], pp. 6–7, Mares and Meyer [2001], pp. 284–286). So, relevance logic uses a different notion of validity than rudimentary logic, with the result, for example, that disjunctive syllogism fails. To keep the discussion of deviant logics within bounds, I won't discuss this added twist.

component in common:[25] 'The gap between the semantic and the syntactic is bridged in part ... by the *variable sharing constraint*' (Mares and Meyer [2001], p. 283). There is some distance from here to the relation of ground and consequent, but perhaps it does represent a first installment on enforcing relevance.[26] There are, however, many ways of meeting this condition: 'Despite some of our hopes and utterances ... the One True Logic *does not exist*' (Mares and Meyer [2001], p. 280). It seems fair to count it an open question whether any of these, or any group of these, can overcome the many advantages of classical logic.

One of the central anomalies that relevance logicians have hoped to block is explosion: (φ and not-φ) implies ψ, no matter what φ and ψ might be; in other words, a contradiction implies anything, no matter how irrelevant! In recent years, this desideratum has taken on a life of its own, in the pursuit of paraconsistent logics.[27] One important paraconsistent logic arises from a simple modification of the strong Kleene truth tables that underlie rudimentary logic:

φ	not-φ
T	F
?	?
F	T

φ ψ	(φ and ψ)	(φ or ψ)
T T	T	T
T ?	?	T
T F	F	T
? T	?	T
? ?	?	?
? F	F	?
F T	F	T
F ?	F	?
F F	F	F

In the Kleene interpretation, ? is taken as a truth value gap, indicating a claim that's neither true nor false. In the paraconsistent system LP, ? means *both* true *and* false. In both cases, validity is the preservation of truth.

[25] This is not, by itself, enough to guarantee a relevance logic, as some paradoxes would remain, e.g., 'if φ, then (if ψ, then φ)'.

[26] As Dunn and Restall note, citing Meyer, 'relevance logic gives, on its face anyway, no separate account of relevance' (Dunn and Restall [2002], p. 6, see also Dunn [1986], p. 124). Rather, the idea is to produce a system that avoids the known 'irrelevances'.

[27] See Priest [2001], pp. 67–69, 122–123, 125–128, Priest [2002], or Priest and Tanaka [2004].

This small change makes a large difference. For example, (φ or not-φ) is a tautology of LP, because where Kleene had a gap, making (φ or not-φ) indeterminate, LP has a glut, making (φ or not-φ) true come what may (though sometimes also false!). On the other hand, (φ and not-φ) implies ψ in Kleene's system, because (φ and not-φ) is never true, but the inference does not hold in LP: (φ and not-φ) can be both true and false, if φ is both true and false, while ψ is just false. So LP is a paraconsistent logic; it blocks explosion.[28]

Formally, this is clear enough, but the interpretation of 'both true and false' is another story. In the face of skepticism, paraconsistent logicians deride the idea that a contradiction implies everything:

> Not only is this highly counterintuitive, there would seem to be definite counterexamples to it. There appear to be a number of... theories which are inconsistent, yet in which it is manifestly incorrect to infer that everything holds. (Priest [2001], p. 67)

Priest gives Bohr's theory of the atom as an example from the history of science:

> To determine the behavior of the atom, Bohr assumed the standard Maxwell electromagnetic equations. But he also assumed that energy could come only in discrete packets (quanta). These two things are inconsistent (as Bohr knew); yet both were integrally required for the account to work. The account was therefore essentially inconsistent. (Priest [2001], p. 67)

In the history of mathematics, Berkeley (in his [1734]) argued persuasively that Newton's original version of the calculus was inconsistent: proofs required *dx* to be non-zero at one point, so it could serve as a divisor, and zero at another, so it could be dropped from a sum. An impressive list of such examples could no doubt be drawn up.

Now surely it is correct to say that we shouldn't go on to infer everything from such a theory—that would be counterproductive. What these cases show, it seems to me, is that great scientists are often able to navigate around severe shortcomings in their theories, using their best instincts for how to proceed and how not to proceed. For example, Wilson discusses the

[28] It also blocks disjunctive syllogism: if φ is both true and false, and ψ is just false, then (φ or ψ) is true (and false), not-φ is true (and false), but ψ is just false. As it happens, disjunctive syllogism is interconnected with explosion (see Priest [2001], pp. 151–152).

theory of 'infinitely near points' from late nineteenth-century geometry, and shows how 'geometrical practice had ... learned, as a silent art of the trade' not to perform certain inferences.[29] Indeed, Euler is revered in some circles for the skill with which he manipulated divergent series, inventing powerful methods that could easily have led him to contradiction,[30] but which he instead wielded to produce 'thousands of results that were later established rigorously'.[31]

From this phenomenon, Priest concludes, 'Clearly, once we admit the existence of such theories, their underlying logics must be paraconsistent' (Priest and Tanaka [2004]). But I'm not sure this *is* clear. The suggestion seems to be that the surprising and sometimes incredible knack these thinkers displayed for avoiding contradictions and striking pay dirt could be codified as applications of a particular deviant logic. This seems to me unlikely. Whatever Newton, Euler, the nineteenth-century geometers, and Bohr were doing, it was closely tied to the complexities of their particular subject matter, and to their deep, if inchoate understanding of its twists and turns.

But leave this aside. It's undoubtedly true that our beliefs are often inconsistent, and even that it can sometimes be rational to hold inconsistent beliefs: Priest's example is the paradox of the preface, whose author believes every statement in her book, but also believes that there must be mistakes somewhere.[32] The hope is that paraconsistent logic will be of help in modeling such systems of belief and in related areas of automated reasoning and belief revision. In all these cases, including the disputed ones from the history of science and mathematics, paraconsistent logic is being proposed as a device for modeling 'inconsistent information/theories from which one might want to draw inferences in a controlled way' (Priest [2002], p. 290).[33] But even if LP or some variant were to perform well in such contexts, this would not in any way compromise the Second Philosopher's account of the logic of the world.[34] Furthermore, though

[29] Wilson [1994], p. 526.

[30] e.g., Euler concluded that the infinite series $1 - 1 + 1 - 1 + \ldots$ sums to 1/2. Kline makes the wry observation, 'there was much confusion in his thinking' (Kline [1972], p. 446).

[31] For a survey, see Kline [1972], pp. 446–453. The quotation is from p. 453.

[32] See Priest and Tanaka [2004].

[33] Priest [2002] calls this 'the major motivation behind paraconsistent logic' (p. 288).

[34] Another example of this phenomenon may be Dummett's interpretation of intuitionistic logic: not as a logic of the world, but as a logic of verification conditions (see Dummett [1973a]). (The

our rudimentary logic tells us that nothing is both true and false, we might recognize that we're sometimes in a position like that of the preface-writer, and it would be handy if a nice paraconsistent logic could show us how to proceed so as to minimize the damage to our system of beliefs.

At this point, then, we've seen deviant logics that reject a restriction of classical logic, that reject each of the idealizations of classical logic, and that change the subject from that of classical logic, but we haven't discussed one that directly disagrees even with rudimentary logic. Of course, there is one, ready to hand, namely intuitionistic logic, the logic of Creator-worlds. We can agree that intuitionistic logic would be appropriate in such a world, but we also feel confident that we don't live in one.[35] Quantum logic may set out to describe a non-KF-world, but so far we've been given no clear idea of what such a world would be like, the kind of idea that would be needed to make quantum logical micro-physics a viable candidate for a successor to our current theories (see III.6).

There is one more candidate in this category, that is, a style of deviant logic that disagrees with rudimentary logic and purports to be the logic of

original Dutch intuitionists also speak of proof conditions in the mathematical context—cf. van Dalen [2002], pp. 4–8—but only because proof is the means by which the mathematician realizes his mental constructions; roughly speaking, in our terminology, giving a proof is how the mathematical Creator 'imagines'.) Dummett holds that our logic *must* be a logic of verification conditions, because a truth-conditional semantics is impossible. Though she would also protest the assumption that these are the only options available (see Part II), the Second Philosopher's first concern is that the whole approach is overly linguistic: she hopes to address the logic of the world directly, not as an offshoot of semantics, while Dummett thinks that without language 'there is no world for us' (Dummett [1973b], p. 311). Where the Second Philosopher thinks we share our most primitive modes of representing the world with animals, Dummett admits only that 'dogs, sharks, etc. ... inhabit *a* world' not that they inhabit 'our world', whose 'general features ... cannot be separated from the understanding of the way in which we express those features' (Dummett [1973b], p. 311). Clearly, Dummett and the Second Philosopher differ so fundamentally that it would take considerable time and effort to trace out and examine their many disagreements. Here I hope it is enough to say that I don't pretend to have refuted or replied to Dummett; I'm only outlining the Second Philosopher's alternative vision.

[35] First, it is clear to each of us that we aren't Creators of this world, that things don't come into being when we imagine them, as we imagine them. Second, if we leave the Creator off-stage, or behind a curtain like the Wizard of Oz, and attend only to the structure of the Creator-world itself, we see that, e.g., the temporal development of our world is very different (things can cease to exist, situations that occur today may not occur tomorrow, etc.). Still, intuitionistic logic is a powerful mathematical tool, not only in constructive mathematics (see Troelstra and van Dalen [1988]), but also, e.g., in the study of topoi (see Fourman [1977]), which includes such attractive items as smooth infinitesimal analysis (see Bell [1998]).

the world. This is dialetheism,[36] the view that there are true contradictions, statements *p* such that both *p* and not-*p* are true. If this is not to imply that all statements are true, the dialetheist must adopt a paraconsistent logic; this is the standard version of the view. Examples of dialetheia, or true contradictions, have been proposed, the most relevant of which involves vague predicates.[37] As the comparison of the Kleene tables with the LP tables suggests, where rudimentary logic sees an object as neither a tadpole nor not a tadpole, the dialetheist sees it as both a tadpole and not a tadpole.

Quite apart from our inability, at least for now, to conceptualize what it would be for a statement to be both true and false, I think this treatment of vagueness misrepresents the underlying structures. If we restrict our attention to tadpoles and frogs, the relevant objects in this case, the dialetheist's account produces two fundamental groupings: the tadpoles and indeterminates; the indeterminates and the frogs. But the biologically significant similarities are those that hold among the tadpoles (and among the frogs), not those that hold between the tadpoles and the indeterminates (or between the indeterminates and the frogs). Much as precisification produces a skewed system of classification—counting the tadpole at t_i as more like the tadpole at t_0 than the tadpole at t_{i+1} (where $[i, i+1]$ is the point of transition from tadpole to frog)—the dialetheist's scheme counts an early indeterminate as like the full-fledged frog, and a late indeterminate as like (in another respect) the tadpole. But these are not the biologically functional divisions. And it seems to me that considerations of this sort carry over to other cases of worldly vagueness.[38]

In sum, then, this glance at deviant logics suggests two morals. We've seen that the efficacy and appropriateness of the restrictions and idealizations

[36] See Priest [2004]. Another terminological disagreement here, this one over spelling. I follow Priest; Routley, co-inventor of the term, omits the second 'e'.

[37] Another example that purports to be 'in the world' (as opposed to cases like conflicts between laws or the liar paradox) comes from quantum mechanics: the electron both goes through one slit and doesn't, as it also goes through the other slit and doesn't (see Priest [2004]). Given the difficulties we've seen with Putnam-style quantum logic (in III.6) and the failure of Reichenbach's previous three-valued proposal (see Putnam [1957], [1965], Gardner [1972]), it seems to me unlikely that this move will resolve the quantum anomalies and provide a viable theory of the quantum world.

[38] This is beside the point, but even if we did embrace a dialetheic account of vagueness, it seems the same pressures that moved us from rudimentary logic to bivalence would apply to the dialetheist as well.

of classical logic are always open to question, depending on the particular context of application. Many deviant logics attempt to provide alternative treatments that forgo these simplifications, and in that way help clarify what's at stake and inform our assessments of the pros and cons. The second moral is more tentative: it seems to me that, at least for now, rudimentary logic has no viable rivals as the underlying logic of the world.

III.8
Caveats

The Second Philosopher's view of logic has been supported here by a wide range of empirical observation, testing, and theory, all of which is subject to revision in the usual way of fallible scientific progress. I suppose it's just barely imaginable that some future Einstein might show conclusively that the apparent KF-structures of the world are illusory; less far-fetched is the possibility that some future Frege will give us a better, deeper analysis of logical form.[1] Either of these might prompt revision of the Second Philosopher's (1′). But clearly the most tentative part of the account must be the psychology, the empirical support given for (2′) and (3′); after all, Piaget's theories were superseded by current thinking within the space of the last few decades. All this can be disheartening to the philosopher in search of certainty, but it is only to be expected by the Second Philosopher, who never undertook to philosophize from a point of view more secure than that of science. Before leaving the subject, let me pause for a brief look at some of the empirical contingencies on which this view of logic rests.

To begin with the more specific, it must be acknowledged that the infant research touched on here is not without its critics. One concern is its heavy reliance on the single experimental paradigm of habituation/preferential looking and its close relatives.[2] Only further research will determine how well the current picture holds up, as researchers develop alternative methods

[1] For what it's worth, it seems to me that plural logic may present an improvement (by extension) on Frege's analysis (see Yi [2005/6]). Given a Henkin-style semantics (which seems to me more plausible for plural logic than for second-order logic), it even has a sound and complete proof procedure, and thus a chance of qualifying as 'epistemically transparent' (see III.1, footnote 1).

[2] Some recent work has addressed this concern, reaching similar conclusions from experiments using manual search and choice designs. See the introduction to Huntley-Fenner et al. [2002] for a survey and references.

and experimental designs for testing infant cognition.[3] But, fascinating as it is, this level of debate is not immediately relevant to the Second Philosopher's current concerns. Though I've tried to present what seems, at the current state of knowledge, to be the most likely story of how we humans come to detect and represent KF-structures, the very general claims (2′) and (3′) could survive considerable modification of the underlying psychology. Suppose, for example,—contrary to current evidence—that Helmholtz or Piaget was right to think that experience of navigating around or manipulating objects is necessary for acquisition of the object concept. From the perspective of the Second Philosopher's account of logic, this major upheaval in psychological theory would only shift the weighting of in-born and early experiential inputs more toward the latter; (2′) and (3′) would be as well supported as before.

Of course, empirical support for the account can't be and isn't immune to all possible transformations in psychology. For example, if—even more contrary to current evidence—it were discovered that the object concept and the rest depend on some linguistic development or other, the case for evolutionary origins would lose the support it now receives from parallels between infant and non-human animal evidence. This might cast doubt on the idea that our tendency to cognize in terms of KF-structures arises purely from our interactions with those structures, not from surrounding cultural phenomena, especially if the linguistic developments required were not cross-cultural. It might still be that we are picking up on real-world structures that these other cultures (and non-human animals) miss, but further evidence would be needed, including a careful look at the alternative cognitive structures found elsewhere.

More globally, the Second Philosopher's account may call to mind the lively recent debate about mental representation in general, be it infant

[3] There are other debates as well: e.g., some challenge the interpretation of these results in terms of 'object representations, object individuation ... suggesting instead that lower level perceptual representations underlie infant performance on these tasks' (Huntley-Fenner [2002], p. 204). See Huntley-Fenner [2002] for discussion and references. There is also a more theoretical debate over whether the infant's conceptions constitute 'core knowledge' that carries forward into adulthood (see, e.g., Spelke et al. [1992], Spelke [2000]) or the first of many theories that change with age (the so-called 'representational-development theory', see Gopnik and Meltzoff [1997], Meltzoff and Moore [1999]). On the Second Philosopher's main contention—that the abilities required to detect and represent KF-structures are present by 9–10 months—these last two schools would seem to agree.

or adult.[4] As the conflict was first characterized, those at one extreme held either that there are no mental representations at all or that all mental representations are primitive, structureless,[5] while those at the other took mental representations to be sentence-like structures in an inner language of 'mentalese'.[6] Jerry Fodor and Zenon Pylyshyn threw down the gauntlet, arguing that only a structured or compositional theory—a language of thought—could account for the productivity and systematicity of thought.[7] Over the years, many in the first camp have come to think that supposedly non-representational connectionist networks both can and should manage complex representations;[8] that the language of thought theorists' insistence that representations be sentence-like, combined by a counterpart to syntactic concatenation, was unnecessarily restrictive; that all that's actually required of structured representations is a general, effective way of getting from the simpler representations to the corresponding complex representation and back.[9] Meanwhile, many adherents of the language of thought came to see Fodor's strong claims—for example, about the innateness of concepts—as both implausible and unnecessary.[10] Some observers began to suggest that both approaches would ultimately be

[4] i.e., the debate over whether or not mental representation works as much of developmental and cognitive psychology supposes, not the debate alluded to in III.3, footnote 3. Thanks to Kent Johnson for his advice on this discussion.

[5] By 'structureless' I mean without discrete, reusable substructures (in contrast, for Second Philosopher presumably a conjunctive representation (*Ca* and *Ha*) is related in some transparent way to independent representations *Ca* and *Ha*). Sterelny [1990], chapter 8, gives an overview of the connectionism debate and references; Case [1999], pp. 39–41, places connectionism in a helpful taxonomy of post-Piagetian child psychology. Some dynamic systems approaches eschew representations altogether (see Thagard [2005], chapter 12, Clark [1997], pp. 148–149, for description and references, van Gelder [1995] for more sustained treatment).

[6] The classic source is Fodor [1975]. See Sterelny [1990], especially chapter 2, for discussion and references.

[7] See Fodor and Pylyshyn [1988]. Thought is 'productive' because we can think indefinitely many, indefinitely complex thoughts; it is 'systematic' because anyone who can entertain the thought that Joe loves Alice can also entertain the thought that Alice loves Joe. Johnson [2004] persuasively challenges the systematicity assumption.

[8] Cf. van Gelder [1990], pp. 355–356: 'There is at least one basic point of agreement among the various parties: in order to exhibit any reasonably sophisticated cognitive functions, a system must be able to represent complex structured items.' That connectionist models can do this is argued, e.g., in Smolensky [1988] (see Fodor and McLaughlin [1990] for reply), van Gelder [1990], Rowlands [1994].

[9] As an example of functional but non-compositional structuring, van Gelder [1990] cites Gödel numbering.

[10] See Sterelny [1990], pp. 27–28.

needed to account for cognition[11] or to doubt that the two schools in fact differ as sharply as it once appeared.[12]

Obviously the view of logic sketched here presupposes mental representations with some minimal structuring—that is, with KF-structuring—though this falls far short of a full-blown language of thought. If the strongest opponents of structured representations turn out to be right, if parsing neural activity in these terms is ultimately not the right way of understanding how we function in the world,[13] then this bit of Second Philosophy will be empirically refuted. But, insofar as evidence from cognitive studies continues to support structured representations in general and KF-structured representations in particular, the Second Philosopher may fairly expect that any viable account of cognitive implementation must support these, one way or another.

This, briefly, is my understanding of some of the ways in which this second-philosophical account of logic is vulnerable to the uncertainties of contemporary science. If the theories it rests on fail, the account of logic fails. But, again, it should be emphasized that the Second Philosopher can't avoid this sort of risk; no matter how good the evidence and the arguments, no scientific theory is infallible, and she is pursuing a variety of science. Notice, for that matter, that much of Quinean naturalism depends on behavioristic theories from the psychology of his day, though Quine himself didn't always put the case quite this way.[14] The issues at stake might have been clearer if he had!

[11] e.g., Clark [1989], p. 175: 'In words that Kant never used: subsymbolic processing without symbolic guidance is blind; symbolic processing without subsymbolic support is empty.'

[12] e.g., Klahr compares 'production systems', which 'focus on symbolically based, rule-oriented, higher cognitive processes', with connectionist systems, which focus on the 'subsymbolic (or non-symbolic)', as candidates for 'cognitive architecture' (Klahr [1999], pp. 132–133), and concludes that 'the two computational approaches are not as distinct as their practitioners have often claimed' (Klahr [1999], p. 150).

[13] See van Gelder [1995] on dynamic systems approaches.

[14] In fact, Quine's behaviorism is more a 'philosophical' stance than a serious attempt to bring current research to bear on the questions he raises. Once again, we're reminded of Fogelin's remark: 'Quine's inspiration comes from the library, not the laboratory' (Fogelin [1997], p. 561).

PART IV

Second Philosophy and Mathematics

IV.1
Second philosophy of science

My goal in this final part is to pull together a few themes from the discussion so far and bring them to bear on a sampling of intertwined issues from the philosophy of science and mathematics. I don't expect to resolve these questions, but I do hope to show how the second-philosophical approach restructures them and to indicate some further directions. One of the most lively contemporary debates in these areas pits various scientific realists against various anti-realists;[1] let's begin here with the boldest and most widely discussed entry on the anti-realist side, namely, Bas van Fraassen's constructive empiricism.[2]

The crux of van Fraassen's position is as dramatic as it is sweeping:

> When the theory has implications about what is not observable, the evidence does not warrant the conclusion that it is true. (van Fraassen [1980], p. 71)

This is not to say that we should give up our theories entirely, but simply that we should refrain from belief in their unobservable posits; we should regard our theories as 'empirically adequate'—as producing truths about observables—while remaining agnostic about the rest. Like Stroud's neo-Cartesian skeptic, who thinks no evidence could ever justify my belief in my hands (see I.2), van Fraassen holds that no evidence could ever justify belief in unobservables. In fact, van Fraassen agrees with the skeptic that our evidence doesn't even warrant belief in 'simple perceptual judgments'—like the one about my hands—not because that evidence consists of sense data (as the skeptic might put it), but presumably because it involves beliefs about matters I haven't actually observed (like what I would see if I were to turn my hand over).[3] By trusting in his senses and in the

[1] See Papineau [1996] for a recent sampling of opinion.

[2] See van Fraassen [1980], [1985]. Churchland and Hooker ([1985]) give a sampling of early reactions.

[3] See van Fraassen [1980], pp. 71–73.

empirical adequacy of his theories, van Fraassen himself goes beyond what he takes the evidence to warrant; he draws this line in conformity not with 'rational compulsion', but with his underlying Empiricism, which instructs him 'to withhold belief in anything that goes beyond actual, observable phenomena' (van Fraassen [1980], p. 202).

What is our Second Philosopher to make of this? The first and most fundamental unobservables are atoms, and we've seen (in I.5 and I.6) that the Second Philosopher takes Perrin's experiments on Brownian motion, confirming Einstein's theoretical predictions, as compelling evidence for their existence. Van Fraassen disagrees, insisting that no evidence can do this job, but unlike Stroud's skeptic, he presents no general argument for this stance. In the absence of such an argument, faced with the bare claim that atomic theory is empirically adequate but not necessarily true, the Second Philosopher might well take heart: van Fraassen appears to occupy the position of Poincaré and Ostwald in 1900, a position she finds sound. Perhaps he can be persuaded, as they were, by a careful review of the new evidence of Einstein and Perrin! If not, perhaps he will explain why her faith in that evidence is misplaced; perhaps there is a weakness she hasn't noticed.

To her surprise, van Fraassen neither admits the force of the Einstein/Perrin evidence nor presents an analysis of its shortcomings. Instead, he grants that from her point of view

> the distinction between [*atom*][4] and *flying horse* is as clear as between *racehorse* and *flying horse*: the first corresponds to something in the actual world, and the other does not. While immersed in the theory, *and* addressing oneself solely to the problems in the domain of the theory, this objectivity of [*atom*] is not and cannot be qualified. (van Fraassen [1980], p. 82)

So for the Second Philosopher, immersed in her science, the Einstein/Perrin evidence *does* provide compelling grounds on which to distinguish atoms from, say, phlogiston, that is, to count atoms as existing, as real. But the 'immersed perspective' is only part of the story; there is also an 'epistemic stance':

> If [the scientist] describes his own epistemic commitment, he is stepping back for a moment, and saying something like: the theory entails that [atoms] exist, *and* not

[4] Van Fraassen uses 'electron' in this quotation and the next, but the same would seem to go for atoms.

all theories do, *and* my epistemic attitude towards this theory is X. (van Fraassen [1980], p. 82)

From this perspective, van Fraassen recommends his Empiricism, with the consequence that we should believe only in the empirical adequacy of atomic theory, not in its truth.

At first blush, the Second Philosopher may think she understands what van Fraassen is talking about when he distinguishes the immersed from the epistemic stance: she might think of Ostwald as taking the epistemic perspective when he counsels readers of his 1904 chemistry textbook to treat atoms as real when explaining chemical phenomena, making predictions, and so on, but to recognize at the same time that their existence hasn't actually been established. In fact, she knows many examples in which a theory is used, taken as true during 'immersion', while the theorist nevertheless retains doubts about certain aspects or entities involved.[5] She agrees that it was reasonable to regard atomic theory as empirically adequate but possibly not true, as Ostwald and others did, before the work of Einstein and Perrin. But she also continues to think that their evidence made the difference, that atomic theory is now properly regarded as true.

Van Fraassen insists that the Second Philosopher has not understood him. Ostwald's 'immersion' and 'stepping back' both take place within what van Fraassen calls 'the scientific world-picture' (van Fraassen [1980], p. 80). In other words, both Ostwald's use of atoms and his reservations about their reality are 'immersed', part of the ordinary scientific project of distinguishing racehorses from flying horses. For a time, he wasn't sure which category atoms belong in, but Einstein and Perrin rightly convinced him that they stand with racehorses. None of this touches van Fraassen's contention that from the epistemic standpoint, atomic theory was, is, and always will be without compelling grounds.

In addition to puzzling over his dismissal of what seems to her good evidence, the Second Philosopher also wonders how van Fraassen can rest content with the bare conclusion that atomic theory is empirically adequate. Scientific curiosity of the simplest sort demands that we ask why this is so, what it is about the world that makes the atomic hypothesis so effective; we've seen this impulse in operation in the case of quantum mechanics (in

[5] As noted in I.5, the use of a mathematical continuum to represent spacetime is often accompanied by agnosticism about its true structure (see my [1997], pp. 143–152, and IV.2.i below).

III.3 and III.6), where we don't know what makes the theory successful yet continue to ask the question. Here, once again, van Fraassen begins by agreeing

> The search for explanation is valued in science because it consists *for the most part* in the search for theories which are simpler, more unified, and more likely to be empirically adequate ... because having a good explanation *consists* for the most part in having a theory with those other qualities. (van Fraassen [1980], pp. 93–94)

Thus the Second Philosopher's impulse to seek an explanation for the empirical adequacy of a given theory is a good one, beneficial for the progress of science. Indeed, van Fraassen insists on the importance of 'learning to find our way around in the world depicted by contemporary science, of speaking its language like a native' (van Fraassen [1980], p. 82). Nevertheless, 'the interpretation of science, and the correct view of its methodology, are two separate topics' (van Fraassen [1980], p. 93). In other words, the Second Philosopher's methodological instincts are impeccable, just the right thing for the pursuit of science, but the interpretation of science is an entirely separate undertaking.

What we have here is yet another two-level view, akin to those of Kant (I.4), Carnap (I.5), and Putnam (I.7). Unlike Descartes,[6] whose ruminations are intended to correct and improve scientific practice, all these thinkers take the methods of ordinary science to be entirely in order. Their goal, then, is not a critique of science, but something else: in Kant's case, to account for synthetic a priori knowledge and block the overreaching of dogmatic philosophy; in Carnap's, to defuse pseudo-questions and put philosophy on its proper syntactic footing; in Putnam's, to combat metaphysical realism and the Correspondence Theory of truth. We've seen that none of these projects holds any appeal for the Second Philosopher, leaving her with no motivation to undertake these various higher-level inquiries.[7] So the question for van Fraassen is: what is the purpose of adopting your epistemic stance?

From van Fraassen's point of view, the trouble with the Second Philosopher is that she's so completely immersed; she doesn't speak the language of contemporary science 'like a native', she *is* a native! He needs to introduce her to his higher level as to a second language, his epistemic

[6] i.e., the Descartes of I.1, not the neo-Cartesian skeptic of I.2.

[7] Not to mention that she would be hard pressed to isolate, much less to validate, any extra-scientific methods that would be needed to conduct those inquiries.

foreign language, and he needs to give her sufficient motivation to undertake the project. His goal, he tells her, is to 'make ... sense of science, and of scientific activity' (van Fraassen [1980], p. 73) and to do so more plausibly than alternative accounts. Of course the Second Philosopher makes her own study of the nature of her inquiries—of its methods, of the explanatory goals of its various undertakings, of its successes and failures and how it might be improved—and she is not inclined to agree with van Fraassen's assessment:

Science aims to give us theories which are empirically adequate; and acceptance of a theory involves as belief only that it is empirically adequate. (van Fraassen [1980], p. 12, italics in original)

That this is untrue seems to her to follow directly from the fact that scientists were not content with the empirically adequate atomic theory, that they wanted to know whether or not it was more than that—and presumably if the atomic hypothesis had failed Perrin's and subsequent tests, they would still have wanted to know why it was empirically adequate, what it was about the world that made atomic theory so successful. Given that none of this seems to be covered by van Fraassen's effort to 'make sense of scientific activity', she sees that effort as inadequate and has no motive to learn in his foreign tongue.[8]

Notice that the Second Philosopher's disagreement with van Fraassen does not align her with those he takes to be his opponents. We've seen that van Fraassen has no quarrel with her belief in atoms on the basis of

[8] The Second Philosopher's reaction to van Fraassen is reminiscent of Fine's: 'faced with such substantial reasons for believing that we are detecting atoms, what, except purely a priori and arbitrary conventions, could possibly dictate the empiricist conclusion that, nevertheless, we are unwarranted actually to engage in *belief* about atoms?' (Fine [1986], p. 146). Fine also rejects the higher level: 'when [the empiricist] sidesteps science and moves into his own courtroom, there to pronounce his judgments of where to believe and where to withhold, he [commits] the sin of epistemology' (p. 147). The position he advocates is the Natural Ontological Attitude (NOA): 'NOA insists that one's ontological attitude towards ... everything ... that might be collected in the scientific zoo (whether observable or not), be governed by the very same standards of evidence and inference that are employed by science itself' (p. 150). It seems, however, that the NOAer begins from a 'humanistic' stance outside science, that he is not simply born native to the contemporary scientific world-view, which raises questions about demarcation criteria and inspires hints of relativism and social constructivism that are foreign to second-philosophical thinking. Also, Fine insists that the NOAer reject all theories of truth, contrary to the second-philosophical inquiry of Part II, perhaps because he assumes these must all take place at a higher level (e.g., he cites the Comparison Problem against what must be a Correspondence Theory of truth, but doesn't seem to consider a correspondence theory). For more, with references, see my [2001b].

the Einstein/Perrin evidence; his concern is with another battle altogether, where his rival is the so-called 'realist'. The terminology is confusing, because the Second Philosopher herself is a realist if this just comes to believing in the existence of unobservables, but in contemporary terms realism often involves more than that. Van Fraassen describes it this way:

> *Science aims to give us, in its theories, a literally true story of what the world is like; and acceptance of a scientific theory involves the belief that it is true.* (Van Fraassen [1980], p. 8, italics in original)

Such a realist may hold that we come to accept a theory because it is the best explanation of phenomena, or because there would otherwise be no explanation of its success, or whatever—these ideas come up again in IV.5[9]—but the salient point for now is that the discussion is taking place at van Fraassen's epistemic level. What's at issue isn't the reliability of the Einstein/Perrin evidence, but the puzzling interpretive question van Fraassen poses, a question that floats free of the ordinary methods of science. Such a realist is not content to agree with the Second Philosopher that the Einstein/Perrin evidence establishes the existence of atoms; he wants to convince van Fraassen, who sets ordinary evidence aside, that atoms *really* exist! The Second Philosopher feels no need for this extra stamp of the foot.[10] (On the typographic conventions introduced in I.7 for the distinction between correspondence theories and Correspondence Theories, we might say that the Second Philosopher is a realist while van Fraassen's opponents are Realists.)

It's worth pondering a moment on what motivates the Realist's foot stamp. Human nature being what it is, even an avowedly naturalistic believer in atoms,[11] when confronted with van Fraassen's agnosticism, may be inclined to insist that they *really* do exist, to try to defeat his constructive empiricism on its own terms with arguments of the sort alluded to above. The trouble with this reaction, as we've seen, is that it grants van

[9] There it becomes clear that the Second Philosopher is no more sympathetic to the Realist's accounts of how science works than she is to van Fraassen's: e.g., atomic theory provided the best explanation of a wide range of chemical and physical phenomena in 1900, but the existence of atoms was not yet established. In general, where the constructive empiricist issues a blanket rejection of all unobservable posits, the Realist issues an equally blanket endorsement; the Second Philosopher faults both for passing over the details of the evidence for each particular posit, for shirking the responsibility to evaluate each case individually.

[10] Fine [1986], p. 129, credits this way of characterizing the Realist to Charles Chastain.

[11] e.g., Boyd [1983].

Fraassen too much at the outset, in particular, it buys into his 'stepping back' to the 'epistemic stance', and as a result, it implicitly grants that the Einstein/Perrin evidence isn't enough by itself, that it stands in need of supplementation.[12] Once this move is made, the game is lost, because the only compelling evidence has been officially set aside as ineffective for these higher purposes. Even if the Realist's effort to answer van Fraassen is couched in purely naturalistic terms,[13] he has betrayed his naturalism the moment he allows that evidence like Einstein and Perrin's is inadequate. The Second Philosopher, obviously, would never make this first move.

What's at work here is the endearing if misguided human tendency to rise to a skeptical challenge. Though van Fraassen isn't a radical skeptic—he believes he has hands—he puts the Second Philosopher in a familiar position: like the neo-Cartesian skeptic of I.2, he asks her to justify something—in this case, her belief in atoms rather than hands—without using any of her tried and true methods for settling such questions. This is a challenge that the Second Philosopher grants she cannot meet—justify *p* without using the methods you have developed and honed for such justifications—but she doesn't take this to undercut her original grounds. (Van Fraassen differs from the neo-Cartesian in moving the debate to the 'epistemic level', so he can allow that her original justification serves admirably, 'for scientific purposes'. This is good enough for the Second Philosopher, who has no other purposes and hasn't yet been convinced that she should.) The Realist, in contrast, feels that *something* must be said in reply, which leads to his distorted picture of scientific method, that is, to his lack of faith in ordinary evidence.

The same basic tendency turns up in many forms. For example, John Worrall presses the point by asking us to consider

> the theory that god created the universe in 4004 B.C. complete with some stuff buried in various places that looks awfully like, but is not, the bones of animals from now extinct species and complete with pretty patterns in various rocks that look awfully like, but are not, the imprints of the skeletons of animals from now extinct species and so on. (Worrall [1999], p. 346)

[12] Of course the Second Philosopher is happy to see the Einstein/Perrin evidence supplemented by evidence from other sources—e.g., from the various detectors touched on in IV.5—but all this is 'immersed' and thus irrelevant from van Fraassen's 'epistemic stance'.

[13] See, e.g., Boyd [1990], p. 227: 'the epistemology of empirical science is an empirical science.' Cf. the Carnap of I.5.

This version of Creationism comes to an Evil Demon hypothesis of restricted scope: the aspects of the world related to evolutionary theory have been arranged (for good or ill)[14] so as to thwart our best scientific methods; we're challenged to justify our rejection of the theory without using any of our best methods. More recent versions that claim to present scientific alternatives to evolution propose their own explanations of such phenomena as the ordering of the fossil record, the geographical distribution of organisms, the age of the earth, most often by evoking The Flood; it isn't hard to imagine a second-philosophical reply in the general tone of Feynman's remarks on astrology quoted in I.1 and at greater length in I.7.[15] Either way, the Second Philosopher remains committed to her own methods; she declines the challenge to defend evolution without them, and she unapologetically employs them in her critique of Flood-based explanations.

Worrall imagines the defender of Creationism insisting on alternative evidential standards of his own: 'the right approach is to apply the standard scientific canons, except were the literal truth of the Bible is at issue' (Worrall [1999], p. 351).[16] In the face of this clash of methods, Worrall insists on 'an unambiguous underwriting of the epistemic specialness of science' (Worrall [1999], p. 341), which would require the Second Philosopher to defend her methods—either against a radical skeptical hypothesis or against an alternative set of methods—without using her own methods.[17] It's understandable, of course, this desire to refute Creationism from a perspective that the Creationist must accept, but the Second Philosopher has no perspective but her own to offer. (Recall that Stroud plays on the analogous desire in the case of the pseudo-Cartesian in I.2.)

[14] Kitcher [1982], p. 127, lists two popular options: '(i) The Devil placed the fossils in the rocks to deceive us; (ii) God put the fossils there to test us.'

[15] See Kitcher [1982], chapter 5, for the raw materials.

[16] Worrall's scientist actually objects to Creationism on the grounds that it violates a methodological principle against ad hoc hypotheses; his Creationist replies with another ad hoc move, i.e., a defense of ad hoc hypotheses when the literal truth of the Bible is at stake. I don't think this move to the level of general methodological maxims is necessary to make the point.

[17] Worrall [1999], pp. 350–351, considers an intermediate possibility: the naturalist might attempt to use principles on which she and the Creationist agree to convince him gradually of the error of his ways. While this 'rhetorical ploy' could conceivably work against a susceptible Creationist, it obviously falls short of a defense of evolution he cannot consistently reject (e.g., by 'claiming that the right approach is to apply the standard scientific cannons, except where the literal truth of the Bible is at issue').

From here, Worrall accuses the naturalist of relativism:[18]

> Philosophers of science are essentially anthropologists with a special interest in the tribe of scientists. ... other tribes—magicians, shamans, ... creationists, scientologists—who also lay claim to creating knowledge could equally well be chosen as the object of study, whereupon different rules, a different epistemology, would obviously be extracted. If we are not ready to assert a special status for rules derived from the anthropological study of scientists compared to those derived from the anthropological study of scientologists then relativism follows. (Worrall [1999], p. 343)

Though the Second Philosopher certainly recognizes the interest and importance of anthropology, her study of scientific method is not the study of some alien tribe but a study of her own methods, with a critical eye to improving them. From her perspective, the methods of the magicians and scientologists can certainly be studied anthropologically, but an inquiry into their reliability as 'knowledge creators' brings their shortcomings into dramatic focus. This is not relativism, but the 'imperialism' of I.7.

There is obviously much more to be said about the realism/anti-realism debate, not to mention broader issues in the philosophy of science,[19] but I hope this much at least suggests how a second-philosophical perspective minimizes the attractions of extra-scientific clashes at some higher level of analysis and of vain sometimes damaging efforts to combat various forms of radical skepticism. Let's now turn to the role of mathematics.

[18] Worrall ([1999], p. 345) also claims that the naturalist 'surrender[s] the normative', but this goes hand in hand with the relativism charge. (See IV.5 for more on normativity.) He sees his disagreement with the naturalist as hanging on the naturalist's rejection and his acceptance of a priori methodological principles, but I don't see that the naturalist must eschew the a priori (see III.6, IV.4, footnote 60) or that believing a principle a priori relieves us of the need to defend its reliability.

[19] Let me note the views of Friedman ([2001], [2002]), DiSalle ([2002], [2006]) and others, who see Kuhn's scientific revolutions (Kuhn [1962]) as transitions mediated not by scientific methods, but by philosophy. On the one hand, I doubt that the Second Philosopher would recognize the required distinction between changes in theory and changes in constitutive principles (the modern-day successors to Kant's forms and categories, Carnap's linguistic frameworks, Kuhn's paradigms, and so on); Kuhn's claim that 'paradigm shifts' are not supported by ordinary scientific rationality has been disputed by many (see, e.g., Kitcher [1993], pp. 272–290, for discussion of the chemical 'revolution' in which Lavoisier's oxygen replaced Priestley's dephlogisticated air) and I have nothing particular to add to that case. On the other hand, though I can't claim any exhaustive analysis, the working parts of the 'philosophical' tools typically called upon—e.g., Helmholtz and Poincaré's meditations on the nature of geometry (Friedman [2001], pp. 108–115), Einstein's 'conceptual analysis' of the difficulties underlying the classical notion of simultaneity (DiSalle [2002], [2006], section 4.2)—would seem to fall well within the Second Philosopher's repertoire.

IV.2
Mathematics in application

The role of mathematics in scientific application raises a host of difficult questions;[1] here I touch on a few of those issues, the ones connected most directly to matters discussed elsewhere in our look at Second Philosophy. In recent philosophical discussion, descended from Quine, the bulk of attention has focused on the claim that the existence of mathematical abstracta can be inferred from their indispensable presence in our best physical theories.[2] A critical examination of this view leads naturally to a broader issue: to what extent is the world mathematically structured? I conclude with a few observations on what Eugene Wigner famously described as 'the miracle' of applied mathematics.[3]

i. Mathematical ontology

Early on, Quine's 'philosophical intuition' dictated that there were no abstract objects whatsoever; he and Nelson Goodman set out to translate away all reference to such things, including 'numbers, functions and other classes claimed as the values of the variables of classical mathematics' (Goodman and Quine [1947], p. 105). It seems a measure of Quine's intellectual integrity, as well as his commitment to some form of naturalism, that when he became convinced of the impossibility of such a translation, he concluded that his deeply held philosophical intuition was in fact incorrect.

[1] I focus here on applications in physical science, but of course mathematics is also applied in other sciences, in non-descriptive uses like cryptography, and in everyday contexts like bookkeeping and batting averages.

[2] For exceptions to this rule—i.e., for philosophical writings on applied mathematics that aren't focused on indispensability considerations—see e.g. the authors represented in the *Monist* issue of April 2000.

[3] Wigner [1960], p. 237. The passage is quoted in (iii) below.

We've seen (in I.6) how his naturalistic study of science led him to confirmational holism and the web of belief; from there, Quine notes that

> A self-contained theory which we can check with experience includes, in point of fact, not only its various theoretical hypotheses of so-called natural science but also such portions of logic and mathematics as it makes use of. (Quine [1954], p. 121)

Given that reference to mathematical entities cannot be eliminated from our best science, the case for the existence of these abstracta runs on exactly the same tracks as the earlier case for the existence of atoms: being part of a theory with the theoretical virtues 'is what evidence is' (Quine [1955], p. 251).

This has come to be known as the indispensability argument for mathematical realism. In a similar spirit, Putnam emphasizes that individual statements of our best confirmed science involve mathematical objects essentially:

> One wants to say that the Law of Universal Gravitation makes an objective statement about bodies ... that bodies behave in such a way that the quotient of two numbers *associated* with the bodies is equal to a third number *associated* with the bodies. But how can such a statement have any objective content at all if numbers and 'associations' (i.e., functions) are alike mere fictions? (Putnam [1975b], p. 74)

The conformational holism involved in this passage emphasizes that the support of empirical evidence extends to the full content of the individual statement as literally understood, not to some purely physical part, as opposed to the more familiar idea that empirical support accrues to large collections of statements. (We might call these vertical and horizontal holism, respectively.) Both are implicit in the original Quinean line of thought.

Recall that the discussion of I.6 raised second-philosophical doubts about holism: the case of atomic theory suggests that a scientific theory is not best regarded as a homogeneous whole, up for confirmation as a unit, that various different types of evidence are at work, that hypotheses can play various different roles in our theorizing.[4] The mere presence, even

[4] The objection to this line of thought raised by Colyvan [2001], pp. 99–101, and discussed in I.6, occurs in the course of his defense of the indispensability argument.

indispensable presence, of a posit in our theory of the world is not enough to warrant the conclusion that its existence has been established; in each case, the particular role of the posit, the specific nature of the available evidence must be taken into account. Turning our attention to the placement of pure mathematical posits, we first notice that they occur most frequently in idealized contexts, where causal factors have been isolated (e.g., friction ignored) or simplifications introduced (e.g., the earth's surface taken to be perfectly flat or the ocean as infinitely deep).[5] We also find outright mathematizations, where a pure mathematical object stands in for a physical item (e.g., distance is replaced by a real number, a planetary orbit by an ellipse, or a particle by a point). Indeed, simplification and mathematization often go together, for example, when we regard a fluid as a continuous substance so as to apply the mathematics of fluid dynamics.[6] In all such cases, the mathematical posits appear in descriptions that we don't regard as true, from which it would be inappropriate to draw ontological morals of any kind.[7]

Furthermore, we sometimes mathematize without any clear sense of what the physical phenomenon is like independently of the mathematization: for example, we represent spacetime as a continuous manifold and the electromagnetic field as taking values over this manifold without being able to describe either one without the mathematics, and hence, without a clear sense of whether or not there is any idealization (by simplification) involved. This happens even in simple cases: time is treated as continuous when we represent motion as a function from real numbers (standing in for times) to real numbers (standing in for locations). What's striking in such cases is that scientists apparently feel free to help themselves to whatever mathematics best suits their purposes, without concern for establishing the existence of, say, the real numbers, and indeed—even more

[5] For more, see my [1997], pp. 143–145.

[6] Set theorists and philosophers of mathematics may be amused to learn that this assumption about fluids is called 'the continuum hypothesis'. See Tritton [1988], pp. 48–51.

[7] Resnik ([1997], p. 44) disagrees—'consider Newton's account of the orbits of the planets ... for this explanation to work it must be true that the type of isolated system (Newtonian model) [exists and] has the mathematical properties that Newton attributed to it' (this even though we don't think the Newtonian explanation itself is literally true)—but it's hard to see why one couldn't, for example, describe the dynamics in a certain family by comparison with a kingdom in which so-and-so, without assuming that such a kingdom actually exists. See IV.4.

surprising—without concern for establishing that time, spacetime, or the electromagnetic field are truly continuous![8]

Returning finally to the question of confirmation, we find that this nonchalance toward mathematical posits and their corresponding assumptions about physical structure has its repercussions: for example, the success of theories that include the continuity of time or spacetime or the electromagnetic field is not taken to confirm that these items are in fact continuous; these questions are considered to be wide open. These physical structural assumptions are apparently regarded as necessary for the use of the most convenient and effective mathematics, perhaps open to investigation and confirmation in the future, but not as having been established.[9] As for the pure mathematical assumptions themselves—the existence of the real numbers or a continuous manifold—we find a lack of concern not only for confirming them, but even for identifying the sort of evidence that would be required to do so. While heated debate and great ingenuity goes into the development of means of establishing physical existence,[10] the introduction of pure mathematical posits passes without comment. As far as mathematical existence goes, I think it's fair to say that natural scientists are happy to let the mathematicians determine both proper standards and results, and to help themselves to whatever the mathematicians come up with whenever it serves their purposes.[11] If this is right, then the methods and results of our scientific description of the physical world tell us nothing about abstract ontology; that topic belongs to mathematics and finds its proper place in IV.4.[12]

[8] See my [1997], pp. 154–156, for more on the topics of this paragraph and the next.

[9] Cf. the quotations from Einstein and Dedekind in I.5, footnote 20.

[10] See IV.5 for discussion.

[11] Sometimes they get out in front of the mathematicians, as with Heaviside's calculus (see II.6). Another such case is Dirac's famous δ function, of which he wrote: 'Strictly, of course, δ(x) is not a proper function of x … All the same, one can use δ(x) as though it were a proper function for practically all the purposes of quantum mechanics without getting incorrect results.' Dirac credits his early engineering training: 'I continued in my later work to use mostly the non-rigorous mathematics of the engineers … The pure mathematician who wants to set up all of his work with absolute accuracy is not likely to get very far in physics' (see Pais [1998], pp. 7 and 3, respectively for discussion and references). Schwartz produced a proper mathematical treatment of the δ function some thirty years later (see, e.g., Zemanian [1965]).

[12] Cf. Burgess: 'I don't think physicists ever are concerned about whether numbers are real in the way they may be concerned about the status of virtual photons. Of course, that's partly because I take

ii. Mathematical structure

Postponing, then, the question of mathematical ontology, we might still wonder to what extent the world is mathematically structured, to what extent it realizes the various configurations described by pure mathematics. Considerations of the sort rehearsed in the previous subsection suggest that, for now, there is no physical phenomenon whose literal continuity has been established.[13] In contrast, we've seen (in III.4) that the world does enjoy a considerable amount of logical structuring, and we now note that this includes the facts underlying such elementary arithmetical claims as $2+2 = 4$: the inference from 'there is an apple on the table and another apple on the table and no more' and 'there is an orange on the table and another orange on the table and no more' to 'there is a fruit on the table and another and another and another and no more' is valid even in the rudimentary logic of Part III.[14] Likewise it appears that any KF-structure validates the worldly instances of simple multiplications, like $2 \times 3 = 6$: 'there are two As' and 'either x isn't A or there are three y's that bear R to x' and 'anything bears R to one thing' implies 'there are six y's that bear R to an A'.[15] Exponentiation is more complex, but still within the range of

it to be clear (to physicists as well as to myself) that numbers, even if real, aren't physical, and so the physicist can just pass the buck to the mathematician' (personal communication 2 May 2002, quoted with permission). Burgess is the leading exponent of what I call 'Thin Realism' (see IV.4).

Might this policy of leaving mathematical truth and existence to the mathematicians be dangerous to science, e.g., if the mathematicians were to decide against items or methods that scientists need? I don't think this danger is real for reasons given in IV.3 and IV.4.

[13] See also my [1997], pp. 146–152. It might be argued that such a claim cannot in principle be established, but we should bear in mind that some said this about the existence of atoms before Einstein and Perrin.

[14] Assuming, of course, that nothing is both an apple and an orange and that there's nothing else on the table. The claim is that the following inference is valid in the KF-world of the table top:

$$\exists x \exists y (Ax \,\&\, Ay \,\&\, x \neq y)$$
$$\exists x \exists y (Ox \,\&\, Oy \,\&\, x \neq y)$$
$$\forall x (Ax \vee Ox)$$
$$\forall x \sim (Ax \,\&\, Ox)$$
$$\overline{\exists x \exists y \exists z \exists w (((Ax \vee Ox) \,\&\, (Ay \vee Oy) \,\&\, (Az \vee Oz) \,\&\, (Aw \vee Ow)) \,\&\, (x \neq y \,\&\, x \neq z \,\&\, x \neq w \,\&\, y \neq z \,\&\, y \neq w \,\&\, z \neq w) \,\&\, \forall u (u = x \vee u = y \vee u = z \vee u = w)))}$$

[15] Using $\exists_n$ to abbreviate the familiar way of saying 'there are exactly n things such that ...' used in the previous footnote, the claim is that the following is valid in any KF-world:

$$\exists_2 \, xAx$$
$$\forall x (\sim Ax \vee \exists_3 y (Ryx))$$
$$\forall y \exists_1 \, xRyx$$
$$\overline{\exists_6 y (\exists x (Ax \,\&\, Ryx)}$$

KF-structuring.[16] Thus, insofar as it is KF-, the world reflects the structure of elementary arithmetical equalities and inequalities.[17]

As a somewhat roundabout approach to the status of higher mathematics, let's return to the cognitive story of III.5 and note that it can be extended to cover the arithmetic of small numbers. Recall, for example, the experiments of Wynn ([1992]):[18] 5-month-old infants were shown a single object, a screen was lowered to block it from view, then a second object was placed behind the screen. Presented afterwards with two test displays, one with one object, the other with two, the infants looked longer at the single object—despite the fact that this was the same display they'd seen at the beginning of the experiment—indicating that they represented two objects and were surprised to see only one. And it seems they expect to see exactly two; in an experiment with the same set-up, they looked longer at a test display with three objects than at one with two objects.

These results were replicated by a number of researchers, and subsequently extended to cases where the objects were replaced by novel objects or the absolute and relative positions of the objects were altered while they were hidden behind the screen. Similar results were obtained with slightly older infants for manual search—when two cookies were placed in a box and one removed, the infants searched for the remaining cookie—and for locomotion—infants seeing two cookies placed in one box and three in another crawled toward the box with the three cookies. As we saw in III.5, this sort of behavior is not found when rigid, coherent objects are replaced by piles of sand. (This is Huntley-Fenner et al. [2002].) And even the abilities with rigid, coherent objects don't extend beyond three (for example, infants won't consistently choose a box in which eight cookies have been placed, one by one, over a box in which four cookies have been so placed).

All this suggests that what's at work here is the infant's object tracking system:

> ...diverse findings provide evidence that infants have a system for representing objects that allows them to keep track of multiple objects simultaneously. The

[16] e.g., an instance of $2^3 = 8$ comes out: 'there are two *A*s' and '*Rxyzw* holds of one and only one triple *xyz* for each *w*' implies 'there are 8 *w*'s such that *Rxyzw* for some *A*'s *x*, *y* and *z*'. Cf. Jeffrey [2002], p. 448.

[17] To say an instance of an equality doesn't hold is to say that the corresponding inference is invalid.

[18] In this and the next two paragraphs, I follow Spelke [2000], especially pp. 1233–1235, which includes references.

system is domain specific (it applies to objects but not to other perceptible entities like sandpiles), it is subject to a set size limit (it allows infants to keep track of about three objects but no more), and it survives changes in a number of object properties, including color, detailed shape, and spatial location. (Spelke [2000], pp. 1234–1235)

These same abilities are found in rhesus monkeys,[19] and in the mid-level object file system of human adults.[20] (One difference is that the set size limitation in monkeys and adult humans is about four, better than the infant's three.) All this suggests a very primitive representational plan, what Spelke calls a 'core knowledge system'.

But there is another system that takes us beyond three or four, a system that was first studied in animals.[21] So, for example, suppose a hungry rat is presented with two levers, and will receive food if and only if he presses the first lever some fixed number of times, followed by the second lever once.

How did rats behave in this rather unusual environment? They initially discovered, by trial and error, that food would appear when they pressed several times on lever A, and then once on lever B. Progressively, the number of times that they had to press was estimated more and more accurately. (Dehaene [1997], p. 19)

The rats did well at this task, but their number of presses remained only approximately correct, and their accuracy declined as the required number of presses increased. It seems pigeons can discriminate the number of pecks up to fairly large numbers—45 from 50, for example[22]—again with inversely proportional accuracy. And so on through many animals, including human adults.[23] Human infants can distinguish 8 dots from 16 dots and 16 dots from 32 dots, but not 8 dots from 12 dots, which suggests at this age a fairly crude detectable ratio of 2 to 1;[24] adults do very well up to a 3 : 2 ratio, with diminishing but always better-than-chance performance as the ratio decreases.[25]

The current theory is that the animals and humans who display these approximate numerical abilities are equipped with an analog magnitude

[19] See also Hauser and Carey [1998].

[20] Spelke cites more evidence (with references) for the identification of the infant's object tracker with the adult's mid-level perceptual system in her [2000], p. 1235.

[21] See Dehaene [1997], chapter 1, for a summary; also Dehaene et al. [1998].

[22] See Dehaene [1997], p. 23.

[23] See Dehaene [1997], pp. 70–77, Dehaene et al. [1998].

[24] See Xu and Spelke [2000].

[25] See Spelke [2000], p. 1237, for discussion and references.

system. The precise physical mechanism is unknown;[26] one proposal for a simple 'accumulator' is due to Meck and Church ([1983]):

Suppose the nervous system has the equivalent of a pulse generator that generates activity at a constant rate, and a gate that can open to allow energy through to an accumulator that registers how much has been let through. When the animal is in a counting mode, the gate is opened for a fixed amount of time ... for each item to be counted. The total energy accumulated will then be an analog representation of number. (Carey [1995], p. 107)

Such an analog system, however it works in detail, would provide an approximate representation that would become less accurate as the numbers increase, just as observed.

So the emerging picture sees two separate systems accounting for the infant's abilities in the small number cases and the large number discriminations:

... when small sets of objects are encountered, parallel individuation occurs and one object-file is created for each object ... Two occasions when an analog-magnitude mechanism is more likely to be deployed are (1) when the number of individuals grossly exceeds the limits of parallel individuation, and (2) when the stimuli are not objects. (Feigenson et al. [2002], pp. 64–65)

This explains why the infant can distinguish 2 from 3, but can't distinguish 4 from 6: they discriminate 2 objects from 3 using the object tracker, 4 and 6 are beyond that system's reach, and the young analog system isn't sensitive to a 3 : 2 ratio. It explains why the infant fails at Wynn's 1+1 task using sand piles: these aren't represented as objects, so they don't reach the object tracker. It explains why the infant doesn't discriminate the four cookies hidden one by one from the eight cookies so hidden: the analog system works when the objects are all present to view, but not when they are occluded one by one. And so on.[27]

[26] See Dehaene [1997], 28–31, for a metaphorical description and more information and references. Also Feigenson et al. [2002], p. 61.

[27] The case for this theory, and the evidence on both sides, is of course more complex than my summary. For example, Wynn [1998a] argues for an exclusively analog model, but her opponent there seems to be an exclusive object tracker model, not the dual system model. See also Xu and Spelke [2000], Spelke [2000], and especially, Feigenson et al. [2002]. The version of the object tracker proposed in the last of these is somewhat more refined (see pp. 61–62 vs. pp. 63–64).

Though both mechanisms support numerical distinctions, it's important to recognize how dramatically the object tracker differs from the analog system:

The difference between the accumulator model and the object-file model is in their representational roles. The accumulator model (and other analog magnitude models of number representations) is a dedicated number representational system ... It has no other role in general cognitive function. Any information about the physical properties of the stimulus, such as size or color, is discarded within this system. The object-file model, on the other hand, has been suggested as a more general cognitive mechanism underlying object-based attention and working memory. While some numerical information is derivable from object-file representations ... its primary purpose is not quantitative ... Rather, it functions to track objects as they move and change. ... property information can be bound to an object-file once it has been opened. (Feigenson et al. [2002], p. 64)

The analog system is a sort of yardstick, designed to measure size; the object tracker is, in the words of one supporter, 'non-numerical'.[28]

In our terms, what the object tracker does in the small number tasks is detect a small fraction of the logical KF-structuring present in the world. Consider, for example, this report

... [in the small number cases,] infants represent objects but not sets with cardinal values. Their ability to discriminate displays of 1 vs. 2 objects therefore does not depend on representations of sets with specific numerosities but rather on representations of 'an object' and 'an object and another object'. (Xu and Spelke [2000], p. B3, see also Carey and Xu [2001], p. 185)

In other words, the infant can represent 'exactly two objects' in purely logical terms—an object, another object, no other object—an early manifestation of its ability to grasp rudimentary logical form.

Given that the distinction between 16 and 32 dots is another aspect of the world's KF-structure, we see that the young infant has at its disposal a second (largely)[29] veridical system of representation—the analog system—which also carries forward (with some improvements) into adulthood. So we have, in this new case, counterparts to our first and second claims of Part III: the

[28] For references, see Feigenson et al. [2002], pp. 61–62.

[29] I insert this because the logical component is still subject to the quantum mechanical limitations discussed in III.4 and III.6. It isn't clear to me that the analog system has corresponding limitations, so I drop the qualifier in the remainder of the paragraph.

world is so structured and our most primitive cognitive mechanisms allow us to detect that structure. Here the third claim—that we are so configured because the world is as it is—seems at least as well supported as the analogous claims in Part III.[30] For example, Wynn writes (of the analog system):

> I propose that there exists a mental mechanism, dedicated to representing and reasoning about number, that comprises part of the inherent structure of the human mind. A range of warmblood vertebrate species, both avian and mammalian, have been found to exhibit numerical discrimination and reasoning abilities similar to those documented in human infants. Because of its adaptive function, this mechanism quite likely evolved through natural selection, either at a point in evolutionary history prior to the branching off of these different species, or separately but analogously within several branches of these species. (Wynn [1998a], p. 297)

Dehaene is somewhat more guarded, admitting the possibility (on present evidence) that learning is involved, but he nevertheless concludes that

> More likely, a brain module specialized for identifying numbers is laid down through the spontaneous maturation of cerebral neuronal networks, under direct genetic control and with minimal guidance from the environment. Since the human genetic code is inherited from millions of years of evolution, we probably share this innate protonumerical system with many other animal species. (Dehaene [1997], p. 62)

Of course these new empirical claims—that humans are configured so as to conceptualize the world in protonumerical terms, that the world is so structured, and that humans are so configured because the world is so structured—are naturally subject to many of the same general caveats explored in III.8, in addition to its own particular fallibilities.

The point of reviewing all this, aside from its inherent interest, is to raise the next question: how do we get from these rudimentary numerical abilities, which we share with many animals, to arithmetic and number theory? The story begins with the predicament of the young child learning to count. She has one system, the object tracker, that tracks up to three objects, distinguishes three objects from two, two from one, but isn't fundamentally numerical. She has another system, the analog system, that is

[30] Notice the richer store of evidence from non-human animals. Also, there is evidence that even newborns can distinguish 2 from 3 dots (see Antell and Keating [1983]).

protonumerical, but which works only approximately on large arrays. She is taught a string of words—one, two, three, four, etc.—and a method of repeating them as she points to a series of objects, one word for each object, one object for each word, no repetitions, and she masters this.[31] But, given the mental machinery she has to work with, it's hard for her to understand what this procedure accomplishes. Where she has precise representations—for one, two, and three things—she isn't representing cardinality; where she does represent cardinality—for large arrays—she doesn't do so precisely. So what's the significance of reaching 'five' at the end of the process of 'counting' her fingers?

Experimental evidence suggests that children beginning to count are in something very like this predicament. In a series of experiments by Wynn on 2½- to 3½-year-olds, all of whom counted with accuracy, many did poorly when asked to give a certain number of items from an available pile. Though all the children succeeded in giving one when asked for one, all but a few of the oldest, when asked for more than one 'tended either to simply grab a handful of items and give them all, or to give an apparently random number of items one at a time' (Wynn [1990], p. 172). The youngest children didn't even approximate the correct number, that is, they didn't grab more for a larger number. In a second study, children successfully gave two objects when they were a few months older, then three objects a few months later, until the 3½-year-olds were accurate to as high as they could count. Wynn concludes:

> This suggests that by the time children learn the meaning of the word 'five', but after they have learned the meaning of 'three', they acquire the meanings of *all* the number words within their counting range. (Wynn [1990], p. 184)

Why do children make these mistakes and how are they able to overcome them?

Spelke tells a compelling story based on the child's progress in coordinating the precise, non-numerical system with the approximate, protonumerical system:

> Children learn to relate the word *one* to their core system for representing objects: They learn that *one* applies just in case there's an object in the scene, and it is roughly synonymous with the determiner *a*. About the same time, children learn to

[31] Gelman and Gallistel [1978]. See Wynn [1990] for a more recent treatment and references.

relate the other number words to their core system for representing numerosities: They learn that the other number words apply just in case there's a set[32] in the scene, and those words are all roughly synonymous with *some* ... (Spelke [2000], p. 1238)

We're now at the state of the youngest children in Wynn's study: one object is given with success; other requests are met with a random handful.

The next and very difficult step requires that children bring their representations of objects and numerosities together. They have to learn that *two* applies just in case there's a set composed of an object and another object. When *two* is mastered, children must learn that *three* also applies to a combination of object and numerosity representations: to a set composed of an object, an object, and an object. (Spelke [2000], p. 1238)

In other words, the child is beginning to connect the logical, object-tracking representation of 'two' and 'three' with the numerosity of arrays via the counting words and the procedure that goes with them. Thus the object-tracking representations acquire number, and the analog representations acquire precision.

The final step described by Wynn would seem to result from a sudden induction: the child has seen that moving from one to two, from two to three, perhaps from three to four, involves adding an object to an array; at some point between three and five, she

Generalize[s] this discovery to all the number words and infer[s] that each word picks out a set containing one more object than the preceding word. (Spelke [2000], p. 1238)

In sum,

The language of number words and the counting routine allow young children to combine their representations of objects as enduring individuals with their representations of numerosities to construct a new system of knowledge of number, in which each distinct number picks out a set of individuals with a distinct cardinal value. (Spelke [2000], p. 1238)

So it is a linguistic and cultural artefact—the system of number words and the counting procedure—that enables the child to combine her

[32] Spelke uses 'set' here (and below) where I've been using 'array', but I think philosophers should resist the temptation to conclude that she means anything significantly different (like an abstract mathematical object!).

two different representational systems to form representations of precise numerical values.[33]

Fascinating studies of adults support this picture of the bridging role of language. We've seen that both the infant's systems seem to be carried forward to adulthood, albeit with slight improvements in extent (from three to four for the object tracker) and accuracy (from 2 : 1 ratios to better than 3 : 2 for the approximations of the analog system). Dehaene, Spelke, and several of their co-workers tested bilingual adults on two-digit addition problems of two kinds: in the exact test, subjects were asked to calculate the sum and select between two close candidate solutions (one ten off, one correct); in the approximate test, they were asked to estimate the sum and select between two approximate candidates (one rounded to the nearest ten, the other off by thirty). Subjects were first trained in the exact or the approximate task, in one or the other of their two languages, before testing. In the exact test,

> Subjects performed faster in the teaching language than in the un-trained language ... This provided evidence that the arithmetic knowledge acquired during training with exact problems was stored in a language-specific format and showed a language-switching cost due to the required internal translation of the arithmetic problem. For approximate addition, in contrast, performance was equivalent in the two languages, providing evidence that the knowledge acquired by exposure to approximate problems was stored in language-independent form. (Dehaene et al. [1999], p. 971)

Brain-imaging experiments with a different group of subjects on similar tasks produced a similar pattern:

> The approximate task showed greater bilateral activation throughout the inferior parietal lobes, including both the areas thought to be involved in representations of objects in multiple-object-tracking tasks and those thought to be involved in representations of sets in numerosity discrimination tasks. In contrast, the exact task showed greater activation on the left side of the inferior frontal lobe ... the area of activation ... typically ... activated in studies requiring retrieval of well-learned verbal facts and word associations. (Spelke [2000], p. 1239)

Thus, it seems that language is what allows us to extend the precision of our logical, small number representations into the realm of larger numerosities.

[33] Cf. Hauser and Carey [1998], p. 75: 'we read the animal literature to be consistent with the claim that [a mental list of symbols to represent number] (widely but not universally expressed in natural languages) is a human cultural construction.'

The child's key discovery is that moving one step forward in the list of number words corresponds to adding one more object to an array, the idea, in other words, of the successor function. At this point children often engage in extended recitations of the sequence of number words, developing the sense that new ones can always be generated, and it is this linguistic conviction that apparently gives rise to the familiar picture of the natural number sequence: 0, 1, 2, 3, 4, 5, ... Neither the logical object tracker nor the analog approximating device—that is, neither of the detectors developed by evolutionary pressures and/or very early experience—includes even implicitly the idea of indefinite extension. The '...' apparently results not from any insight into the world, but from the childhood conviction that we will never run out of number words; this ground-breaking mathematical idea appears to be essentially linguistic—not logical, not empirical.

This is the point I want to highlight from this long digression. It suggests, I submit, that number theory, the mathematical investigation of the sequence of natural numbers, of the so-called 'standard model of arithmetic', is an elaborate study of this linguistically generated idea of the '...'. It's tempting to respond—isn't there an infinite sequence of number words? Aren't we studying that?[34]—but a moment's reflection reveals that there's a limit to how complex understandable expressions for numbers can be, and thus that there are only finitely many potential 'number words'. Our tendency to think otherwise results from a common pattern of thought: we replace the vague and amorphous structure of actual language with an idealized model of words and recursive rules that does indeed generate indefinitely long expressions, but which does so only by virtue of a mathematization; we use our theory of the natural numbers to define this model, then draw our conclusion about the abundance of number words from an artefact of the modeling. The '...' appears to be present in the sequence of number words only because we've built it into our idealized theory of the number words.[35]

Still the fact that our two known detectors don't encompass the '...' doesn't rule out the possibility that an infinite sequence is realized

[34] e.g., see Wiese [2003], p. 79: 'counting words ... do not refer to ... numbers, they *are* numbers.'

[35] This happens in metalogic, too: we imagine that formulas and proofs can be indefinitely long because we define them recursively, building in the '...' so to speak. This is one of several reasons I resisted formalizing the rudimentary logic of Part III.

somewhere in the physical world; it just means that our access to that physical realization would have to come from elsewhere, presumably from our theorizing. Indeed, in natural science we do sometimes treat a large finite collection as infinite,[36] but in such cases the infinite obviously isn't literally realized. Whether or not such prime candidates as time and space are infinite in extent depends on whether or not there will be a Big Crunch to accompany the Big Bang, a difficult cosmological question that remains open.[37] Of course the non-existence of a physical realization wouldn't affect mathematics: even if application of the natural numbers turned out to be invariably an idealization, it's an idealization that's taken on a life of its own; we would still want to know if Goldbach's Conjecture is true,[38] and if so, if it can be proved. For that matter, as in the case of continuous mathematics, we would probably still have good reason to persist in those idealized applications.

The upshot is that we cannot presently feel confident that any mathematics beyond the rudimentary logic and arithmetic ratified by KF-structuring is literally realized in the physical world.

[36] See e.g. Narens and Luce [1990], p. 124: 'Many kinds of generalizations used in science require us to ignore certain properties that are inherent in finite models ... For example, in finite, ordered domains, maximal and minimal elements necessarily exist ... a desirable generalization may exclude such elements, as for example: "For each American middle class individual there is a slightly richer American middle class individual".' Another example is the 'thermodynamic limit' in which the number of particles in a sample of gas increases to infinity (see Sklar [1993], pp. 78–81).

In fact, idealizations from finite to the size of the natural numbers are relatively rare: 'Intuitively, the idealization of a finite domain should be to some denumerably infinite one. However, mathematical science routinely employs nondenumerable domains (e.g., continua ...) for idealizations, because these have special, desirable modeling properties that are not possible for denumerable domains' (Narens and Luce [1990], p. 124). (For the contrast between the countable or denumerable infinity of the natural numbers and the uncountable or non-denumerable infinite of the reals and other continua, see Enderton [1977], pp. 132–133.) We've seen this in the case of fluid dynamics, where a finite collection of molecules is replaced by a continuous 'ideal fluid', and I'm told something similar happens with mass in cosmology.

[37] Cf. Wald [1992]: 'the universe started with a "*big bang*" a finite time ago! The best observational evidence indicates that the "big bang" took place between 10 and 20 billion years ago' (p. 50); 'What is going to happen in the future? ... the most important question in this regard is, Is the universe open or closed? ... if the universe is open ... it will continue to expand forever. If the universe is closed ..., the expansion will eventually come to a halt; the universe will recontract and will again approach a singular state within a finite time' (p. 68); 'if the presently available observational evidence is taken at face value, the best guess is that the universe in open ... and will, therefore, continue to expand forever. However, there is still considerable room for doubt in this conclusion ... Perhaps within the next few decades, further observations will be made which will conclusively tell us the eventual fate of our universe' (p. 72).

[38] Goldbach's conjecture asserts that every even number greater than 2 is the sum of two primes. It has neither been proved nor refuted since it was first proposed in 1742.

iii. The 'miracle' of applied mathematics

In his well-known meditation on the applicability of mathematics, Wigner writes:

> the miracle of the appropriateness of the language of mathematics for the formulation of the laws of physics is a wonderful gift which we neither understand nor deserve. (Wigner [1960], p. 237)

Similar sentiments have been expressed by many physicists, including Hertz, Einstein, and Feynman.[39] Of course explaining why a particular bit of mathematics applies in a particular case is a central problem in all areas of applied mathematics. Consider, for example, fluid dynamics, with its

> assumption of the applicability of continuum mechanics [[40]] ... We suppose that we can associate with any volume of fluid, no matter how small, those macroscopic properties that we associate with the fluid in bulk.... Now we know that this assumption is not correct if we go right down to molecular scales. (Tritton [1988], p. 48)

Tritton goes on to derive conditions under which the assumption is harmless. Similarly, Wilson ([2006], pp. 214–217) discusses how Euler's method for computing the path of a cannon ball is effective as long as the so-called Lipschitz condition is satisfied. Or, to take an example of a different sort, Keller [1986] explains why a coin flip can reasonably be treated as a random device. But this is not the sort of answer sought by those who see miracles in applied mathematics. What impresses them isn't individual instances of successful use of mathematical methods, but more general phenomena that might be sorted roughly under three headings: Origins, More Out Than In, and Transfers.

Origins. Until recently, the relation between mathematics and the natural sciences was understood to be quite simple: the two were hardly distinguished at all.[41] The historian Morris Kline describes the situation this way:

> the Greeks, Descartes, Newton, Euler, and many others believed mathematics to be the accurate description of real phenomena ... they regarded their work as the uncovering of the mathematical design of the universe. (Kline [1972], p. 1028)

[39] See Steiner [1989], pp. 449–450, for a list of quotations with references.

[40] See footnote 6. [41] See IV.3 for more on this point.

Over the course of the nineteenth century, this picture changed dramatically: 'gradually and unwittingly mathematicians began to introduce concepts that had little or no direct physical meaning' (Kline [1972], p. 1029). Citing the rise of negative numbers, complex numbers, *n*-dimensional spaces, and non-commutative algebras, he remarks that 'mathematics was progressing beyond concepts suggested by experience', but that 'mathematicians had yet to grasp that their subject ... was no longer, if it ever had been, a reading of nature' (Kline [1972], p. 1030). By mid-century, the tide had turned:

> after about 1850, the view that mathematics can introduce and deal with rather arbitrary concepts and theories that do not have immediate physical interpretation but may nevertheless be useful, as in the case of quaternions, or satisfy a desire for generality, as in the case of n-dimensional geometry, gained acceptance. (Kline [1972], p. 1031)

This movement continued with the study, for example, of abstract algebras, pathological functions, and transfinite numbers.

The heady new view of mathematics that accompanied this change is perhaps best expressed by Georg Cantor: 'Mathematics is entirely free in its development ... The essence of mathematics lies in its freedom' (as quoted in Kline [1972], p. 1031). This sentiment appears in the thinking of many of the most innovative mathematicians of the late nineteenth century; today, it is standard orthodoxy. Mathematics progresses by its own lights, independent of ties to the physical world. Legitimate mathematical concepts and theories need have no direct physical interpretation. Still, mathematical structures identified and studied for purely mathematical purposes have been notably effective in applications. One conspicuous example is group theory.

The physicist Freeman Dyson begins the story this way:

> In 1910 the mathematician Oswald Veblen and the physicist James Jeans were discussing the reform of the mathematical curriculum at Princeton University. 'We may as well cut out group theory,' said Jeans. 'That is a subject which will never be of any use in physics.' It is not recorded whether Veblen disputed Jeans's point, or whether he argued for the retention of group theory on purely mathematical grounds. All we know is that group theory continued to be taught. (Dyson [1964], p. 249)

Some years later, Wigner took up the problem of the interaction of more than two identical particles. The physicist and historian Abraham Pais recounts that

> He rapidly mastered the case n=3 (without spin). His methods were rather laborious; for example, he had to solve a (reducible) equation of degree six. It would be pretty awful to go on this way to higher n. So, Wigner told me, he went to consult his friend the mathematician Johnny von Neumann. Johnny thought a few moments then told him that he should read certain papers by Frobenius and by Schur which he promised to bring the next day. As a result Wigner's paper on the case of general n (no spin) was ready soon and was submitted in November 1926. It contains an acknowledgement to von Neumann, and also the following phrase: 'There exists a well-developed mathematical theory which one can use here: the theory of transformation groups which are isomorphic with the symmetric group (the group of permutations)'. (Pais [1986], pp. 265–266)

Pais concludes: 'Thus did group theory enter quantum mechanics' (Pais [1986], p. 266). Dyson echoes the momentous tone of this remark in the epilogue to his account of Jeans's gaffe:

> By an irony of fate group theory later grew into one of the central themes of physics, and it now dominates the thinking of all of us who are struggling to understand the fundamental particles of nature. (Dyson [1964], p. 249)

Examples could be multiplied: exploring their physical problems, physicists are often surprised to find that the mathematical tools they need are ready and waiting.

This oddity drives Wigner's worries. Mathematical concepts, he writes,

> are defined with a view of permitting ingenious logical operations which appeal to our aesthetic sense ... [they are chosen] for their amenability to clever manipulations and to striking, brilliant arguments. (Wigner [1960], pp. 225, 229)

Mathematician and computer scientist Richard Hamming makes a similar point:

> *Artistic taste* ... plays ... a large role in modern mathematics ... we have tried to make mathematics a consistent, beautiful thing, and by doing so we have had an amazing number of successful applications to the physical world. (Hamming [1980], pp. 83, 87)

The obvious question is why theories devised to satisfy purely aesthetic desiderata, constrained only by human standards of taste, should end up applying so aptly to the physical world. I argue (in IV.3) that mathematical concept formation is guided by more substantive factors than mere aesthetic preference, but given that the considerations involved seem equally far removed from any concern with potential applications, the troublesome question remains: why do theories devised to satisfy purely mathematical values and goals end up applying so aptly to the physical world?

More Out Than In. Beyond the 'miracle' that pure mathematics turns out to apply so often, there are surprises in the ways it does so: 'When you get it right ... more comes out than goes in' (Feynman [1965], p. 171). One much discussed example comes from the history of electromagnetism. When Maxwell set out (in 1873) to develop his mathematical theory of the phenomenon, he ran into difficulty; in order to preserve conservation of charge, he had to add an extra factor, the 'displacement current'. This led to the prediction of radio waves, which were detected in the laboratory by Hertz in 1887.[42] Hertz was moved to declare that

> One cannot escape the feeling that these mathematical formulae have an independent existence and an intelligence of their own, that they are wiser than we are, wiser even than their discoverers, that we get more out of them than was originally put into them. (As quoted by Dyson [1964], p. 249)[43]

How could this be?!

Another striking example comes from quantum mechanics:[44]

> The Dirac equation necessarily implies that at each space-time point there are four wave-functions—not two ... along with the two electron states à la Pauli, there appears another pair, also spin-doubled, but with negative kinetic energy. That is very bad: kinetic energy is essentially by definition always positive or zero. (Pais [1991], p. 355)

Pais reports the general dismay this generated, leading Heisenberg to remark that 'the saddest chapter of modern physics is and remains the Dirac theory'

[42] Cf. Hamming [1980], p. 82: 'Many of you know the story of Maxwell's equations, how to some extent for reasons of symmetry he put in a certain term, and in time the radio waves that the theory predicted were found by Hertz.' See Steiner [1998], pp. 77–79, for discussion and references.

[43] This quotation also appears in Steiner [1998], p. 13, and Bell [1937], p. 16, but none of these authors provides a reference in Hertz. (Steiner cites Dyson.) No Hertz expert myself, I confess I haven't managed to locate the passage.

[44] See Steiner [1998], pp. 82–83, 159–161, for discussion.

(Pais [1991], p. 355). Eventually Dirac proposed 'A new kind of particle, unknown to experimental physics, having the same mass and opposite charge of the electron' (as quoted by Pais [1991], p. 356). In other words, the positron. This prediction, springing entirely from the mathematics, was confirmed the next year, in 1932:

The positron ... was first seen by Carl Anderson from Cal Tech, in a cloud chamber exposed to cosmic radiation. In early 1933 the first creation of electron-positron pairs was observed, also in cosmic radiation. (Pais [1991], p. 356)

In such cases, pure mathematics not only provides models of empirical structures by some mechanism more or less mysterious; these models contain further physical information hidden in their equations![45] Or so the worry goes.

Transfers. Finally there is the odd phenomenon of a mathematical treatment that works well in one context successfully transferring to an entirely unrelated context. In his introductory lectures, Feynman observes:

There is a most remarkable coincidence: *The equations for many different physical situations have exactly the same appearance.* Of course, the symbols may be different—one letter is substituted for another—but the mathematical form of the equations is the same. (Feynman et al. [1964], p. 12-1)

He goes on to show how the equations he's just derived for electrostatics reappear in accounts of heat flow, distortion of a stretched membrane, neutron diffusion, irrotational fluid flow, and uniform illumination of a plane.

Why are the equations from different phenomena so similar? We might say, 'It is the underlying unity of nature' ... everything is made out of the same stuff, and therefore obeys the same equations. That sounds like a good explanation, but let us think. The electrostatic potential, the diffusion of neutrons, heat flow—are we really dealing with the same stuff? Can we really imagine that the electrostatic potential is *physically* identical to the temperature or to the density of particles? ... The displacement of a membrane is certainly *not* like a temperature. Why, then, is there 'an underlying unity'? (Feynman et al. [1964], p. 12-12)

[45] Another popular example is Heisenberg's application of matrix mechanics; see Wigner [1960], p. 232: 'Surely in this case we "got something out" of the equations that we did not put in.' See also Steiner [1998], pp. 146–149, and Liston [2000] for a rebuttal of Steiner's analysis.

Why indeed?![46] This phenomenon turns up frequently in the literature on the mystery of applied mathematics:

Wigner also observes that *the same mathematical concepts* turn up in entirely unexpected connections. For example, the trigonometric functions which occur in Ptolemy's astronomy turn out to be the functions which are invariant with respect to translation (time invariance). They are also the appropriate functions for linear systems. The enormous usefulness of the same pieces of mathematics in widely different situations has no rational explanation (as yet). (Hamming [1980], p. 82)

Why should what works here also work there?!

Many variations on these themes have been essayed over the years[47] and the conviction that mathematics is 'unreasonably' successful in application remains strong. This isn't the place to attempt a full treatment, even if I were capable of providing one, but I would like to pick up one thread from these complex and multifaceted discussions, perhaps the most general to be found in the case for miracles, namely, the appeal as in Feynman to 'a remarkable coincidence'. We hear this repeatedly. After treating several individual cases, Jody Azzouni writes:

It may be felt that this way of putting the matter has neatly evaded the real puzzle about applicable tractable mathematics. For the above sorts of explanation handle each application separately. But surely there is something global still to be explained here: Why has *so much* tractable mathematics proven successfully applicable? ... Isn't this something of a weird coincidence? (Azzouni [2000], p. 221)

It turns up again in Michael Liston's version of 'Origins':

Complex numbers, analytic functions, and Hilbert spaces ... were developed and evolved in pure mathematical contexts, guided solely by aesthetic criteria. Yet they are surprisingly effective in physics. The coincidence is unreasonable, surprising, mysterious, bizarre. (Liston [2000], p. 195)

[46] In some cases, the mathematics is transferred to a new context where its original justification no longer applies: see, e.g., Wilson [2006], pp. 158–159, 175–176, on the Navier–Stokes equation (cf. II.6, footnote 15) or Wigner [1960], p. 232, on Heisenberg.

[47] e.g., Steiner ([1998], p. 5) argues that reasoning by mathematical analogies depends on an unnaturalistic belief 'that the human species has a special place in the scheme of things'. See Liston [2000], Simons [2001], Azzouni [2004], pp. 175–180, Bangu [2006], for a range of skeptical responses.

Let's consider the matter from this perspective.[48] Is there a coincidence here, in need of explanation?

In a series of writings and lectures, the statistician and magician Persi Diaconis has used statistical methods and a conjuror's eye to debunk many claims of significant coincidence, in contexts ranging from Carl Jung's synchronicity to contemporary ESP research (see Diaconis [1978], Diaconis and Mosteller [1989]). I offer here the modest observation that many of the considerations Diaconis brings to bear on his examples have counterparts for the case of applied mathematics; indeed, these counterparts often turn up—though not under this description—in the writings of those atypical authors who sometimes tentatively, sometimes apologetically, suggest that perhaps applicability is not as miraculous as it first appears. Of course nothing follows conclusively from this observation, but it might well give us pause: perhaps at least some of our amazement is due to tempting errors of judgment that are common in a wide range of human dealings.

The most straightforward way of explaining away an apparent coincidence is to uncover a hidden cause.[49] I sit quietly at home and suddenly remember a friend I haven't seen or thought of in years; a knock comes at the door announcing that very friend! I tell myself this must be synchronicity, some weird, but meaningful coincidence—unless I realize that without noticing it consciously I most likely heard the characteristic sound of his jalopy in the driveway or caught her favorite perfume wafting in the open window, and that this clue must have triggered the recollection. In such cases, the true explanation is causal, not synchronic.[50]

For comparison, consider the 'miraculous' case of non-Euclidean geometry. One hears tell of a lively research area pursued well into the late 1800s (part of this development is sketched in IV.3). Kline reports that

> none of the mathematicians who worked in the later period believed that these basic non-Euclidean geometries would be physically significant ... the thought that

[48] For the record, Azzouni ([2000], pp. 221–222) ends up denying that there is in fact any global question to answer. Liston ([2000], p. 195) elaborates a brief remark of Steiner's to touch on one of the themes developed below (see footnote 58).

[49] See Diaconis and Mosteller [1989], p. 859.

[50] As Diaconis and Mosteller note ([1989], p. 853), Jung regarded synchronicity as 'a hypothetical factor equal in rank to causality as a principle of explanation'.

physical space ... could be non-Euclidean was dismissed. In fact, most mathematicians regarded non-Euclidean geometry as a logical curiosity. (Kline [1972], p. 921)

Euclid's was held to be the true geometry; the others were studied for their purely mathematical interest.

The story picks up in 1912, when Einstein moved to Zurich. By then he had described the statics of gravitation, but the dynamics eluded him. In June and July of that year, he wrote:[51]

The further development of the theory of gravitation meets with great obstacles ... The generalization [of the static case] appears to be very difficult. ... it cannot yet be grasped what form the general space-time transformation equations could have. I would ask all colleagues to apply themselves to this important problem!

The supreme difficulty was that Euclidean geometry would not do.

If all [accelerated] systems are equivalent, then Euclidean geometry cannot hold in all of them. To throw out geometry and keep [physical] laws is equivalent to describing thoughts without words. We must search for words before we can express thoughts. What must we search for at this point?

Early in 1912, Einstein realized that 'Gauss's theory of surfaces holds the key for unlocking this mystery ... I realized that the foundations of geometry have physical significance.' But despite these insights, the problem remained so troublesome that Einstein, on arrival in Zurich, told his mathematician friend Marcel Grossman, 'You must help me or else I'll go crazy.'

As history records, Grossman did help. Pais, Einstein's co-worker and biographer, reports that

he told Grossman of his problems and asked him to please go to the library and see if there existed an appropriate geometry to handle such questions. The next day Grossman returned (Einstein told me) and said that there indeed was such a geometry, Riemannian geometry. (Pais [1982], p. 213)

Einstein himself describes arriving in Zurich

without being aware at that time of the work of Riemann, Ricci, and Levi-Civita. This [work] was first brought to my attention by my friend Grossman when I posed to him the problem ... (as quoted by Pais [1982], p. 212)

[51] This paragraph and the next follow Pais [1982], chapter 12. The quotations in this paragraph come from pp. 211–212 where references can be found.

Grossman and Einstein immediately set to work on their collaboration, during which Einstein wrote:

At present, I occupy myself exclusively with the problem of gravitation and now believe that I shall master all difficulties with the help of a friendly mathematician. (as quoted by Pais [1982], p. 216)

This was premature, as the theory of general relativity didn't reach its final form until a few years after the collaboration with Grossman, but the introduction of Riemannian geometry was nevertheless a fundamental breakthrough.

The story as told so far might appear an impressive example of the phenomenon we've labeled 'Origins': a bit of mathematics pursued as a 'logical curiosity' in Kline's phrase (Kline [1972], p. 921, quoted above) ends up playing a fundamental role in physics. But in fact Kline is describing classical synthetic non-Euclidean geometry, not Riemannian geometry, and a fuller history reveals that Riemann himself, faced with earlier non-Euclidean geometries, was motivated by the physicist's question:

'Since these geometries differ from each other what are we really sure is true of physical space?' This question was Riemann's point of departure. In answering it he created more general geometries, now known as Riemannian geometries, which, because of their very nature and in view of our limited physical knowledge, could be as instrumental in representing physical space as Euclidean geometry. Is it then to be wondered that Einstein found Riemannian geometry useful? The physical relevance of Riemannian geometry does not detract from the ingenious use Einstein made of it; nevertheless its suitability was a consequence of work on the most fundamental physical problem that mathematicians have ever tackled, the nature of physical space. (Kline [1968a], p. 234)[52]

So Riemann's motivation was not 'purely aesthetic' or in any sense 'purely mathematical'; he was concerned, rather, with the needs of physical geometry and his efforts were successful. The apparent coincidence disappears when the hidden cause is revealed.

Some instances of 'More Out Than In' permit similar dissolution.[53] Consider again the case of Maxwell. The historian Daniel Siegel writes:

[52] A similar paragraph appears in Kline [1980], p. 293.

[53] Cf. Oldershaw [1990], p. 146: 'These triumphs [which include the case of Maxwell noted above] are important and very impressive, but are they "mysterious" or "miraculous"? The author thinks not. Is it not more reasonable to believe that our successful physical theories originate from the recognition of an apparent pattern that approximates the underlying order of nature, ... and if the applied mathematics

According to the standard account, Maxwell began with a set of four field equations ... Maxwell realized ... that this is not yet a complete and consistent set of equations ... Ampère's law is in need of modification. The difficulty could be resolved, Maxwell realized, by adding to the right hand side ... another term. ... What was required for mathematical consistency ... was an additional current. (Siegel [1991], pp. 87–89)

Siegel goes on to show that the standard account is incorrect, that Maxwell saw himself as developing Faraday's purely physical theory:

Faraday had carried out experiments to demonstrate ... forces distributed in space ... and he was convinced of the primacy of these ... 'powers' in space in determining magnetic phenomena. Faraday used the method of iron filings to visualize the patterns of force in space, and later he began to refer to these spatially distributed powers as 'lines of force', together constituting a 'field' in space. In the full development of Faraday's theory ... electric charges were to be regarded as epiphenomena ... having no independent or substantial existence ... In Faraday's approach, then, it was the field that was truly primary. (Siegel [1991], p. 9)

When Maxwell added the displacement current, he was in fact attempting to provide a mechanical model for Faraday's field:

In sum, Maxwell did basically what the standard account says he did: he modified Ampère's law ... in a manner consistent with the equation of continuity and Coulomb's law. His goal in that, however, was not a complete and consistent set of electromagnetic equations for its own sake, but rather a complete and consistent mechanical model of the electromagnetic field. (Siegel [1991], p. 97)

It was the detailed requirements of this mechanical model that prompted Maxwell's addition, not a purely mathematical concern.[54]

Finally, after presenting his example of 'Transfer' cited above, Feynman provides this demystifying explanation:

As long as things are reasonably smooth in space, then the important things that will be involved will be rates of change of quantities with position in space.

accurately embodies that pattern, then it should not be the least bit surprising that the pattern or its mathematical embodiment holds good beyond the empirical limits that are applicable at a given time and can thus be used to make successful predictions.'

[54] Cf. Siegel [1991], chapter 4, Harman [1998], chapter V. See Steiner [1998], pp. 77–79, for an effort to preserve the impression of mathematical coincidence in this case.

That is why we always get an equation with a gradient. The derivatives *must* appear in the form of a gradient or a divergence; because the laws of physics are *independent of direction*, they must be expressible in vector form. The equations of electrostatics are the simplest vector equations that one can get which involve only the spatial derivatives of quantities. Any other *simple* problem—or simplification of a complicated problem—must look like electrostatics. What is common to all our problems is that they involve *space* and that we have *imitated* what is actually a complicated phenomenon with a simple differential equation. (Feynman et al. [1964], p. 12-12)

In other words, many different phenomena share high-level structural features that allow them to be treated using the electrostatic equation.

But even if many cases of surprising application can be explained away by underlying causes, there will presumably remain a reserve of inexplicable and thus 'miraculous' examples, and it's here that Diaconis's distinctive statistician's and magician's thinking comes into play.[55] For a characteristic example, consider the 'new word' phenomenon:

Here is a coincidence every person will have enjoyed: On hearing or reading a word for the first time, we hear or see it again in a few days, often more than once. (Diaconis and Mosteller [1989], p. 858)

Diaconis and Mosteller give a number of explanations for such cases; the one that concerns us here involves 'heightened perception':

Very likely this word has been going by your eyes and ears at a steady low rate, but you have not noticed it. You are now tuned to it, and where blindness used to be sits an eagle eye. (Diaconis and Mosteller [1989], p. 858)

As a mathematical analog, I suggest that we tend to notice those phenomena we have the tools to describe. There's a saying: when all you've got is a hammer, everything looks like a nail. I propose a variant: if all you've got is a hammer, you tend to notice the nails. As Hamming notes, 'almost all our experiences in this world do not fall under the domain of science or mathematics' (Hamming [1980], p. 89).[56] Armed with the tools we happen to have, perhaps we tend to fix on those aspects of the world that seem

[55] In popular lectures, he speaks of viewing a phenomenon through 'coincidence spectacles'.
[56] See IV.4, footnote 47, for an example.

likely to admit of mathematical treatment.[57] Perhaps our impression that mathematical science is unreasonably successful in describing the world arises in part from selective attention to amenable cases. Just as some odd coincidence seems to bring the new word under our eyes more frequently than before, some happy miracle seems to predispose the world to mathematical description. In particular, this mental quirk might help explain our surprise at 'Transfers'.

Another variety of cases arises from selective memory: a passing thought of Thelonius Monk flits through our jazz fan's mind fairly often, but if she's thinking of him when one of his recordings comes on the radio, she forgets all other occasions and finds a remarkable coincidence: 'What are the chances of that?!' (Depending on how often her thoughts turn to Monk, how often she listens to the radio, and how often the radio station plays Monk, the probability of this 'coincidence' might in fact be fairly high.) As Diaconis and Mosteller note:

> Frequency of forecasting the same dire event improves the chances of simultaneity of forecast and outcome. Forgetting many failed predictions makes success seem more surprising. (Diaconis and Mosteller [1989], p. 859)

The counterpart in the case of applied mathematics is our forgetfulness of its many failures:

> it seems a little strange to claim, after very lengthy trial-and-error ... to find appropriate mathematical models of natural phenomena ..., that the *successful* results (and let's not forget the far more numerous examples of less happy results) of mathematical endeavors are 'unreasonably' effective. (Oldershaw [1990], p. 145)[58]

As we've seen (in II.6), Wilson's work is filled with examples of just how difficult it is to apply mathematics successfully: 'Let us begin with the simple observation that applied mathematics can be very tough!' (Wilson [2000a], p. 296).

[57] Cf. Wilson [2000a], p. 297: 'it is the job of the applied mathematician to look out for the *special circumstances* that allow mathematics to say something useful about physical behavior.'

[58] Cf. Steiner [1998], p. 46: 'what about all the failed attempts to apply mathematics to nature? Are not, in fact, most such attempts doomed to failure?' Also Liston ([2000], p. 195): 'If the ratio of successful to failed application is small enough, the coincidence can be attributed to mere chance.'

Distortions from selective memory would seem to be at work in our reaction to particular cases falling under all three categories of mathematical coincidence: how many mathematical concepts devised for purely mathematical purposes are never successfully applied (origins)? How many predictions based on purely mathematical manipulations turn out to be inaccurate (more out than in)? How many attempts to treat a new phenomenon with the same mathematics as some known phenomenon end in failure (transfers)?

A related phenomenon is the 'Law of Truly Large Numbers':

> With a large enough sample, any outrageous thing is likely to happen. ... truly rare events, say events that occur only once in a million ... are bound to be plentiful in a population of 250 million people. If a coincidence occurs to one person in a million each day, then we expect 250 occurrences a day and close to 100,000 such occurrences a year. Going from a year to a lifetime and from the population of the United States to that of the world (5 billion at this writing), we can be absolutely sure that we will see incredibly remarkable events. When such events occur, they are often noted and recorded. (Diaconis and Mosteller [1989], p. 859)

The analogous point for our case doesn't rest on the sheer number of mathematical structures—an infinity too high to count!—but on the large finite number of roughly distinct structural theories available to provide physical models. We've seen how pure mathematics diverged from natural science in the nineteenth century; since then there has been an explosion of sometimes bewildering mathematical diversity. With the vast warehouses of mathematics so generously stocked, it's perhaps less surprising that a bit of ready-made theory can sometimes be pulled down from the shelf and effectively applied.

Finally there's a more subtle statistical effect that Diaconis and Mosteller ([1989], p. 859) call 'multiple end-points and the cost of "close" '. Diaconis illustrates the idea with an example from ESP testing: a subject B.D. asks two subjects to name two different playing cards—they pick the ace of spades and the three of hearts—after which each is to draw cards one at a time from two well-shuffled decks, one red, one blue.

> The red-backed three of hearts appeared first. At this point, B.D. shouted, 'Fourteen', and we were instructed to count down 14 more cards in the blue pack.

We were amazed to find that the 14th card was the blue-backed three of hearts. Many other tests of this kind were performed. Sometimes the performer guessed correctly, sometimes not.

... B.D. was a skilled opportunist. ... Suppose that, as the cards were turned face upwards, both threes of hearts appeared simultaneously. This would be considered a striking coincidence and the experiment could have been terminated. The experiment would also have been judged successful if the two aces of spades appeared simultaneously or if the ace of spades were turned up in one deck at the same time the three of hearts was turned up in the other. There are other possibilities: suppose that, after 14 cards had been counted off, the next (15th) card had been the matching three of hearts. Certainly this would have been considered most unusual. Similarly, if the 14th or 15th card had been the ace of spades, B.D. would have been thought successful. What if the 14th card had been the three of diamonds? B.D. would have been 'close'.

A major key to B.D.'s success was that he did not specify in advance the result to be considered surprising. The odds against a coincidence *of some sort* are dramatically less than those against any prespecified *particular one* of them. For the experiment just described, including as successful outcomes all possibilities mentioned, the probability of success is greater than one chance in eight. This is an example of exploiting multiple endpoints. (Diaconis [1978], p. 132)[59]

Likewise, in the case of applied mathematics, the scope of what counts as 'success' is not specified in advance: do we require literal application (as with elementary arithmetic)? Is idealized and/or mathematized description allowed (as with our account of motion)? What about cases where computational complexities require a patchwork account (as with Sommerfield's treatment of light reflecting off a razor blade discussed in II.6) or where the methods are employed are entirely illogical and inexplicable (like Heaviside's, again in II.6)? Recall Wilson's hard-won 'lesson of applied mathematics':

The universe in which we have been deposited seems disinclined to render the practical description of the macroscopic bodies around us especially easy. ... Insofar

[59] Imagine how the chances of the jazz fan's 'coincidence' increase if she is also inclined to be surprised to hear a recording of someone else playing a Monk tune, or a recording of Charlie Rouse, Monk's long-time saxophonist, or ... Diaconis notes that B.D.'s chances of success were further increased because 'several such ill-defined experiments were often conducted simultaneously, interacting with one another' and because he made use of sleight of hand and other magician's tricks (Diaconis [1978], p. 132).

> as we are capable of achieving *descriptive successes* within a workable language ... we are frequently forced to rely upon unexpectedly roundabout strategies to achieve these objectives. It is as if the great house of science stands before us, but mathematics can't find the keys to its front door, so if we are to enter the edifice at all, we must scramble up backyard trellises, crawl through shuttered attic windows and stumble along half-lighted halls and stairwells. (Wilson [2006], p. 26; see also p. 452)

The fact that our partial successes are often gained by a hodge-podge of cobbled-together approximations and tricks may be enough by itself to make the results seem less miraculous, but the point here is that these makeshift solutions also greatly increase the likelihood that a given bit of mathematics will count as successfully applied or a given phenomenon as subject to mathematical treatment.

Let me stop here. This brief discussion of applied mathematics is aimed at a modest point: perhaps the impression of a 'miracle' in the relation of pure mathematics to natural science is at least partly due to patterns of thought that tend to lead us astray in more familiar settings. We forget how much of even the purest mathematics has its roots in physical sources and how many structural similarities hold between diverse physical situations; we forget how many phenomena can't be described in mathematical terms and how much pure mathematics has no application; we forget what a wide range of pure mathematics there is to choose from in our efforts to describe the world; and we don't take into account the widespread fudging that's involved in successful applications. Human beings are susceptible to apparent coincidences—as pattern-seeking creatures perhaps we tend to see connections where none exist—and some portion of Wigner's 'miracle' may well spring from this source.

IV.3
Second methodology of mathematics

Having looked at mathematics in application—having concluded that it tells us nothing about mathematical truth or ontology and that only the mathematics closely tied to KF-structures can be confidently viewed as literally instantiated in the world—let's now turn to pure mathematics, and in particular to the conduct of pure mathematics. How are its proper methods to be adjudicated? What makes for a good definition, an acceptable axiom, a dependable method of proof? It's clear that the practice of pure mathematics is highly constrained—this is hardly a realm where anything goes!—but what is the nature of those constraints and are they rationally defensible?

A Second Philosopher in an earlier era would have had an easier time with these questions, because (as noted in IV.2.iii) mathematics and natural science weren't always as sharply distinguished as they are today.[1] Early in the seventeenth century, Galileo famously remarked that Nature

> is written in that great book which ever lies before our eyes—I mean the universe—but we cannot understand it if we do not first learn the language ... in which it is written. The book is written in mathematical language. (as elsewhere) (As quoted by Kline [1972], pp. 328–329)

Galileo and later Newton believed that 'nature is mathematically designed' (Kline [1972], p. 331); indeed, they both held that science should aim for mathematical descriptions rather than physical explanations of phenomena. By the end of the century, after the development of the calculus by Newton and Leibniz,

[1] For more on the developments in this paragraph, see Kline [1972], chapters 16, 18, §2, and 26, §4.

the boundary between mathematics and science became blurred ... as science began to rely more and more upon mathematics to produce its physical conclusions, mathematics began to rely more and more upon scientific results to justify its own procedures. The upshot of this interdependence was a virtual fusion of mathematics and vast areas of science. (Kline [1972], p. 395)

This 'belief in the mathematical design of nature' (Kline [1972], p. 620) continued through the eighteenth century, when Euler inaugurated the study of partial differential equations with his work on the wave equation, and into the early nineteenth century, when Fourier developed his heat equation:

Profound study of nature is the most fertile source of mathematical discoveries ... it is ... a sure method of forming analysis itself, and of discovering the elements which it concerns us to know, and which natural science ought always to preserve: these are the fundamental elements which are reproduced in all natural effects. (Fourier [1822], p. 7)

Indeed Fourier agreed with Galileo and Newton that

Primary causes are unknown to us; but are subject to simple and constant laws, which may be discovered by observation, the study of them being the object of [science]. (Fourier [1822], p. 1)

For a Second Philosopher in this climate, the study of mathematical methods would be an inseparable part of her general investigation of scientific methods.

By the end of the nineteenth century, however, the advent of non-Euclidean and algebraic geometries, abstract algebras, higher set theory, and the rest had divided mathematics and natural science into two inquiries; mathematicians broke away from the necessity of application, asserting their freedom to pursue their own goals, to investigate whichever mathematical structures they found interesting and fruitful. Thus the contemporary Second Philosopher faces a vast body of pure mathematics pursued for its own reasons, using its own methods,[2] quite independent of her well-honed arsenal of observation, experiment, theory formation and so on. Of course pure mathematics isn't the only stretch of discourse satisfying that description; we've touched earlier on

[2] This is what generates the worry labeled 'Origins' in IV.2.iii.

such cases as astrology and creationism. Insofar as these others make claims about the causal structure of the world—for example, when the astrologer insists that a particular behavior was caused by the position of the stars or the creationist holds that certain fossils were deposited by the Flood—we've seen that the Second Philosopher subjects those challenges to her usual tests and dismisses them when they're found wanting. But what about a version of astrology, call it 'pure astrology', that doesn't make such claims, that contents itself with describing astral forces that don't interact with the physical world? From the Second Philosopher's point of view, is pure mathematics on a par with pure astrology?

Notice first how different this hypothetical pure astrology is from actual astrology; we're to imagine that it presents an elaborate account of these arcane forces, perhaps bearing on the personal auras connected to individual people, but that none of this has any causal efficacy. We could imagine a similar 'pure theology' that makes no claims about intervention of God or angels in the natural order, but instead describes their interactions in a realm entirely isolated from the world we live in. These activities would run parallel to the workings of pure mathematics, as it lays out the relations between its acausal abstracta. All these, as human activities, would be apt subjects for the Second Philosopher's sociological or anthropological study of the role they play in our culture, her psychological study of their role in the individual psyche, perhaps biological or evolutionary studies of their basis and origins, and so on.

What sets pure mathematics apart, obviously, is its apparently essential contribution to our scientific description of the world.[3] Because of this added feature, the Second Philosopher's treatment of pure mathematics in sociology, anthropology, psychology, biology, and so on—the treatment that runs parallel to her approach to pure astrology or pure theology—will not be enough.[4] In the course of her examination of her own best methods, she will need an account of how and why pure mathematics plays the role

[3] Philosophers love bizarre hypotheticals, e.g., what if it turned out that pure theology could be used in application? Then pure theology would be functioning as a dressed-up version of what we call pure mathematics—however the people in the hypothetical situation might be inclined to refer to it—and the Second Philosopher would properly so treat it.

[4] This is not a reversion to a Quinean indispensability argument, because the conclusion is only that mathematics is different from pure astrology, not that mathematics is confirmed. Cf. Dieterle [1999], p. 131, Tappenden [2001], pp. 496–497.

it does.[5] Indeed, given its importance, she comes to see that one way of pursuing our understanding of the world is to pursue pure mathematics; she might well be inclined to participate in the practice herself. At this point, she needs answers to the questions that opened this section; she needs to know which are the right mathematical methods and why.

It's been objected that this answer is only effective for that portion of pure mathematics that actually figures in applications.[6] Or, to put it another way, perhaps the Second Philosopher should confine her pursuit of pure mathematics to those portions most directly connected to natural science. This isn't, as we've noted, how mathematicians actually go about their business, but perhaps—so the objection goes—it would be rational for the second-philosophical mathematician not to follow her mathematical impulses wherever they might lead, but to direct her efforts more narrowly. I think the example of group theory from IV.2.iii is enough to cast serious doubt on the wisdom of this strategy: though it was developed for purely mathematical reasons, it eventually became a central mathematical tool in quantum mechanics; had the attitude now recommended to the Second Philosopher prevailed (the attitude actually represented at the time by Jeans), physics would have suffered. The needs of natural science seem better served by the free flowering of contemporary pure mathematics, by its lush diversity, than by any policy of containment.[7]

So assuming the Second Philosopher should reject any call to curtail the free pursuit of pure mathematics, how is she to resolve its methodological questions? It's often suggested that the answer lies in metaphysics, in the nature of the abstract subject matter of mathematics: for example, a

[5] Cf. Tappenden [2001], p. 497: 'the central role of mathematics in our scientific world view demands an account of the distinctive value of mathematics in any complete epistemology'.

[6] See, e.g., Dieterle [1999], Rosen [1999], p. 472, Tappenden [2001], p. 497.

[7] Cf. the well-stocked warehouses of IV.2.iii. See Kline's less purely historical writings ([1968a], [1980], chapter XIII) for a spirited dissent: the 'confidence that a mathematics freed from bondage to science will produce richer, more varied, and more fruitful themes that will be applicable to far more than the older mathematics is not backed by anything but words' (Kline [1980], pp. 300–301). Still he admits that 'the great mathematicians often transcended the immediate problem of science' (Kline [1980], p. 279) to good effect and concludes that 'in the final analysis, sound judgment must decide what research is worth pursuing' (Kline [1980], p. 306). Much of Kline's concern seems to be that 'the freedom to create arbitrary structures [is] being abused' (Kline [1980], p. 288) or that too many if not all the resources of contemporary mathematics are being turned away from applied problems (see the image of the tree deprived of soil in Kline [1980], pp. 298–299). It is no part of my claim here that any consistent concept or axiom system is as good as any other (this is the 'Glib Formalism' rejected in my [1997], p. 202) or that the health and worth of mathematics would not be irreparably harmed if issues of application were ignored entirely (see below).

constructivist might insist that mathematical entities are formed by our mental operations, so a proof of existence must give a recipe for such a construction; another might hold that mathematical entities exist only insofar as they are defined, and thus that impredicative definitions—which define an object in terms of a collection to which that very object belongs—should not be allowed; or finally a set theoretic realist might insist that a proper axiom must be true in the objective, independently existing world of sets. Indeed the Second Philosopher's investigation of various historical episodes indicates that theories of mathematical truth or existence or knowledge do in fact appear in most mathematical debates over proper methods, alongside more typically mathematical considerations. She wants to know which factors have actually shaped the mathematics we now have, so as to approach contemporary debates more effectively.

What she finds, for example, in the case of impredicative definitions is revealing. During the early years of the twentieth century, Russell and Whitehead attempted to reproduce classical mathematics while adhering to the so-called Vicious Circle Principle, which prohibited impredicativity. This effort was partly motivated by the hope of avoiding the paradoxes of early set theory, though opposition to an ontology of sets was also present in the allied no-class theory. The Second Philosopher notes that the controversy was eventually resolved in favor of allowing impredicative definitions and that ontological debates over the existence and nature of sets remain unresolved to this day. This strongly suggests that metaphysical agreement did not underlie this methodological outcome. On closer examination, more typically mathematical factors reveal themselves; for example, Gödel writes that the Vicious Circle Principle

> makes impredicative definitions impossible and thereby destroys the derivation of mathematics ... effected by Dedekind and Frege, and a good deal of modern mathematics itself. ... classical mathematics ... [implies] the existence of real numbers definable ... only by reference to all real numbers. ... I would consider this rather as a proof that the vicious circle principle is false than that classical mathematics is false. (Gödel [1944], p. 127)[8]

[8] Cf. Gödel [1944], pp. 132, 134: 'The classes ... introduced in this way do not have all the properties required for their use in mathematics ... the theory of real numbers in its present form cannot be obtained.'

The reasoning here is straightforward: if the goal is to reproduce the full range of classical mathematics, then impredicative definitions should be allowed.[9] Apparently it was this consideration that carried the day.

After uncovering corresponding methodological argumentation in a range of cases, the Second Philosopher concludes that though metaphysical theories on the nature of mathematical truth and existence undeniably do turn up in such debates, they are not in fact decisive, they are in fact distractions from the underlying purely mathematical considerations at work.[10] Actual methodological decisions, she sees, are based on a perfectly rational style of means-ends reasoning: the most effective methods available toward the concrete mathematical goals in play are the ones endorsed and adopted.[11] Acting on her assumption that the actual methods of mathematics are the ones that should be followed, she resolves to apply such typically mathematical methodological reasoning to any contemporary debates she might face.[12]

One might worry that this leaves the well-being of science at the mercy of the mathematician's whim. Consider, for example, the foundational difficulties that plagued the calculus from Newton's time till the late nineteenth century; what if[13] the mathematical community hadn't eventually come up with a solution—the Cauchy/Weierstrass definition of limit,

[9] This obviously doesn't rule out the development of predicative mathematics in pursuit of different goals, e.g., that of determining how much can be done with these more limited tools. See Weyl [1918], Feferman [1964]. For more on this case, see my [1997], pp. 8–14, 172–174.

[10] By way of contrast, theories about the nature of physical things *are* an integral part of the practice of natural science. See IV.5.

[11] Notice that the methodological upshot of the constructivist's position might be adopted on means-ends rather than metaphysical grounds: it might be argued that for certain mathematical purposes—e.g., in contexts where algorithms are needed—a non-constructive existence proof is useless. The same might be said for the requirement that all mathematical entities be definable (see the discussion in my [1997], II.4.ii).

[12] For more on the line of thought in the last two paragraphs, see my [1997], III.4. There I phrase the conclusion as recommending 'mathematical' as opposed to 'philosophical' considerations, an unfortunate terminological choice that led to unproductive debate over what counts as 'philosophy' (e.g., is the goal of deriving classical mathematics a 'philosophical' one?). The Second Philosopher—bless her!—doesn't talk this way; just as she employs no demarcation criteria for science vs. non-science, she has no litmus test for philosophy vs. non-philosophy. Instead, she notes, as in the text, that considerations of existence and truth and knowledge, of ontology and epistemology, do not in fact play an instrumental role in settling questions of mathematical method. Just as we describe her methods as 'scientific', we might describe those considerations as 'philosophical', but these are just our rough-and-ready way of describing the Second Philosopher's deliberations. (In my [2005b], without the resources of Second Philosophy at my disposal, I use the terms 'metaphysical' vs. 'methodological philosophy', hoping for the best.)

[13] See footnote 3.

set theoretic constructions for the real numbers, the theory of sets on which these were ultimately based—but had instead decreed on some grounds or other that the calculus was irredeemably flawed, that it should be expunged from mathematics (and hence, presumably, from science)?[14] I think this concern mistakes the attitudes of pure mathematicians. Though they may not be primarily motivated by physical applications, providing tools for natural sciences remains one among the overarching goals of the practice of mathematics; history demonstrates the tenacity with which pure mathematicians have sought rigorous versions of the physicist's rough-and-ready improvisations.[15] Contemporary mathematicians—like contemporary scientists—take for granted that a tool doesn't work well for no reason; they tirelessly pursue explanation.

This response invites a more far-fetched version of the original worry: what if mathematicians were to decide that the goal of providing tools for natural science should be outweighed by some other worthy objective, whatever that might be?[16] What if the entire community were to wander off in pursuit of this new goal, leaving science bereft? One unhappy thought seems to me unavoidable: if the Second Philosopher couldn't somehow persuade these hypothetical mathematicians in terms of other shared goals and values—from among those currently in play—she would have no extraordinary means by which to convince them that they were wrong. In the case of the astrologers' star-based explanations of human behavior, she can show by her methods why they are misguided, but she cannot do so, as the astrologer would insist, without appeal to her methods (see IV.1); similarly, she can show by her mathematical methods why these hypothetical wayward mathematicians are wrong, but she cannot do so, as would be required to return them to the fold, without appeal to the very methods they have forsaken.[17]

But I think this conclusion is not as dire as it may first appear. There's nothing in the strange tale told so far to determine whether or not the practice of these wayward souls would continue to be called 'mathematics',

[14] I'm grateful to Peter Railton for this formulation of the objection.

[15] e.g., Heaviside's operational calculus from II.6 or Dirac's delta function from IV.2.iii.

[16] I have only myself to blame for this wild imagining, as I entertained it in my [1997], p. 198, footnote 9. Cf. Tennant [2000], p. 329.

[17] In neither case does she have access to a show-stopper of the form '*x* is science iff...' or '*y* is mathematics iff...'.

and of course the word doesn't matter. What is clear is that the new practice, whatever it's called, wouldn't play the same role in the Second Philosopher's investigation of the world as the discipline we call 'mathematics' now plays. Presumably the evolved practice would end up more or less comparable to 'pure astrology' or 'pure theology' and the Second Philosopher would have no interest beyond the sociological, anthropological, biological, etc. Furthermore, for purposes of her ongoing investigation of the world, she and her fellows would need to reinvent a practice more or less the same as what we now call 'mathematics', and that practice would command precisely the attention previously awarded to the discipline that wandered off. If the word 'mathematics' were retained by the wayward practice, she would need a new one, but again, the word isn't what's at issue.

Returning from the land of imagination, let me illustrate the Second Philosopher's approach with a few cases. Suppose for example that she becomes interested in the set theoretic problem of Cantor's Continuum Hypothesis (CH): is there an infinite cardinality between that of the natural numbers and that of the real numbers?[18] The CH arose as a conjectured answer to the first non-trivial question in the exponentiation table for infinite numbers, it exercised Cantor himself and many of his illustrious successors, and the reason for its recalcitrance was eventually established by Gödel and Cohen: it can't be proved or disproved from the standard assumptions of set theory, the axioms of Zermelo-Fraenkel with Choice or ZFC (assuming ZFC is itself consistent).[19] Some regard this outcome as the end of the story; others continue to think the CH is the sort of question that ought to have an answer and that some extension of ZFC should be sought to answer it.[20] Here the Second Philosopher confronts a serious methodological question: should an answer to the CH be pursued, and if so, what methods are appropriate for settling it?

This is obviously a big question,[21] but to illustrate the Second Philosopher's way of thinking, I'd like to consider two arguments often given in support of a negative answer to the first part, that is, against the opinion that CH needs an answer. Both are so commonplace as to count as folkloric; both rely on analogies with other parts of mathematics. The first goes like this: trying to decide the CH is like trying to decide if groups are

[18] See my [1997], pp. 63–66, for a bit of history and references.

[19] See Enderton [1977] for a textbook treatment of ZFC.

[20] See Gödel's remarks quoted in IV.4.

[21] See IV.4 for more.

commutative; in both set theory and group theory, we study all models of a certain collection of assumptions; some models of the group axioms will be commutative, some not; some models of ZFC satisfy CH, some not; trying to figure out which models are the 'real' groups or the 'real' sets is a nonsensical undertaking. The second compares set theory to geometry instead, but to similar effect: trying to settle the CH is like trying to settle the parallel postulate; there are Euclidean geometries that satisfy the parallel postulate and non-Euclidean geometries that don't, just as there are models of ZFC that do and don't satisfy CH; unless geometry is interpreted physically, there's nothing more to be said; the same goes for set theory.

Now the Second Philosopher sees the appropriateness or inappropriateness of a given mathematical method as determined by its effectiveness toward the goals of the practice for which it is proposed, so she will assess the aptness of these analogies by attending to the underlying aims of the three mathematical theories involved: set theory, group theory, and geometry. Taking group theory first, she inquires into those purely mathematical considerations—so often noted above!—that led to its development. She finds the first glimmerings of the group notion in the work of Galois (around 1830) on the solvability of equations by algebraic means, which she and her fellows now understand in terms of substitution groups and their subgroups. She notes, however, that Galois himself never isolated the group concept; this was left to Cayley, inspired by Galois, some twenty years later. Cayley found the abstract group structure in multiplication on matrices and addition on quaterions, as well as Galois's substitution groups. The surprise is that Cayley's idea passed unnoticed. Kline explains:

> Cayley's introduction of the abstract group concept attracted no attention at this time, partly because matrices and quaternions were new and not well known and the many other mathematical systems that could be subsumed under the notion of groups were either yet to be developed or were not recognized to be so subsumable. (Kline [1972], pp. 769–770)[22]

About ten years later, Dedekind derived the abstract notion from his work on permutation groups, again without much influence.

[22] Cf. Wussing [1969], p. 233: 'Cayley's premature abstraction of 1854 failed, in the first place, through the lack of a body of concrete representations accepted in mathematical practice'.

It wasn't until the 1870s that the group concept found its audience. During that decade, Dedekind identified finite groups in his work on algebraic number theory, Kroneker isolated the group concept while working on Kummer's ideal numbers, Cayley wrote several more papers emphasizing that the abstract structure goes beyond substitution groups, Frobenius and Stickelberger extended the concept to congruences and Gauss's composition of forms, Netto introduced the notions of isomorphism and homomorphism between groups, Klein used infinite transformation groups to classify geometries, and Lie considered infinite transformation groups in connection with differential equations. Kline summarizes:

> By 1880 four main types of groups were known. There are the discontinuous groups of finite order, exemplified by substitution groups; the infinite discontinuous (or discrete) groups, such as occur in the theory of automorphic functions; the finite continuous groups of Lie exemplified by the transformation groups of Klein and the more general analytic transformations of Lie; and the infinite continuous groups of Lie defined by differential equations. (Kline [1972], p. 1140)

At this point, von Dyck, influenced by Cayley and by his teacher Klein, brought together the threads of 'theory of equations, number theory, and infinite transformation groups' under the abstract notion of a group, and the theory flourished from there (Kline [1972], p. 1141).[23]

The Second Philosopher concludes from this story that the notion of group only came into mathematical prominence when it began to serve a particular mathematical goal. Speaking of abstract algebra in general, Kline puts it this way:

> Its concepts were formulated to unify various seemingly diverse and dissimilar mathematical domains as, for example, group theory did. (Kline [1972], p. 1157)

Given this understanding of the role of the group concept, it's clear why Galois didn't bother to draw it out, why Cayley's initial definition fell on deaf ears, and why group theory leapt to the fore in the 1880s. The purpose of the notion of group is to call attention to similarities between a broad range of otherwise dissimilar structures. In doing so, it not only provides an elaborate and detailed general theory that can be applied again and again; it also more accurately isolates the features responsible for the

[23] See also Wussing [1969], pp. 233–245.

particular phenomena ('that x has feature y isn't due to its idiosyncrasies z or v or w, but only to its group structure').[24] Until sufficiently many such structures had been examined and explored, the concept wasn't doing any work, wasn't furthering any mathematical goal. Only when it began to do so was it embraced by the mathematical community.

Now what about set theory? No doubt a mathematical practice as rich and varied as contemporary set theory functions in service of a range of different goals and subgoals, but for present purposes, the Second Philosopher isolates one fairly uncontroversial motivation: set theory hopes to provide a foundation for classical mathematics, that is, to provide a dependable and perspicuous mathematical theory that is ample enough to include (surrogates for) all the objects of classical mathematics and strong enough to imply all the classical theorems about them. In this way, set theory aims to provide a court of final appeal for claims of existence and proof in classical mathematics: the vague question 'is there a mathematical object of such-and-such description?' is replaced by the precise question 'is there a set like this?'; the vague question 'can so-and-so be proved mathematically?' is replaced by the precise question 'is there a proof from the axioms of set theory?' Thus set theory aims to provide a single arena in which the objects of classical mathematics are all included, where they can be compared and contrasted and manipulated and studied side-by-side.[25] Given this foundational goal, the Second Philosopher recognizes that set theoretic practice must strive to settle on one official theory of sets, a single, fundamental theory. This is not to say that alternative set theories could not or should not be studied, but their models would be viewed as residing in the one true universe of sets, V.

Given this understanding of the mathematical forces that shaped them, the Second Philosopher returns to the purported analogy between group theory and set theory: trying to settle CH is like trying to decide if a

[24] e.g., von Dyck writes, 'I wish to emphasize here ... that this approach does not aim to surrender the *individual* advantages that may derive from a particular formulation of each particular [group] ... For each specific [group] we have at our disposal a treasure of specific information ... *But it is precisely these special connections that call for a discussion of the extent to which they are based on purely group-theoretic as against other properties*' (as translated and quoted by Wussing [1969], p. 242, italics in the original).

[25] The Second Philosopher is not committed to various over-strong interpretations of this idea, e.g., that all mathematical objects are really sets, or that all mathematical knowledge arises from proof from set theory, or even that it is impossible that some other theory, like category theory, could play a similar role. See my [1997], I.2, or [2001a], pp. 18–19, for elaboration of these caveats.

group must be commutative. Reconsidering this argument in light of the contrasting goals of set theory and group theory, it no longer appears so persuasive. Given that group theory is *designed* to bring together a wide range of disparate mathematical structures, it would make no sense to try to rule out some of those structures as groups. (We might go on to consider rings and fields, of course, but this doesn't undercut the importance of the underlying group structure.) On the other hand, given that set theory is (at least partly) *designed* to provide a foundation for classical mathematics, to provide a single arena for mathematical existence and proof, it does make sense to try to make our theory of sets as decisive as possible, to try to choose between alternative axioms, to try to rule out models that do this foundational job less well than others. There's nothing wrong with either group theory or set theory, but they're aimed at different mathematical goals.

Even as she rejects this common analogy between set theory and group theory, the Second Philosopher notes a different parallel at a higher remove: the two practices are both aimed at identifiable—though different—mathematical goals and the large-scale structure of their efforts to meet those goals are analogous. Her historical story shows how various mathematical and conceptual developments gradually converged on the 'right' formulation of the group concept. In the early days, for example, when only finite groups were known, Cayley required only associativity and cancellation laws, from which the existence of the identity and of inverses could be derived.[26] But a notion with such limited application was mathematically idle, as we've seen, and when infinite groups came into consideration, this was not enough:

> Geometry was historically the most important source of infinite groups ... It was in extending Cayley's abstract group theory to cover symmetry groups of infinite tessellations that Dyke made first mention of inverses in the definition of group. (Stillwell [2002], p. 367)

There are, of course, many perfectly consistent concepts in the general vicinity of the group concept, but presumably none that would serve the mathematical purposes so well. The real mathematical work came

[26] If G is finite, a in G, then the powers of a are all in G and must eventually reach a point at which $a^m = a^n$, where $m < n$. Canceling a^m from both sides, a^{n-m} is the identity and a^{n-m-1} is the inverse of a. See Stillwell [2002], p. 367, for discussion, and all of chapter 9 for more on the history of group theory.

in isolating precisely the underlying structure responsible for so many important features of the particular examples at hand, and in codifying that structure in the concept of a group. The Second Philosopher notes that the same can be said of set theory: set theoretic axiomatics is involved in a similar process of zeroing in on the best notion of set, that is, the notion best suited to the mathematical goals that it's intended to serve.

In any case, the Second Philosopher sees good grounds on which to dismiss the argument against inquiry into the CH based on this purported analogy with group theory: because the goals of set theory and group theory are different, their appropriate methods should be expected to differ as well. To assess the second analogy, with geometry, she turns her attention once again to the development of the theory,[27] to better understand the goals that shaped it. She first discovers that dissatisfaction with the parallel postulate pre-dated any concerns about the truth of Euclidean geometry. Kline describes the situation this way:

> The axioms adopted by Euclid were supposed to be self-evident truths about physical space ... However, the parallel axiom ... was believed to be somewhat too complicated. No one really doubted its truth and yet it lacked the compelling quality of the other axioms. Apparently even Euclid himself did not like his own version of the parallel axiom because he did not call upon it until he had proved all the theorems he could without it. (Kline [1972], p. 863)

Generations of geometers undertook to prove the parallel postulate from the other axioms without success. One common method was indirect: consider the alternatives to the parallel postulate—that there is no line through a given point parallel to a given line or that there are many lines through a given point parallel to a given line—and try to rule them out by deriving contradictions from them. Sacchieri employed a version of this strategy and eventually, in 1733, published a book called *Euclid Vindicated from All Faults*. What this book in fact contains is a proof that one of the options does lead to contradiction, while the other leads to conclusions so unacceptable as to rule it out as well.

Thirty years later, Klügel suggested that the parallel postulate is known to be true only by experience, and that the conclusions repugnant to Sacchieri were only contrary to experience; soon Lambert envisioned a range of

[27] Here I follow Kline [1972], chapter 36.

logically possible geometries that might have little to do with physical reality. By 1818, Schweikart proposed an alternative to Euclid called 'astral geometry', which he thought might be the correct geometry for the stars. Taurinus also pursued astral geometry, and though he considered Euclidean geometry to be true, he insisted that astral geometry was logically consistent. All these mathematicians believed that the parallel postulate could not be proved, that alternative geometries were logically consistent, and that Euclidean geometry was the true theory of physical space (with the possible exception of the stars).

The turning point came with Gauss. In the 1790s, he too was engaged in the attempt to prove the parallel postulate, but in 1799, he wrote to a fellow mathematician:

> the path I have chosen does not lead at all to the goal which we seek [a proof of the parallel postulate] ... It seems rather to compel me to doubt the truth of geometry itself. (As quoted in Kline [1972], p. 872)

This sentiment grew stronger as he developed his non-Euclidean geometry; in 1817, he wrote to another colleague:

> I am becoming more and more convinced that the [physical] necessity of our [Euclidean] geometry cannot be proved ... Perhaps in another life we will be able to obtain insight into the nature of space ... Until then we must place geometry not in the same class with arithmetic, which is purely a priori, but with mechanics. (As quoted in Kline [1972], p. 872)

A sometimes-disputed anecdote has it that Gauss went so far as to measure the sum of the angles of a triangle formed by three nearby mountains, to see if it differed from the Euclidean 180°, only to conclude that the disparity fell within the margins of experimental error. It seems beyond dispute that Gauss considered alternative geometries to be candidates for application to the physical world.

This line of thought reached its full flowering when Gauss set the foundations of geometry as the topic for the qualifying lecture of his student, Riemann.[28] The result (touched on in IV.2.iii) was the famous work of 1854, finally published in 1868. There Riemann presents a general formalism for a range of geometries: parameters for determining the local

[28] For this paragraph, see Kline [1972], chapter 37, §3.

metric, and hence the local curvature, of the space are left open; Euclidean geometry then appears as a special case among an abundance of non-Euclidean possibilities. Which of these best represents physical space is left as a matter for empirical study. Perhaps most striking to the contemporary eye is Riemann's suggestion that local variations in the metric of physical space might depend on the actual spatial distribution of mass! This idea appears again in Clifford, in 1870, but otherwise rests unexplored until 1912, when Einstein arrived in Zurich and consulted Grossman.

Clearly non-Euclidean geometry became, in the hands of Gauss and Riemann, an investigation of the possibilities for the structure of physical space. Given that the goal, quite explicit in Riemann, is to provide a range of models, leaving the choice between them to empirical science, it would be counterproductive for the mathematician to limit that range by coming to a prior decision on the parallel postulate. Once again, as in the case of group theory, the goals of geometric practice differ so dramatically from those of set theory that it is unreasonable to expect the same methods to be rational in both cases. While the demand for a single unified theory makes perfect sense for the set theoretic community in its efforts to provide a foundation, it would be madness for the geometric community in its efforts to provide a broad range of models for the scientist. Once again the Second Philosopher concludes that the purported analogy is not apt.

In the course of showing that these various efforts to undermine the problem of the CH are ineffective, the Second Philosopher has uncovered set theory's foundational goal, which in turn provides rational grounds for the search for a single axiom system that is as decisive as possible and in particular for one that settles CH. How does she then approach this problem, the second part of our original methodological question? There is, in fact, an available new axiom candidate that would settle not only the CH but also most other open questions in the field—namely, Gödel's Axiom of Constructibility ($V = L$)—but it seems to offend against a second methodological moral to be drawn from set theory's foundational goal. The idea this time is that if set theory is to serve as arbiter of mathematical existence, it should be as generous as possible, so as not to constrain the free pursuit of pure mathematics. The trouble for $V = L$ is that there are other available axiom candidates, so-called large cardinal axioms (LCs), that seem better suited to this maximizing goal: there is a natural interpretation of ZFC + $V = L$ in ZFC + LCs—that is, every theorem of ZFC + $V = L$ can be proved from ZFC + LCs when relativized to the submodel L—but ZFC + $V = L$ cannot recapture ZFC + LCs in a similar way. Furthermore, ZFC + LCs can prove the existence of mathematically interesting structures that can't exist in L. Thus it seems that ZFC + $V = L$ isn't as generous as possible, that it restricts the range of structures available to the pure mathematician (and hence, to the natural scientist), and thus, that it ought to be rejected.[29]

Large cardinals themselves are clearly more congenial to the maximizing objective than $V = L$, but they're not sufficient to settle the CH.[30] Recently, Woodin has presented an innovative case against CH, innovative because it doesn't take the usual form of proposing and defending a new axiom. Instead Woodin argues, in rough outline, that it is possible to extend ZFC to a theory with certain 'good' properties and that any such extension implies not-CH.[31] A serious second-philosophical investigation of this intriguing idea would involve examining the justification for classifying these specific properties as 'good', but what's of central interest for present purposes is that Woodin's case is couched entirely in the means-ends terms that the Second Philosopher believes will properly carry the day in the long run: for example, the 'good' properties require the theory to be decisive in certain ways.

In sum, then, the Second Philosopher sees fit to adjudicate the methodological questions of mathematics—what makes for a good definition, an acceptable axiom, a dependable proof technique?—by assessing the effectiveness of the method at issue as means toward the goals of the particular stretch of mathematics involved. Straightforward examination of the historical record suggests that theories about the nature of mathematical existence and truth don't play an instrumental role in these determinations,

[29] See my [1997], pp. 206–234, and Steel [2000], p. 423, for discussions of the methodological principle *maximize*. The differences between my version and Steel's are largely due to my not entirely successful effort to spell out what counts as a 'natural' extension of ZFC and a 'natural' interpretation. See Löwe [2001] and [2003] for more. For present purposes, the details or even the ultimate viability of the argument are less important than the style of argument it represents.

[30] See Levy and Solovay [1967].

[31] Woodin's proof of this result is complete except for an outstanding piece called the Omega Conjecture. See Woodin [2001] or Koellner [2006] for an overview. Steel advocates a conflicting program for settling CH that depends on the failure of the Omega Conjecture. See my [2005b], Koellner [2006], for discussion and references. I come back to Steel's thinking in IV.4.

but this is not to say that such metaphysical questions evaporate entirely from the second-philosophical point of view.[32] Mathematics remains a distinctive human practice that needs to be explained—what is the nature of mathematical discourse? how does it manage to play the role it does in science? and so on—and the answers to these questions will require the Second Philosopher to take on traditional problems of mathematical ontology and epistemology.

[32] Because my [1997] focuses on methodology, these typically philosophical questions are set aside there—but they are not dismissed (see pp. 200–203).

IV.4

Second philosophy of mathematics

We've seen that the Second Philosopher, in her effort to understand the world, may well turn to the pursuit of pure mathematics, and that when she does so, her assessment of proper methods rests on weighing their efficacy toward her mathematical goals. But if traditionally philosophical questions of ontology and epistemology are irrelevant to the methodology of mathematics, they must still be faced when the Second Philosopher turns her attention to understanding mathematics as a human practice. She confronts a highly disciplined discourse whose history stretches back centuries and whose products are uniquely useful in scientific application. Does mathematics have a subject matter, like physics, chemistry, or astronomy? Are mathematical claims true or false in the same sense? If so, by what means do we come to know these things? What makes our methods reliable indicators of truth? The answers to these questions will not come from mathematics itself—which presents a wonderfully rich picture of mathematical things and their relations, but tells us nothing about the nature of their existence[1]—but they are constrained by the methodological autonomy of mathematics; they must sustain its methodological verdicts. To engage these issues, my plan is to describe three general styles of response and to assess the second-philosophical potential of each.

Let's begin with the metaphysical questions about the subject matter of mathematics, on the theory that the outlines of an appropriate epistemology will depend on their answers. We've seen (in IV.2.ii) that elementary

[1] If the metaphysical considerations surrounding methodological decisions were playing a functional role there, they would lay claim to being part of mathematical practice, but they aren't (see IV.3).

arithmetical claims like $2 + 2 = 4$ are answerable[2] to the logical structure of the world—which means they have the same status as rudimentary logic[3]—but that the '...' of mathematical number theory is another matter. It isn't clear that the world contains any literally infinite structures, and even if it did, our confidence that every number has a successor presumably doesn't hinge on that fact, but on some underlying quasi-linguistic conviction. If the full theory of numbers has a subject matter, it isn't to be found in the physical world; as soon as the '...' comes into play, we've entered the realm of higher mathematics. As metaphysicians, we want to know whether or not the abstracta described there, including the infinity of natural numbers, can be said to exist.

Serious doubts about mathematical existence first arose in mathematics itself with the introduction of negative and complex numbers: Kline reports that 'opposition ... was expressed throughout the [eighteenth] century', and the controversy continued into the first half of the nineteenth (Kline [1972], pp. 592, 596).[4] Opinion began to shift with Gauss's introduction of geometric interpretations in 1831. By identifying a complex number with a point in the plane, Gauss claims the

> intuitive meaning of complex numbers [is] completely established and more is not needed to admit these quantities into the domain of arithmetic. (As quoted in Kline [1972], pp. 631–632)

Kline continues:

> He also says that if the units 1, −1 and $\sqrt{-1}$ had not been given the names positive, negative, and imaginary units but were called direct, inverse, and lateral, people would not have gotten the impression that there was some dark mystery in these numbers. (Kline [1972], p. 632)

[2] 'Answerable' in the sense that our effective use of simple arithmetic rests on its correlation with KF-structures.

[3] i.e., contingently true where the required KF-structures are present, believed independently of experience (see III.6).

[4] e.g., in 1759, Masères writes that negative roots 'serve only, as far as I am able to judge, to puzzle the whole doctrine of equations, and to render obscure and mysterious things that are in their own nature exceedingly plain and simple'. As late as 1831, DeMorgan wrote 'the imaginary expression $\sqrt{-a}$ and the negative expression $-b$ have this resemblance, that either of them occurring as the solution of a problem indicates some inconsistency or absurdity. As far as real meaning is concerned, both are equally imaginary, since $0 - a$ is as inconceivable as $\sqrt{-a}$'. Both are quoted in Kline [1972], pp. 592–593, with references.

Unfortunately, the dependability of geometry as arbiter of existence was soon undercut by increasing awareness of the logical shortcomings of Euclid's arguments[5] and the rise of non-Euclidean geometries (see IV.3). Against this backdrop, efforts to found analysis turned to numerical notions instead, including, for example, the now-standard Bolzano–Cauchy–Weierstrass definition of limit.[6] This line of development led to the set theoretic notion of a function as a set of ordered pairs,[7] to real numbers understood as sets (most prominently by Cantor and Dedekind),[8] and ultimately to our contemporary orthodoxy—to show that there are so-and-sos is to prove 'so-and-sos exist' from the axioms of set theory—and set theory's foundational goal (see IV.3).[9] So existence in set theory seems a reasonable place to begin an inquiry into mathematical existence in general.

Set theory is often regarded as an essentially 'realistic' or 'Platonistic' theory, as if a metaphysics of objectively existing abstracta is straightforwardly presupposed in its axioms and theorems. So, for example, Paul Bernays ([1935]) sees a brand of Platonism as implicit in the combinatorial notion of set,[10] the Axioms of Infinity and Choice,[11] and the use of the law of the excluded middle and impredicative definitions. More recently, Solomon Feferman writes that in set theory

> Sets are conceived to be objects having an existence independent of human thoughts and constructions. Though abstract, they are supposed to be part of an external, objective reality. (Feferman [1987], p. 44)

[5] See Kline [1972], pp. 1005–1007.

[6] See Kline [1972], chapter 40. For more, see Boyer [1949] or Grabiner [1981].

[7] See my [1997], pp. 116–128, and the references cited there.

[8] See Kline [1972], pp. 983–987, Dedekind [1872].

[9] Of course, this is only the 'contemporary orthodoxy' for mainstream classical mathematics; various schools of constructivism or predicativism, for example, would disagree. Furthermore, set theory only claims to provide surrogates for the objects of classical mathematics; those sympathetic to the idea that mathematical objects like numbers and functions have an existence (or not!) quite independent of their set theoretic surrogates should regard what follows as a discussion of truth and existence in set theory proper.

[10] On the combinatorial picture, the existence, e.g., of a set of integers does not depend on there being a definition or rule picking out its elements; rather, it is viewed as the result of 'an infinity of independent determinations' (Bernays [1935], p. 260) of membership or non-membership for each integer. See my [1997], pp. 127–129, for discussion.

[11] The Axiom of Infinity asserts the existence of an infinite set. The Axiom of Choice posits a choice set for every family of non-empty, disjoint sets (that is, of a set with one element from each of the family's members), but provides no method for defining or constructing such a thing. See my [1997], pp. 51–52, 54–57, for discussion and references, or Enderton [1977], pp. 68, 151–158, for a textbook treatment.

This underlying Platonism 'reveals itself most obviously in such principles as ... the Axiom of Choice' (Feferman [1987], p. 44). Indeed, as we've seen (in IV.3), defenses of impredicative definitions do often involve metaphysical pictures of this kind—mathematical things exist objectively, they aren't created by our definitions—and the same is true for the Axiom of Choice.[12]

The most well-known position of this type is due to Kurt Gödel:

> For someone who considers mathematical objects to exist independently of our constructions and of our having an intuition of them individually ... there exists, I believe, a satisfactory foundation of Cantor's set theory. (Gödel [1964], p. 258)

He goes on to describe what we now call the iterative hierarchy of sets.[13] If this conception of set is 'accepted as sound, it follows that the set-theoretical concepts and theorems describe some well-determined reality' (Gödel [1964], p. 260). From this Platonistic perspective, Gödel gains his objective, that is, a defense of the CH as a legitimate question (see IV.3): in this reality

> Cantor's conjecture must be either true or false. Hence its undecidability from the axioms being assumed today can only mean that these axioms do not contain a complete description of that reality. (Gödel [1964], p. 260)

The prescription, then, is a search for new axioms touched on toward the end of IV.3.

Such realistic views have come under serious philosophical attack, both epistemological and referential: it seems that these abstract, eternal, objective things fall outside the range of our human cognitive powers, so it's difficult to see how we could know or talk about them.[14] Some have hoped to improve the prospects for a viable theory of set theoretic knowledge by insisting that the relevant subject matter is no mysterious world of mathematical things, but the concept of set. The axioms of set theory, on this view, are not descriptions of a set theoretic reality, but explications of this concept; set theoretic statements are true if they are part of the concept (or follow from it). Obviously, views of this sort vary dramatically

[12] See my [1997], pp. 54–57, for a summary of various considerations raised for and against Choice.

[13] See Enderton [1977], pp. 7–9, for a textbook presentation, or my [1997], I.3, for its role in defending the various axioms of set theory.

[14] The classic source is Benacerraf [1973]; for more recent discussions, see my [1990], pp. 36–48, Field [1989], pp. 25–30, Burgess [1990]. This challenge is discussed briefly in II.5.

depending on the notion of concept involved, but if the resulting set theory is to be objective, if the truth or falsity of its statements is to be independent of our thought, of our ways of knowing, then it seems the concept of set must itself be objective.[15] So, though eliminating abstract objects also eliminates the problem of how we can know about them, this realistic version of conceptualism faces the equally baffling problem of how we know about objective concepts. Another thought is to reject sets as objects in favor of an objectively existing set theoretic hierarchy structure,[16] which again only succeeds in shifting the epistemic problem. Let me lump these positions together as Robust versions of realism.

Advocates of Robust Realism are often drawn to it by the promise of a determinate truth value to independent statements like the CH, and familiar forms of object realism like Gödel's do take the universe of sets to be entirely precise. Still, we should note the possibility of a robust reality that isn't fully determinate. So, for example, a concept realist's concept of set might exist objectively, independently of our inquiries, and so on, without deciding every particular: it might be vague or indefinite in places. Likewise, there are forms of object realism that allow abstract objects, like sets, to be indeterminate in various respects.[17] The defining feature of Robust Realism is that it sees set theory as the study of some objective, independent reality; that reality might dictate that the CH is true, that the CH is false, or that there is no fact of the matter about the CH.

Despite its popularity and its distinguished pedigree, Robust Realism will not serve as a template for a Second Philosophy of mathematics—and not in the first instance because of its epistemological woes. To see this, recall the case against the axiom candidate $V = L$ that was sketched toward

[15] Both Frege and Gödel apparently thought of concepts in this way (see Frege [1884], section 47, Gödel [1944], p. 128). Martin [2005] examines Gödel's conceptual realism in some detail.

[16] e.g., Shapiro's 'ante rem structuralism' (Shapiro [1997]).

[17] Akiba [2000] defends such a view; see also Field [1998c], pp. 398–399. As an example, Field suggests versions of structuralism that take a mathematical object to be a position in a structure, with no properties other than those conferred by its being that position; Akiba also compares his position with this variety of structuralism. Both writers focus on numbers, which may be neither identical to nor distinct from, e.g., the von Neumann ordinals, but Parsons deals directly with the possibility of sets as 'incomplete objects' (see his [1990], [1995], and [2004]). He allows that structures can have non-isomorphic instances (Parsons [2004], p. 69), so perhaps sets (as positions in the set theoretic structure) are not determinate enough to settle CH. For a hint of this, see Parsons [1995], p. 79: 'looking at the structuralist view itself, one might ask whether it concedes to set theory the degree of objectivity that many set theorists are themselves inclined to claim, following the example of Gödel in "What is Cantor's Continuum Problem?" [i.e., Gödel [1964]]'.

the end of IV.3: $V = L$ should be rejected because it conflicts with the maximizing goal of set theory. We have here an argument against adding $V = L$ to the list of standard axioms. Now the Robust Realist holds that there is an objective world of sets (or an objective concept of set or an objective set theoretic structure or ...) and that our set theoretic statements aim to assert truths about this world; in particular, the axioms of our theory of sets should be true in this objective world and (given that logic is truth-preserving) our theorems will be, too. So from a Robust Realist's point of view, an argument against adding $V = L$ to the list of standard axioms must be an argument that $V = L$ is false in the world of sets.

That's the trouble. The argument based on our maximizing goal shows why adding $V = L$ leads to a theory of the sort we dislike, a restrictive theory, but how does its being a theory we dislike provide compelling evidence that it's false? To put the problem more generally: granting that selecting for maximization generates theories we like, what reason do we have to think it likely to generate theories that are true? As any casual observer of the scientific process knows, the physical world has all too often failed to conform to scientists' preferences;[18] surely this is one of the characteristic hazards of any attempt to describe an objective, independent reality. Until he has reason to believe that the preference for maximization tends to produce true theories—and it's devilishly hard to see how that claim could be defended—the Robust Realist must look elsewhere for evidence against $V = L$.[19] Thus Robust Realism injects a metaphysical component, a theory of the nature of sets, into a methodological debate—a move the Second Philosopher eyes with suspicion given the track record of irrelevance racked up by such considerations—and worse, this metaphysical component purports to undercut a purely mathematical argument with impeccable means-ends credentials.

Under the circumstances, it's worth considering a form of realism weaker than the Robust variety, a realism that begins from the Second Philosopher's

[18] Tracing the history of the wave theory of light, Einstein and Infeld ([1938], p. 117) ask whether light waves are longitudinal (like sound waves) or transverse (like water waves): 'Before solving this problem let us try to decide which answer should be preferred. Obviously, we should be fortunate if light waves were longitudinal. The difficulties in designing a mechanical ether would be much simpler in this case. ... But nature cares very little for our limitations. Was nature, in this case, merciful to the physicists attempting to understand all events from a mechanical point of view?' Of course the answer was no. 'This is very sad!', Einstein and Infeld conclude (p. 119).

[19] e.g., to confirmed predictions of hypotheses inconsistent with it. See Martin [1998].

conviction that the choice of methods for set theory is properly adjudicated within set theory itself, in light of their effectiveness as means to its various goals. By this realist's lights, the axioms and theorems of set theory, as generated by those methods, are taken to be true, including the existential claims, so sets exist. But what are these sets like? The answer, according to our realist, is that the methods of set theory tell us what sets are like—and indeed, set theory does tell us an impressive story of what sets there are, what properties they have, how they are related, and so on. Still, to answer our metaphysical questions we need information of a different sort about these sets: do they exist objectively or are they mental constructions or even fictions? Are they part of the spatiotemporal, causal universe? Is their existence contingent on something or other, or do they exist necessarily? And so on.

Here our new realist must tread lightly, because set theory itself is silent on these topics, and set theory, on this account, is supposed to tell us all there is to know. There may be some temptation to dismiss the metaphysical questions as unscientific pseudo-questions—perhaps in the spirit of Carnap (see I.5)—but whatever its pros and cons in general, this is not an option here. As we've seen, beyond questions of mathematical method, the Second Philosopher is faced with perfectly legitimate and challenging scientific questions about the nature of human mathematical activity: how does mathematical language function? Does it relate the world in the same ways as the language of natural science? What happens when human beings come to understand mathematical theories? How does mathematics work in various kinds of applications? And so on. To answer these questions, she must face many of the metaphysician's concerns: do mathematical entities exist, and if so, what is the nature of that existence? Are mathematical claims true, and if so, how do humans come to know this? These are not detached, extra-scientific pseudo-questions, but straightforward components of our scientific study of human mathematical activity, itself part of our scientific investigation of the world around us.

To see how our new realist's answers to these metaphysical questions might go, let's turn to the comments of two contemporary thinkers whose positions seem to me to exemplify this weaker style of realism. Consider first the set theorist John Steel, who writes:

> To my mind, Realism in set theory is simply the doctrine that there are sets ... Virtually everything mathematicians say professionally implies there are

sets. ... As a philosophical framework, Realism is right but not all that interesting. (FOM posting 15 Jan. 1998, quoted with permission)

'there are sets' ... is not very intriguing. 'There are sets' is, by itself, a pretty weak assertion! Realism asserts that there are sets, and hence ... that 'there are sets' is true. (FOM posting 30 Jan. 1998, quoted with permission)

Here the existence of sets, a version of realism, is being deduced from what mathematicians say 'professionally', that is, when applying the methods and principles of mathematics in general and set theory in particular. Steel admits that

Both proponents and opponents sometimes try to present it as something more intriguing than it is, say by speaking of an 'objective world of sets'. (FOM posting 15 Jan. 1998, quoted with permission)

Steel remarks that 'such rhetoric adds more heat than light', and he seems uncertain that his 'not very interesting' realism is Robust: 'whether this is Gödelian naïve realism I don't know' (FOM posting 30 Jan. 1998, quoted with permission). I suggest that it is not, and it may be that Steel himself has more recently come to agree:

The 'official' Cabal philosophy has been dubbed *consciously näive realism*. This was an appropriate attitude when the founding fathers were first laying down the new large cardinals/determinacy theory ... It may be useful now to attempt a more sophisticated realism, one accompanied by some self-conscious, metamathematical considerations related to meaning and evidence in mathematics. (Steel [2004], p. 2)

Perhaps maximization arguments count as such self-conscious metamathematical considerations: we talk about what sorts of theories are to be preferred and why.

So what about the metaphysical questions? Steel addresses only one of them directly:

sets do not depend causally on us (or anything else, for that matter). Virtually everything mathematicians say professionally implies there are sets, and *none of it is about their causal relations to anything*. (FOM posting 15 Jan. 1998, italics added, quoted with permission)

Set theory, again, tells us all there is to know about sets, and it says nothing about their being causally related to anything, so they aren't. Addressing mass rather than causation, John Burgess (our second exemplar) extends this way of thinking to include what natural scientists say and do not say about sets:

> I think the fact that ... sets ... don't have mass can be inferred from the fact that when seeking to solve the 'missing mass' problem in cosmology, physicists may speculate that neutrinos have mass, but never make such speculations about ... sets. (personal communication 24 Apr. 2002, quoted with permission)

Similarly, set theory does not tell us that sets are located in space, or that they have beginnings or endings in time, or that they are involved in causal interactions, and natural scientists do not seek them out or appeal to them as causal agents in explanations, so we should conclude that they are not spatiotemporal or causally active or passive. Summing up, Burgess writes:

> One can justify classifying mathematical objects as having all the negative properties that philosophers describe in a misleadingly positive-sounding way when they say that they are abstract. ... But beyond this negative fact, and the positive things asserted by set theory, I don't think there is anything more that can be or needs to be said about 'what sets are like'. (personal communication, 24 Apr. 2002, quoted with permission)

In sum, sets are taken to have the properties ascribed to them by set theory and to lack the properties set theory and natural science ignore as irrelevant. There is nothing more to be said about them. Such posits are sometimes called 'thin', so let's call this Thin Realism.[20]

Given that the mathematical things of Thin Realism are non-spatiotemporal and acausal, it might seem to share the Achilles heel of Robust Realism, that is, it might seem to impede a reasonable account of how human beings come to know mathematical facts. This is where 'thinness' does its work: sets just are the sort of thing that can be known about by careful application of the methods of set theory. To see how we come to know what we do about sets—assuming we're not concerned about the logical connections at the moment—we need only examine how various mathematical and set theoretic considerations led us to accept the axioms, a study, Burgess notes, that is 'given in the standard histories'.[21]

[20] The term 'Thin Realism' is reminiscent of Azzouni's 'thin posits' (Azzouni [1994]), but various features of his position rule out straightforward comparison with the alternatives under consideration here. I come back to Azzouni's views in IV.5.

[21] Burgess is writing here with Rosen, in their [1997], pp. 45–46. The Second Philosopher would emphasize that these considerations include assessment in the means-ends terms of IV.3.

So, for example, the maximization arguments touched on in IV.3 are just the right sort of thing for defending methods and axioms of set theory. The Robust Realist's conviction that some further justification is needed strikes the Thin Realist as akin to the radical skeptic's challenge to the Second Philosopher's scientific beliefs (see I.2): yes, your claims meet the most stringent of scientific standards, but are they *really* justified? The Thin Realist is unmoved.[22]

Though it escapes the familiar difficulties, Thin Realism also fails to deliver the metaphysical pay-offs that put many Robust Realists on this path in the first place: for example, it cannot play the role of Robust Realism in the debate over independent statements like CH, in particular, in assessing whether or not CH has what the Robust Realist calls 'a determinate truth value'. The Thin Realist is perfectly willing to assert 'CH or not-CH', or 'CH is either true or false', or 'Either CH is true or not-CH is true'—as these are all straightforward assertions of set theory—but as we have seen, the Robust Realist wants something more than an appeal to classical logic. Thus he posits an objective reality that the ordinary methods of set theory, including classical logic, do (or do not) allow us to track.

In stark contrast, the Thin Realist sees no such gap between the methods of set theory and sets: sets just are the sort of thing that can be known about in these ways; the Robust Realist is wrong to interpose some elusive extra layer of justification between the two. Various considerations lead the Robust Realist to his concern over determinacy: the independence of CH, the existence of assorted models of set theory, the inability of large cardinal axioms to settle CH, the lack of intuitive force behind the strong axiom candidates currently available, and so on. These same considerations may lead the Thin Realist to fear that we will never come up with an acceptable theory that decides the question, but this possibility does not change the fact that CH is either true or false.[23]

It's sometimes suggested that the Thin Realist's evasion of the difficulties of the Robust Realist hinges on her acceptance of a disquotational as

[22] This is not to say that the Thin Realist's reasons for being unmoved are precisely the same as the Second Philosopher's in I.2. See below.

[23] Speaking of apparent indeterminacy in vague contexts, Field accuses epistemic theorists—those who hold that Borderline Joe is either bald or not-bald, we just can't tell which (see III.7)—of failing to distinguish between 'factual but unknown' and 'non-factual' (see Field [2000], p. 284). For the epistemicist, everything is factual. The same might be said of the Thin Realist with regard to set theory.

opposed to a correspondence theory of truth.[24] To see why this view is tempting, consider what a disquotationalist might say about CH. Assuming she adheres to classical logic, she'll endorse 'CH or not-CH', and hence 'CH is true or not-CH is true', because 'CH' and 'CH is true' come to the same thing. A correspondence theorist might introduce caveats about the nature of set theoretic truth and reference, about the means of connection between set theoretic language and sets, that could undercut this bald conclusion, but the disquotationalist can't take this route without admitting that there's more to truth and reference than the T- and R-sentences. If this is right, then for the disquotationalist, the CH has a determinate truth value as an immediate consequence of classical logic. As we've seen, the Thin Realist draws the same conclusion from the same source, so perhaps it's natural that the two positions are often thought to be linked.

The claim, then, is that a disquotational theory of truth and reference is what frees realism from the challenges facing the Robust Realist (and hence, that we would all be better off to embrace disquotationalism). To keep the issues simple, let's focus on the referential challenge:[25] how do we humans manage to refer to sets? Here the correspondence theorist must provide an account of a substantive connection between sets and our use of the word 'set', and the acausal, non-spatiotemporal nature of sets makes this sound like a potentially difficult task. In contrast, the disquotationalist need only remark that 'set' refers to sets.[26] Thus the disquotationalist avoids a serious challenge to realism in set theory. Or so it seems.

[24] Burgess (with Rosen in their [1997], p. 33) takes naturalism to include a deflationary theory of truth more or less by definition. (Azzouni [1994], [2004] is also a deflationist.) In contrast, as we saw in Part II, the Second Philosopher regards the correct theory of truth as a question open to debate; if she sides tentatively with a version of disquotationalism in II.4, this is on scientific, not definitional grounds.

[25] In Burgess's treatment of the epistemic challenge (with Rosen, in their [1997], pp. 47–49), a deflationary theory is used to replace 'is belief in the truth of set theory justified?' with 'is belief in set theory justified?', on the grounds that the deflationist takes '...is true' to be interchangeable with '...'. Because sets are the kind of thing that set theory tells us about—an appeal to what I'm calling Thin Realism—the reformulated question is then answered positively. But it seems to me that even a correspondence theorist would approve the first move (Field [1994a], §5, argues that a correspondence theorist will need a deflationary notion of truth in contexts like this); what's at issue here isn't about word–world relations, but the more basic question of whether or not we have epistemic access to abstracta. The work of dissolving the problem of knowledge is done by Thin Realism (or in terms closer to Burgess's: by his version of naturalism), not by deflationism.

[26] If there aren't any sets, this is an empty claim, but we're concerned for now with Robust and Thin Realists, who agree that sets exist.

In fact, I think this advantage is illusory. Though the observation that 'set' refers to sets is all the disquotationalist requires as an account of reference, we've seen (in II.4) that many questions about word–world relations remain legitimate even for the disquotationalist: for example, how did the word 'set' come to be used to refer to sets? How is it that the set theorist's beliefs involving 'set' are often good indicators of the facts about sets? If our disquotationalist is also a Thin Realist, these challenges are easily met by appeal to thinness. To understand how we came to use 'set' to refer to sets, we need only examine how the word came to be used as it is.[27] Burgess suggests that this story would include attention to the development of Cantor's work, to how and why it led to

> the transition between thinking of 'the points of discontinuity of the function f' in the plural and thinking of 'the set of points of discontinuity of the function f' in the singular, and how this transition contributed to the development of analysis in the 19th century. (personal communication 24 Apr. 2002, quoted with permission)[28]

Once again, sets just are the sort of things that can be referred to by using the set theoretic vocabulary as we do. Similarly, my beliefs are good indicators of the facts about sets because sets just are the sort of things that can be learned about in the ways I've come to have those beliefs.

But—and here's the twist—what's to keep a correspondence theorist who's also a Thin Realist from regarding the same historical/methodological story as a substantive account of the facts by virtue of which the word 'set' refers to sets? The correspondence theorist who's also a Robust Realist may well worry that reference requires a causal connection between sets and our use of 'set', just as she might well worry that knowledge of sets requires a causal connection between sets and our cognitive machinery, but a Thin Realist will have set aside such concerns in light of her confidence that sets just are the sort of things that can be referred to and known about by the actual language and methods of set theory. Causal requirements—either for a disquotational account of how we came to refer or for a correspondence account of the facts by virtue of which we refer—may strike the Thin

[27] Burgess (with Rosen) points out that when the deflationist Quine takes up the question of reference (in his [1974]), he is 'concerned solely or mainly with how the usage of our terms got to be as it is' (Burgess and Rosen [1997], p. 59).

[28] Here Burgess is alluding to the origins of Cantor's set theory in his investigation of representations of functions by trigonometric series. See my [1997], pp. 15–17, for discussion and references.

Realist as appropriate to the case of concreta, but by her lights, only a dogmatic partisan of a causal theory of reference would see this as a problem for the word 'set'.[29] As Burgess (with Rosen) writes

> Why should one have more confidence in ... a theory of reference than in such ordinary judgments ... as 'the symbol "π" has been used to refer to the number pi since the 18th century?' (Burgess and Rosen [1997], p. 51)

The point here is that the referential challenge is overcome, just as the epistemic challenge is overcome, not by disquotationalism, but by Thin Realism.

I claim then that the Thin Realist can avoid the epistemic and referential challenges to realism by appeal to thinness, regardless of which theory of truth she adopts, in other words, that Thin Realism does not require disquotationalism. But those who link the two may have the converse in mind: they might hold that the only form of realism open to the disquotationalist is Thin Realism, that disquotationalism requires that any realism be Thin. On this way of thinking, adopting a disquotational theory of truth would be a way to avoid the challenges to realism, because it would force the realism in question to be Thin. On this approach, the question becomes: can the disquotationalist be a Robust Realist? Or: must a Robust Realist be a correspondence theorist?

We've seen a hint of why Robust Realism might be thought to be incompatible with disquotationalism: the Robust Realist worries that the CH might have no determinate truth value; in familiar cases, like Gödel's, he provides metaphysical evidence—there is an objective world of sets—designed to show that it does. The disquotationalist, on the other hand, might simply follow the inference from 'CH or not-CH' to 'either

[29] Leeds ([1995], pp. 11–15) gives an in principle argument against the very possibility of a correspondence theory of reference, based (partly) on the grounds that whatever such theory is proposed, we would be willing to give it up if a new term in our language didn't satisfy it. Such a theory would '*describe* the way our current language connects ... with the objects we speak about now' (p. 14), without constraining our future language. This is precisely how my Thin Realist correspondence theorist has reacted: faced with the authority of set theoretic methods, she revises her theory of reference to include the customary use of set theoretic language as a legitimate way of referring to sets. Why does Leeds find this unacceptable? Because 'it is hard to see why anyone would find this project of much interest, given the philosophical uses to which we wish to put the correspondence theory' (p. 15). A footnote gives exactly one example of these 'philosophical uses': in the terms used here, it is the referential challenge to Robust Realism. Unlike Leeds, then, my correspondence theorist isn't interested in a normative theory to serve such philosophical purposes; as a Second Philosopher, she rests content with a descriptive theory of how reference in fact works.

CH is true or not-CH is true', from the use of classical logic to the conclusion that CH has a determinate truth value. So, this line of thought continues, the Robust Realist must be a correspondence theorist, so as to allow for semantic failures of a sort that would allow for indeterminacy of truth value; perhaps, for example, 'set' is referentially indeterminate.

If this is right, it seems to me that the disquotationalist is in deep trouble, trouble that extends far beyond his treatment of set theory. In II.4, we considered the case of Priestley's use of 'dephlogisticated air' and the likelihood that our current scientific language suffers from similar difficulties. If the disquotationalist is unable to recognize these phenomena, if his understanding of how our language connects to the world is limited to R-sentences, then disquotationalism is simply descriptively inadequate. But matters are not so grim: what our problematic contemporary cases most likely have in common with Priestley's is that our beliefs, like his, are good indicators of something in the world, but not of their own truth conditions. We've seen that the disquotationalist can even mimic the correspondence theorist's claims about indeterminacy of reference, though for her, these are secondary, derivative, and not-fully-objective consequences of the underlying distortion in the indication relations.

We've also seen that the Robust Realist and the Thin Realist disagree over the indication relations in set theory: for example, the Thin Realist thinks her belief that $V = L$ is false—based on a maximization argument—is a good indicator of the facts about sets, because sets just are the sort of thing that can be learned about by application of such set theoretic methods; the Robust Realist sees no grounds for this conclusion in the maximization argument alone, because the reliability of such set theoretic methods requires further defense. Indeed, the Robust Realist's gap between set theoretic methods and the facts about sets is what allows him to take various considerations—again, the independence of CH, the existence of assorted models of set theory, the inability of large cardinal axioms to settled CH, the lack of intuitive force behind strong axioms now under consideration—as evidence that CH may, in fact, be indeterminate, that there may be no fact of the matter to uncover. For the Thin Realist, to think this way is misguided, as it improperly separates sets from the straightforward application of set theoretic methods, which tell us that CH is either true or false.

The question, once again, is whether or not the disquotationalist is forced to follow the Thin Realist here, and I think we can now see that the answer is no. Disquotationalism by itself, unlike Thin Realism, makes no claims about the evidential relations of set theory. Recall that the question—why are a knowledgeable person's beliefs involving 'Hume' such good indicators of the facts about Hume?—is a perfectly legitimate question for the disquotationalist to ask; and the question—why are the set theorist's 'set' beliefs such good indicators of the facts about sets?—is just as legitimate. The Thin Realist will answer this question one way: sets are just the kind of thing that can be known about using the methods of set theory. The Robust Realist doesn't accept this answer; he requires something more, which is what generates the epistemological challenge to his position. What matters for our purposes is that nothing stops the disquotationalist from siding with the Robust Realist here; nothing blocks the combination of Robust Realism with disquotationalism.[30]

The suggestion, then, is that the disquotationalist can speak with the Robust Realist. As in Priestley's case, contemporary set theoretic beliefs may be good indicators of some features of sets, but not quite of their own truth conditions. As in contemporary physics, there are hints that something like this may be happening, hints to be found in the very phenomena that inspire the Robust Realist's concern. Though set theory tells us that CH is true or false, and we continue to use classical logic while doing set theory, it could be that set theory is wrong about this; perhaps, for example, our 'set' talk connects to sets via a concept of set and perhaps that concept isn't as well defined as our set theoretic methods presuppose. As Field puts it:

> [a] view ... I find ... plausible ... is that our set-theoretic concepts are indeterminate: we can adopt any one of the above claims about the size of the continuum we choose, without danger of error, for our prior set-theoretic concepts aren't determinate enough to rule the answer out. Prior to our choice there is no determinate fact of the matter. (Field [2000], p. 304)[31]

Though our practice reflects an assumption that CH is 'factual, but unknown', the disquotationalist can admit the possibility that it is 'non-factual'.[32]

[30] Field agrees that the legitimacy of the 'Hume'–Hume question shows that the epistemological problem for Robust Realism 'isn't dissolved by a disquotational theory of truth' (Field [1994a], p. 118, footnote 15).

[31] Notice how close Field comes here to a conceptual strain of Robust Realism.

[32] At one point, Field believed that 'standard mathematical reasoning can go unchanged when indeterminacy in mathematics is recognized: all that is changed is philosophical commentaries on

By this point, it should be obvious why this move open to the disquotationalist is not open to the Thin Realist. To say that our set theoretic methods may be wrong because of... well, whatever... is to interpose an intermediary between set theoretic methodology and sets, something the Thin Realist will not do. The illusion that the two positions are linked is no doubt encouraged by the fact that both regard

'set' refers to sets

as, in some sense, laid down by fiat. But the fiats have different sources. For the Thin Realist, that source is the authority of set theoretic methods—sets just are the sort of thing that can be referred to by customary use of set theoretic language—and the R-sentence is a conceptual truth about sets. For the disquotationalist, on the other hand, the source of the fiat is our notion of reference; the R-sentence is a trivial truth about reference. The two positions are entirely independent.

The possibility of this Thin Realism[33] suggests that set theory is not, in and of itself, committed to Robust Realism. Contrary to the claims of Feferman and others,[34] the methods and axioms of set theory—like

mathematics' (Field [1998b], p. 337). Reacting in part to Leeds's criticisms (in Leeds [1995]), Field [2000], p. 304, writes that there is 'something fishy in the idea: if the philosophical commentaries about determinateness are as divorced from mathematical practices as [suggested in Field [1998b]], then they seem suspiciously idle.' He concludes that whether or not a mathematician regards CH as determinate will be reflected in practice after all: one who does will look for evidence one way or the other; one who doesn't will think that 'each possibility is worth developing... [though] practical and aesthetic considerations may make some answers more interesting than others' (Field [2000], p. 304). I agree that a set theorist who regards CH as determinate will look for evidence one way or the other, but a Robust Realist who believes this and a Thin Realist (who must believe this) will disagree crucially on what counts as evidence. Furthermore, the Thin Realist's evidence consists of characteristically set theoretic means-ends reasons for preferring one theory to another; to dismiss these considerations as merely 'practical and aesthetic' overlooks the deep mathematical motivations driving this practice.

[33] Let me note here a variant of Thin Realism that would identify truth in set theory with following from some eventual defensible theory of sets (i.e., a theory generated by effective methods towards set theoretic goals). Such a Thin Realist-2 (unlike our official Thin Realist-1) could say, with the Robust Realist, that the CH may turn out not to have a determinate truth value, though she would mean that CH might turn out not to be provable or disprovable from any defensible theory of sets (=absolutely undecidable in the sense of Koellner [2006]), not what the Robust Realist means (i.e., that the objective metaphysics to which set theory is answerable has shortcomings). This variant seems to me somewhat less appealing because it involves the adoption of an epistemically constrained notion of truth for set theory (see I.7). (As we've seen, the Thin Realist-1 doesn't define truth for set theory in terms of set theoretic methods; the reliability of those methods is an ontological fact about the nature of sets themselves.)

[34] These 'others' apparently don't include Bernays, who distinguishes between 'platonistically inspired' methods of set theory, which he calls a 'restricted platonism', and an 'absolute' or 'extreme

maximization and the Axiom of Choice—can be justified internally, in terms of set theory's goals and values; they do not require appeal to

> an external, objective reality... [of] independently existing entities [in which] statements such as the Continuum Hypothesis... have a definite truth value (Feferman [1987], pp. 44–45)

The Thin Realist will hold, in Burgess's negative way, that sets are not created by our thoughts or definitions, that they are acausal and non-spatiotemporal, but he will regard the Robust Realist's further worry over whether or not CH has a determinate truth value as misguided. CH is either true or false because our best theory of sets includes 'CH or not-CH'; there is no more to it than that. Our coming to know which will depend on whether or not there will one day be a mathematically well-motivated way of settling it—perhaps an extension of ZFC, perhaps something else, along the lines of Woodin's recent efforts (noted in IV.3)—and that question remains open. But set theory tells us what there is to know about sets and it reveals no further problem.

With the Robust Realist and the Thin Realist on stage, let me now introduce the third and final player in our little drama. We've seen (in IV.2.i) that the mathematical entities present in our best scientific theories are not among the posits of those theories whose existence has been confirmed. It's possible, of course, that the truth of mathematical claims, the existence of mathematical entities will be required to explain how mathematics manages to work in application, but let me postpone this question for a moment and consider the hypothesis that mathematical things do not exist, that pure mathematics is not in the business of discovering truths. Let's call this position Arealism.[35]

So what is set theory doing, according to our Arealist, if it's not advancing our store of truths, not telling us what sets there are and what they are like? Here philosophers seem occupationally predisposed to analogies:

platonism', which interprets the methods of set theory 'in the sense of conceptual realism, postulating the existence of a world of ideal objects containing all the objects and relations of mathematics' (Bernays [1935], pp. 259, 261).

[35] I avoid the more usual term 'anti-realism' that appears, e.g., in the realism/anti-realism debate discussed in IV.1 and often marks a principled or a priori resistance to Realism (to recall the typographical convention of I.7 and IV.1). My Arealist isn't set against mathematical entities any more than she is against unicorns; she just sees no evidence for the existence of either.

mathematics is like a game,[36] or mathematics is like fictional story-telling,[37] or mathematical language is like metaphorical language.[38] (Even the Robust Realist often appeals to an analogy between mathematics and natural science.[39]) Let's imagine that our Arealist instead undertakes to characterize mathematics directly, as itself; instead of trying to understand mathematics by analogy with something more familiar, she tries to make mathematics itself more familiar. The pursuit of set theory, as we've seen (in IV.3), is a process of devising and elaborating a theory of sets, prompted by certain problems (recall its beginnings in Cantor's analysis of sets of singularities[40]), guided by certain values (power, consistency, depth, ...), in pursuit of certain goals (a foundation for classical mathematics, a complete theory of reals and sets of reals, ...). This particular mathematical process of theory formation is directed by these internal mathematical problems, norms, and goals just as other processes of mathematical concept formation are directed by their own constellation of considerations, for example, the development of the concept of function or group or topological space. With care, these processes can be described and analyzed, and their underlying rationality assessed.

This Arealism recalls some versions of Formalism—both deny that mathematics is in the business of seeking truths about certain abstracta[41]—but the dissimilarities are at least as dramatic. Let me note just a few. Arealism doesn't limit its attention to formalized theories: the natural language discussions of goals and methods that surround ZFC are fully part of set theoretic practice, and even proofs from ZFC and its extensions are rarely formalized. In addition, the Arealist is obviously far from judging, as some Formalists do, that any consistent axiom system is as good as any other:[42] one of the central aims of her methodological study of set theory is to understand and assess arguments for and against axiom candidates. Finally, the Arealist may well agree that some strictly finitary statements—like $2 + 2 = 4$—are different in kind from those of infinitary mathematics, indeed that they are true (see IV.2.ii), but she does not view its ability to generate true statements of this 'contentful mathematics' as grounding the

[36] See Resnik [1980], pp. 55–66, or Shapiro [2000], pp. 144–148, for discussion.

[37] See Burgess [2004] for criticism of the fictionalist analogy.

[38] See Burgess and Rosen [2005] for criticism and references.

[39] e.g., in Gödel [1944], [1964], or my [1990].

[40] See footnote 28.

[41] Formalism comes in many varieties. See Resnik [1980], chapter 2, or Shapiro [2000], chapter 6, for an overview and references.

[42] This is the 'Glib Formalism' discussed on p. 202 of my [1997].

full weight and worth of 'ideal mathematics'.[43] Pure mathematics has a wide range of uses, from facilitating various internal mathematical goals to providing methods and structures for applications outside mathematics—uses which the Arealistic methodologist is keen to investigate and understand.

Let's now return to the question postponed: does a viable explanation of how mathematics manages to work in application require the existence of its entities? Consider for illustration a case from elementary textbook physics, that of projectile motion.[44] Suppose we have a cannon equipped with a cannon ball and some gunpowder, and we want to know where the ball would hit the ground if we were to light the fuse. We begin by implementing a policy of representing the times, distances, velocities, accelerations, and angles involved (in other cases, lengths, volumes, masses, charges, energies, and so on) as if they vary continuously. In some cases this is an explicit simplification (for example, we know charge is quantized), in others (space and time) we aren't sure (see IV.2.i), but we do have good reason to believe that the idealization is harmless for present purposes. Here simplification comes as it often does in service of mathematization: we now replace times, distances, and angles with real numbers, velocities and accelerations with functions from reals to reals, so that we can apply the machinery of the calculus.[45] Let's suppose we've already determined by experiment the acceleration due to gravity (at sea level, where the cannon is), the independence of the horizontal and vertical components of projectile motion, and the negligibility of air resistance (in the case of a cannon ball, as opposed to a baseball). We then measure the angle θ between the cannon and the ground (treating both as perfectly straight).[46] This leaves the question of what velocity will be generated in this cannon ball fired from this cannon at this angle by the explosion of this amount of

[43] See Detlefsen [1986], especially, pp. 3–24, for this instrumentalist reading of Hilbert's Formalism.

[44] See Halliday et al. [2005], pp. 21–26, 64–69, especially problem 4-7(b).

[45] It may seem odd to regard the move to a full continuum as a simplification, given the higher-order infinity involved, but in fact the mathematics is much streamlined—opening the way to the all-important use of differential equations. (See Koperski [2001], §4, for a related discussion.)

[46] It's often noted that our measurements are inaccurate due to shortcomings in our instruments: 'to say that the temperature of a body is 10°, 9.99° or 10.01° is to formulate three incompatible [mathematized] facts, but these three incompatible facts correspond to one and the same practical fact when our thermometer is accurate only to a fifth of a degree' (Duhem [1906], p. 134). But the indeterminacy goes deeper, into the facts themselves: an actual physical object doesn't have a precise length because its boundaries are fuzzy; the cannon has no precise angle to the ground because of irregularities in both.

gunpowder, but let's put these matters in a black box and assume the initial velocity is known.

With all this in place, we execute a course of purely mathematical reasoning on those functions and real numbers: we want to know $R = x(t) - x(0)$ (where t is such that $y(t) - y(0) = 0$); the horizontal velocity is constant, so $R = v_x(0) \cdot t = v(0) \cdot \cos\theta \cdot t$. Meanwhile the vertical velocity is undergoing a constant acceleration $-g$, so $\frac{dv}{dt} = -g$. Integrating twice, we get $y(t) = y(0) + v_y(0) \cdot t - \frac{gt^2}{2}$. This means that $0 = y(t) - y(0) = v_y(0) \cdot t - \frac{gt^2}{2} = v(0) \cdot \sin\theta \cdot t - \frac{gt^2}{2}$. Substituting $t = \frac{R}{v(0)\cos\theta}$ gives us $v(0) \cdot \sin\theta \cdot (\frac{R}{v(0)\cos\theta}) = \frac{g}{2} \cdot (\frac{R}{v(0)\cos\theta})^2$. A little algebra plus the trigonometric identity $2\sin\theta\cos\theta = \sin(2\theta)$ yields $R = v(0)^2 \cdot (\frac{\sin 2\theta}{g})$. We now feed in the values we've measured—$\theta = 45°, g = 9.8 \text{ m/s}^2$, and $v(0) = 82\text{m/s}$—and calculate that R is a little over 686 m.[47] I belabor this simple example to illustrate the general pattern: we replace physical items with mathematical ones, engage in a sometimes extended stretch of pure mathematics,[48] then draw conclusions for the physical situation based on the assumption that the two are sufficiently similar; in support of that assumption, we rely on the results of experimental tests and on evidence that the explicit and potential idealizations involved are both harmless and helpful. The use of mathematics in more elaborate cases like Maxwell's theory of electromagnetism or Einstein's

[47] Notice that if we'd asked where the cannon ball would come to a stop rather than where it would first touch down, an accurate prediction would have been hard to come by, since this would depend on the texture of surrounding terrain, which is subject to odd local variations (bumps and depressions), wind and weather (if the grass is or isn't wet), and countless miscellaneous accidents (if the grass was or wasn't recently mowed). This recalls one thread from the morals of IV.2.iii: our mathematized science can answer some questions, not others; we tend to notice the phenomena we can treat and to forget the many we can't.

[48] Duhem ([1906], pp. 155–157) describes this process: 'When a physicist does an experiment, two very distinct representations of the instrument on which he is working fill his mind: one is the image of the concrete instrument that he manipulates in reality; the other is a schematic model of the same instrument, constructed with the aid of symbols supplied by theories; and it is on this ideal and symbolic instrument that he does his reasoning, and it is to it that he applies the laws and formulas of physics'; for example, a manometer is 'on the one hand, a series of glass tubes, solidly connected to one another ... filled with a very heavy metallic liquid called mercury by the chemists; on the other hand, a column of that creature of reason called a perfect fluid in mechanics, and having at each point a certain density and temperature defined by a certain equation of compressibility and expansion'. Unfortunately this view of Duhem's is wedded to a stark fictionalism about natural science that's served as an inspiration for contemporary anti-realists, but appears misguided to the Second Philosopher (cf. I.6, footnote 17, II.6, footnote 16, IV.5, footnote 35).

theory of relativity is more complex but apparently much the same in principle.

The question for the Arealist is: to do this, must we assume that the passage of pure mathematics involved is true, that the abstracta involved exist? Given that the pure mathematical representation is used as an object of comparison—the physical situation is enough like this mathematical one to suit our purposes—it's hard to see why it needs to exist in order to do the job. What matters is that we have a well-articulated theory describing an appropriate mathematical structure. It's unclear what's accomplished by adding—and by the way that mathematical structure exists!

Recall Putnam's exclamation from IV.2.i:

> One wants to say that the Law of Universal Gravitation makes an objective statement about bodies—not just about sense data or meter readings. What is that statement? It is just that bodies behave in such a way that the quotient of two numbers *associated* with the bodies is equal to a third number *associated* with the bodies. But how can such a statement have any objective content at all if numbers and 'associations' (i.e., functions) are alike mere fictions? (Putnam [1975b], p. 74)

Here as before we measure the masses and the distance, replacing them with real numbers; the objective content of the Law is that the masses of the bodies, the distance separating them, and the gravitational force between them are related to each other as these numbers are related in the equation $F = Gm_1m_2/r^2$. (I've stated this claim informally. When Putnam speaks of functions from objects to real numbers, he imagines a mathematization of the informal statement in a mathematical theory that includes physical objects with mass properties as urelements[49] and formally specifies the purported structural similarity between the two situations.) Putnam continues:

> It is like trying to maintain that God does not exist and angels do not exist while maintaining at the very same time that it is an objective fact that God has put an angel in charge of each star and the angels in charge of each of a pair of binary stars were always created at the same time! (Putnam [1975b], p. 74)

Not quite. We may hold that God and angels don't exist, but this needn't prevent us from articulating a rich vision of the heavenly hosts and their

[49] In set theory, urelements are non-sets. Imagine the usual iterative hierarchy beginning from some collection of objects. (See Enderton [1977], pp. 7–9, where urelements are called 'atoms'.)

relations. Nor would it prevent us from making an objective claim of the form: this man and his two sons are related much as God is related to these two angels (he favors the weaker of the two).[50] Likewise for our description of the physical world making use of mathematical abstracta.[51]

More would be needed to fill out this picture in detail, but assuming at least some plausibility for the claim that the Arealist can explain how mathematics is applied in natural science, let me touch briefly on another concern: some would argue that standard mathematical claims must be true if we are to apply logic there, because logic is a way of proceeding from truths to truths.[52] The Arealist need not be troubled by this: she need only agree to treat mathematical claims as if they were claims about a KF-world of abstracta; she need only stipulate that this is how one goes about constructing mathematical theories.[53] In fact, given that the properties of mathematics have sharp boundaries, that the objects of mathematics are subject to none of the spatiotemporal blurriness of ordinary objects, that the only relevant ground-consequent is the material conditional, the idealizations of classical logic present none of the potential headaches here that they do in ordinary physical cases.[54]

Assuming, then, that Arealism is at least prima facie viable, let's return to our exercise in compare and contrast. Earlier, we illuminated Thin Realism from the right by contrasting it with Robust Realism: the 'thinness' of its mathematical entities serves to undermine any challenge to the effectiveness of set theoretic methods for securing knowledge of sets; the price paid is the loss of any Robust claims about 'determinate truth values' for independent statements. Let's now examine Thin Realism from the left, in contrast this time with Arealism. At first blush, the difference here is more stark than with Robust Realism—the Thin Realist takes set theoretic theorems to be

[50] Indeed, see IV.3, footnote 3. Also IV.2, footnote 7.

[51] Bearing in mind that we often aren't able to state the objective fact about the physical world without comparing it to the abstracta.

[52] e.g., for Frege, logic concerns relations between thoughts, that is, between 'things for which the question of truth can arise' (Frege [1918], p. 328).

[53] So set theory is to be conducted as if it were describing the KF-world V, the universe of sets.

[54] It's worth recalling that Frege invented his logic expressly for the case of mathematics: 'when I came to consider the question to which of these two kinds [analytic or synthetic] the judgments of arithmetic belong, I first had to ascertain how far one could proceed in arithmetic by means of inferences alone ... to prevent anything intuitive from penetrating here unnoticed, I had to bend every effort to keep the chain of inferences free of gaps ... I found the inadequacy of language to be an obstacle ... This deficiency led me to the idea of the present ideography' (Frege [1879], pp. 5–6).

true and sets to exist; the Arealist does not—but in fact it seems to me that the actual differences between Thin Realism and Arealism are considerably less significant than those separating Thin Realism from Robust Realism.

To see this, consider a simple case: the justification of the Axiom of Replacement. History and mathematical considerations suggest that it was adopted as a fundamental axiom for a number of reasons:[55] it is needed to prove the existence of cardinal numbers like $\aleph_\omega$ that were present in informal set theory; it squares with limitation-of-size, one of the traditional guidelines for avoiding paradox;[56] it implies a number of important theorems that make set theory more tractable (every set is equinumerous with an ordinal,[57] transfinite recursion, Borel determinacy); and so on.[58] We've seen that the Robust Realist must explain why a principle with all these welcome features is also likely to be true; the Thin Realist dismisses this worry by characterizing sets as entities we can learn about by the exercise of set theoretic methods like these. But how will the Thin Realist differ from the Arealist? Each will tell the same story on Replacement, citing the same turns of history, the same mathematical consequences, the same set theoretic problems, norms, and goals. Only at this point, once all relevant theorems and facts of methodology are in place, does the question of truth arise: the Thin Realist counts the facts on the table as justifying a belief in the truth of Replacement, while the Arealist views them simply as good reasons to include Replacement among the axioms of our theory of sets.

What exactly is added in moving from 'this is a good axiom for reasons *x*, *y*, *z*' to 'this is a good axiom for reasons *x*, *y*, *z*, and therefore likely to be true'? Setting aside formal truth predicates, which are equally available to the Thin Realist and the Arealist, the word 'true' turns up in various contexts in the practice of set theory: 'Cantor's theorem is true'; 'So-and-so's conjecture turned out to be true, much to my surprise'; 'I think the Axiom of Measurable Cardinals is true'; 'CH does (or does not) have a truth value'. Different set theorists might attach different cognitive content to these claims, depending on their metaphysical

[55] Replacement is no different from the other standard axioms in this respect; similar lists can be given for each of them (see my [1997], pp. 36–62). For Replacement in particular, see my [1997], pp. 57–60.

[56] See, e.g., Fraenkel et al. [1973], pp. 32, 50.

[57] Assuming the Axiom of Choice.

[58] See Mathias [2001] for more on the poverty of set theory without Replacement. (I'm grateful to Akihiro Kanamori for this reference.)

leanings (realistic, formalistic, nominalistic ...), but I submit that for all such statements involving truth—by inspecting the role they play in practice, what underwrites their assertion, what is taken to follow from them, and so on—we can easily isolate what they come to for the practice of set theory, a sort of methodological core or 'cash value' if you will, that is available to the Arealist as well as the Thin Realist. I have in mind such simple readings as: 'such-and-such is provable from appropriate axioms'; 'such-and-such is a good axiom for reasons x, y, z'; 'I think we will (or won't) find compelling mathematical reasons for adopting a theory in which such-and-such turns out to be provable or disprovable.' On this much the Arealist and the Thin Realist will completely agree; the difference only comes in the way the word 'true' is then applied: the Thin Realist will find in all this good evidence for truth, while the Arealist takes talk of 'truth' as inessential and sticks to the methodological facts unadorned. As far as the practice of set theory is concerned, it is hard to see what's lost on the Arealist's approach.

Returning at last to the Second Philosopher, where do her sympathies lie? At issue is what happens when she concludes that she has good reason to pursue pure mathematics (see IV.3). We've seen how she studies and adopts the actual methods she finds in the practice, gradually refining and improving them[59] and tentatively setting aside appeals to metaphysical views about the nature of mathematical things. This will be enough to sour her on Robust Realism, but what of Thin Realism and Arealism? In Gideon Rosen's terms, the choice between the two can be put this way: does the Second Philosopher take mathematics to be an 'authoritative practice'?

> A practice is *authoritative* if, whenever we have reason to accept a statement given the proximate goal of the practice, we thereby have reasons to believe that it is true. (Rosen [1999], p. 471)

The Thin Realist answers yes: set theoretic claims generated by our most effective methods should be regarded as true and the sets whose existence is thereby asserted should be taken to exist. The Arealist answers no: set theoretic claims, including existence claims, generated by its most effective methods should be adopted as appropriate means toward theories that serve our goals, but natural science is the final arbiter of truth and existence, and it confirms neither the truth of mathematics nor the existence of sets.

[59] e.g. see the rise and fall of Definabilism in my [1997], II.4.ii.

The dispute between them can be posed this way: does the Second-Philosopher-turned-mathematician view her new undertaking as continuous with her original inquiry, as part of one complex effort to figure out what the world is like, or does she see it as a fundamentally distinct enterprise, pursued for its instrumental value? From a Thin Realist point of view, she has discovered, in the course of her inquiry, a new type of entity for which her methods must be expanded, a type of entity whose connection to its proper forms of evidence is conceptual.[60] (She realizes that others may also propose entities with conceptual connections of an analogous sort—for example, a theist might propose that the reliability of revelation is conceptually connected to his notion of God—but these other concepts, unlike the concept set, present no attractions that motivate her to adopt them.) Speaking instead from an Arealist point of view, she would say that she has good reason to employ new methods to develop a new kind of theory, a theory describing a vast array of structures, but unconcerned with truth or existence. Which of these options should the Second Philosopher embrace?

It's tempting to feel that there's a right and a wrong answer here, but I want to suggest that the situation bears a disquieting resemblance to that of Wilson's Druids in II.6: each group of natives stoutly defends its usage, but we can see that there is no predetermined fact to which the decision—bird or house?—is answerable, that the natives' strong convictions, one way or the other, are due to mere historical contingency. As an imaginary parallel, suppose first that our Second Philosopher begins with her fundamental uses of 'true' and 'exists', based on observation, experiment, confirmation of hypotheses, and so on. She recognizes an array of human activities that use different methods, from pure mathematics to astrology, but she also comes to realize that she has good reason to pursue pure mathematics, given its usefulness to her scientific theorizing. Still, this is not enough to convince her that an extension of the terms 'true' and 'exists' is appropriate; her original methods are the arbiters of truth and existence and they don't confirm either in the case of mathematics. Now picture instead the Second Philosopher beginning from a different perspective, this time from a contemplation of the full range of human endeavor with the terms 'true'

[60] This would presumably be another candidate for a priori knowledge, though of a different sort from the rudimentary logical truths of III.6.

and 'exists' liberally sprinkled throughout. She quickly sees good grounds on which to remove these terms from many places, like astrology, but they seem unobjectionable in their mathematical occurrences. So there they remain.

In our rough-and-ready terms, we might describe this contrast in yet another way: the Arealist holds that mathematics is distinguished from other extra-scientific enterprises by its role as a handmaiden to science; this very handmaiden role prompts our Thin Realist to assume that mathematics *is* a science, alongside the various natural sciences. For the Thin Realist, then, the methods of science—in particular, the methods of mathematical science—lead us to assert the existence of sets and the truth of set theoretic claims. In other words, in terms the Second Philosopher doesn't use, the question comes down to whether or not mathematics should count as a science. One might try to give this question bite by requiring a science to describe something objective, to be constrained by something outside ourselves, but this won't help. For the Thin Realist, set theory is answerable to the way sets are, and the way sets are doesn't depend on us, but for the Arealist, set theory is also answerable to matters independent of us, namely, to logic and to the facts about which mathematical concepts effectively serve which mathematical goals. When we recall that the Thin Realist's 'the way sets are' is conceptually tied to the methods of set theory, and that those methods come down to logic and the facts cited by the Arealist, the sense of any real distinction here begins to slip away.

Stripped to fundamentals, the disagreement is over the classification of methods. Are the methods of mathematics of a piece with observation, experimentation, theory formation and testing, etc., or are they of a different sort entirely? Our Arealist is struck by the stark differences between the methods of natural science and the methods of mathematics: for example, whatever the details of her conversation with the radical skeptic (see I.2), she never claims that his challenge is incoherent, but the Thin Realist's reply to a set theoretic skeptic is that there is no gap at all between sets and set theoretic methods, that the connection there is conceptual, and thus that the challenge is ill formed. The Arealist sees this as a sharp divide—her methods for finding out about the world are empirical, not conceptual—but the Thin Realist replies with complete composure that some aspects of the world can be uncovered empirically and some conceptually; for the Thin Realist, the sharp divide is between those methods that reliably investigate the world and those that don't. It's hard to see that this is more than a matter of emphasis.

It might seem that the possibility of an eventual bifurcation or fragmentation in set theory reveals a disagreement here that's more than verbal or cosmetic: the Thin Realist's link between truth and provability in set theory depends on the fact that there is, so far, a single best theory—ZFC plus large cardinals—that set theorists are out to explore and extend; if set theory were to split into a range of equally fundamental theories, the Thin Realist's treatment of 'existence' and 'truth' would be compromised. In the course of exploring his alternative to Woodin's approach to solving the CH,[61] Steel devotes considerable attention to a speculative scenario of this sort.[62]

Notice first that competing theories of sets could conceivably arise in a variety of different ways. The only sort we've actually encountered is exemplified by the opposition between $V = L$ and LCs, and here the response is straightforward, because, as we've seen (in IV.3), the structures available from $V = L$ can be recaptured in a theory with LCs as the theory of the inner model L. From here, we enter the realm of speculation, but perhaps we can discern some general patterns, even without the concrete factors that can be expected to do the real work in particular cases. So, for example, suppose the set theoretic community were faced with two attractive theories—attractive, that is, in terms of set theoretic norms and goals—neither of which is capable of recapturing the other. If no particulars of the case interfere, it seems the large-scale goals of maximization and unification, of providing a single theory as rich as possible, would recommend a concerted search for a more powerful theory that could encompass both existing theories in a sense relevantly comparable to LCs ability to encompass $V = L$.[63]

Steel's speculative scenario presents another variety: he imagines the possibility of opposing options, each of which could encompass the other. As both would then be equally maximizing, the concern is that there

[61] See IV.3. Steel's approach is only viable if the Omega Conjecture is false.

[62] See my [2005b] and Koellner [2006] for discussion and references.

[63] Analogous considerations seem to me to count against metaphysical positions that posit various universes of sets 'existing side-by-side' (Balaguer [1998], p. 59). As soon as those universes are posited, we find ourselves wanting to talk about the relations between them—about isomorphisms and embeddings and so on—which requires a larger arena containing both.

would be no ground on which to choose between them. Peter Koellner ([2006]) disputes this claim, suggesting additional considerations that might weigh one way or the other; under certain, again-speculative conditions, he details how one theory might miss a significant structural parallel that the other captures:

> we have here a case where there is further structure that we can leverage to provide reason for favouring one theory over the other. (Koellner [2006], p. 185)

Koellner may be right about special features of Steel's case, but I suspect that the force exerted by the foundational goal toward a single fundamental theory would be felt even if such otherwise persuasive features were absent. We're imagining the set theoretic community faced with two attractive theories, either of which is capable of encompassing the other: practitioners could see themselves as working in the first theory and recapturing the interesting consequences of the second in a natural interpretation, or as working in the second and so recapturing the first; methodologically, it would make no significant difference. Still, the foundational goal would motivate an Arealist to fasten on some consideration, however small, to tip the balance one way or the other, even if it ultimately came down to such matters as familiarity or historical contingency. And—consistent with the thought that Arealism and Thin Realism are in some sense cosmetic variants of the same position—it seems the Thin Realist would have good reason to use the very same consideration that the Arealist used in order to make her own choice; she holds, after all, that the methods of set theory are reliable indicators of the truth about sets, and she and the Arealist agree on which methods are proper, including this one. So it seems to me that the Thin Realist is as well equipped as the Arealist to deal with a Steel-like situation.

Thus mounting a threat to Thin Realism requires us to imagine the development of two attractive theories of sets, neither of which can recapture the other, and for which the promise of an overarching theory encompassing them both is somehow blocked. At this point, we've progressed so far into the realm of imagination as to raise doubts that anything sensible can be said: set theorists would be faced with a choice between maximizing the available mathematical structures and maintaining a single fundamental theory; either way, a basic goal would have to be sacrificed. If the goal of unification were to win out, there would be no problem for the

Thin Realist. If the goal of maximizing were to win out, the Arealist would say we now have good reason to explore alternative, equally fundamental theories of sets. The Thin Realist would say this, too, and presumably add that there turned out, to her surprise, to be different kinds of sets, that truth and existence in mathematics turned out to be relative to a particular universe of sets. But perhaps it's more likely that all bets would simply be off, that mathematicians would come up with a new line of development we can't begin to foresee. In any case, it isn't obvious that the Thin Realist would be any worse off than the Arealist in this dire eventuality.

If this is a fair description of the state of debate between the Thin Realist and the Arealist, then it's hard to see that there is any fact of the matter here about which we can be right or wrong, just as nothing in the natives' usage before they saw the airplane predetermined what they should call it: the decision between Thin Realism and Arealism appears to hinge on matters of convenience, taste, and preference in the bestowing of these honorific terms (true, exists, science, knowledge).[64] This tentative conclusion could hardly be further from Rosen's view of the matter; for him, whether or not mathematics is classified as authoritative is the central issue.[65] Of the Arealist option, he writes:

> In this case [the Second Philosopher's position] can only be the modest view that mathematics is ... immune to external criticism *only in its judgments about what makes for good mathematics* ... (Rosen (1999), p. 473)

While this may sound trivial, notice that the Robust Realist insists that 'making for good mathematics' is not enough to justify adopting an axiom or method. The Second Philosopher's deceptively modest-sounding suggestion in fact makes a significant difference to the evaluation and adjudication of methodological disputes, freeing them from metaphysical disputes over the nature of mathematical existence and truth.[66] Meanwhile, Rosen sees Thin Realism

> as a radical epistemology, according to which the intra-mathematical case for an axiom is automatically an adequate justification for believing it to be true. (Rosen (1999), p. 474)

[64] Perhaps this is why mathematicians are often reluctant to take the question seriously.

[65] Cf. Tappenden [2001], p. 496.

[66] After all, the question of how new axiom candidates should be judged is the central problem of the book Rosen is reviewing.

Perhaps this is radical epistemology, but I hope enough has been said to defend it against Rosen's 'Authority Problem':

> to give some sort of principle for telling the authoritative practices from the rest ... [from] theology, pure astrology or anything else from the anthropologist's chamber of curiosities. (Rosen (1999), pp. 471, 474)

The Second Philosopher, either Arealist or Thin Realist, distinguishes mathematics as essential to her investigation of the world.[67]

In sum, then, I see no significant difference between Arealism and Thin Realism; they are alternative descriptions of the same underlying facts.[68] Each way of speaking has its own advantages and disadvantages. The awkwardness of Arealism is easy to spot: though the existence claims of set theory are rational and appropriate moves in the development of set theory, sets do not in fact exist; the theorems of set theory are not true. Burgess ([2004], p. 19) would disparage this as 'taking ... back ... in one's philosophical moments what one says in one's scientific moments', but it seems to me that a better description would be taking back in one's scientific moments what one says in one's mathematical moments. Still, it would be more comfortable, admittedly, never to take anything back at all.

The awkwardness of Thin Realism, on the other hand, is more diffuse: in various areas, its ontology of abstracta allows uncomfortable questions to be asked. In fact, if Arealism and Thin Realism are identical under the skin, these questions can all be answered much as the epistemological problem—how do we come to know about sets?—and the referential problem—how can we refer to sets?—are both answered by the declaration of thinness: sets just are the kind of thing that can be referred to in these ways and known about by these methods. Questions about the nature of set theoretic existence are to be answered negatively—not causal, not

[67] Rosen ends his review by comparing the naturalism of my [1997] with van Fraassen's position, because 'for the constructive empiricist ... intra-scientific norms for acceptance are a matter for philosophical description, not legislation'—likewise, for the Second Philosopher, intra-mathematical norms for acceptance (Rosen [1999], p. 474). As we've seen (in IV.1), the Second Philosopher doesn't take up van Fraassen's external take on science, and Rosen wonders if she 'leaves it entirely open whether we have any reason to believe what set theory says' (ibid.). The answer is no. For the Second Philosopher, there is no perspective external to natural science, but natural science itself provides a perspective external to mathematics, a perspective from which she embraces Thin Realism/Arealism. In other words, we have the Thin Realist's reasons to believe in set theory, but recognize that this 'belief' is just a dressed-up version of Arealism.

[68] Those bothered by the Arealist's account of how mathematics applies might prefer to go back and restate it in the Thin Realist's idiom.

spatiotemporal—but here we might worry that it is not always obvious which of a pair of opposing properties is the negative one. For example, do sets exist necessarily or contingently? Presumably the answer should be that they exist necessarily, that existence in all possible worlds, despite appearances, is a negative feature because it comes to a denial that the existence in question is contingent on anything—but one might rather avoid this topic altogether. Or, to take a fresh example, are we referring to the same sets as Zermelo did? As Cantor did? Our methods of establishing their existence are different, as are some of the problems we're trying to solve, some of our goals, and so on. Does this matter? Again, there are no doubt ways of answering these questions, but a point of view from which they do not arise in the first place has its obvious attractions.

To sum up: the Second Philosopher undertakes pure mathematics as an essential part of her investigation of the world. She resolves any methodological questions that arise by assessing the efficacy of the alternatives for the realization of her mathematical goals. In her extra-mathematical study of mathematical practice, she recognizes that attributions of mathematical existence, truth, knowledge, reference, and so on are inessential cosmetic add-ons to the purely methodological analysis. In such discussions, she feels free to speak as an Arealist or as a Thin Realist, confident that the one mode of expression is effectively interchangeable with the other.

IV.5
Second metaphysics

Let me conclude this long journey into Second Philosophy with one last return to Quine. We've seen (in I.6) how Quine proposed that epistemology be naturalized and how the Second Philosopher follows his lead while dispensing with the empiricist and behaviorist trappings. Metaphysics, too, was to be naturalized, though here again the Second Philosopher jettisons such central Quinean doctrines as holism and the indispensability argument (see I.6 and IV.2.i). We've examined some particular aspects of Second Metaphysics (in IV.1 and IV.4 as well as the aforementioned), but it's worth stepping back for a look at the overarching question: what do the Second Philosopher's inquiries tell her about what there is? The demise of holism, the various subtleties of idealization and mathematization, leave us without clear instruction on how to evaluate the ontological morals of a given scientific theory. I'm hardly the only generally Quine-sympathetic philosopher to have considered this question, so I hope to illuminate the charge of the Second Metaphysician by tracing the development of metaphysics naturalized in our post-Quinean age.

From the start, the indispensability arguments tended to overshadow the rest. The notion that a hard-nosed naturalist should condone the existence of abstracta was so dramatic, so mesmerizing, that it held sway in the philosophy of mathematics for decades: nominalists rallied to argue that the key assumption was false, that mathematics is dispensable in natural science despite appearances;[1] realists used the argument to support their position, focusing their attention on the puzzle of how we come to know about non-spatiotemporal entities.[2] Meanwhile, ontological debates in the philosophy of science centered on such topics as the reliability of inference to the best explanation or the proposal that only the causal posits of our best

[1] e.g., Field [1980], Chihara [1973], [1990].

[2] e.g., my [1990].

theories are in fact confirmed.[3] In all this, the viability of Quine's picture as an account of physical ontology went surprisingly unexplored.

One contemporary school of thought in the philosophy of mathematics directly addresses the question of post-Quinean metaphysics, presenting a new proposal for judging the ontology of a given scientific theory. Consider, for example, Mark Balaguer's Fictionalist[4] who rejects holism[5] and holds that:

> Empirical science has a purely nominalistic content that captures its 'complete picture' of the physical world... It is coherent and sensible to maintain that the nominalistic content of empirical science is true and the platonistic content of empirical science is fictional. (Balaguer [1998], p. 131)

And this is true, the Fictionalist continues, even if we can't actually express this nominalistic content without appeal to fictional mathematical objects (Balaguer [1998], section 3.2). The purported causal isolation of mathematical entities is marshaled to support the claim that they appear in science only as 'descriptive aids' (Balaguer [1998], p. 137):

> The nominalistic content of empirical science is its picture of the physical world, whereas its platonistic content is the canvas... on which this picture is painted. (Balaguer [1998], p. 141)

Similar suggestions occur in the writings of other Fictionalists.[6]

As we've seen (in IV.2.i), there's considerable evidence for a position of this type in the scientist's practice of adopting whatever mathematical tools are handy and effective, without concern for the ontological status of the abstracta involved, for example, when real-valued functions are used to model motion without worries over whether or not the existence of

[3] e.g., van Fraassen [1980], Boyd [1983], Cartwright [1983].

[4] I put it this way because Balaguer distances himself from his Fictionalist in two ways. First, in addition to Fictionalism, he also sees a version of Platonism as viable, and indeed he thinks there is no fact of the matter concerning the disagreements between the two (see Balaguer [1998], Part III). For the second, see footnote 7.

[5] See, e.g., Balaguer [1998], p. 40: '... confirmational holism is, in fact, false. Confirmation may well be holistic with respect to the *nominalistic* parts of our empirical theories, but the mathematical parts of our empirical theories are *not* confirmed by empirical findings'. We've seen (in I.6) that the Second Philosopher would go further, suggesting that holism fails for physical hypotheses as well.

[6] See, e.g., Melia ([2000]): mathematics allows us to express nominalistic claims that we couldn't otherwise, but it's acceptable to adopt a mathematized scientific theory, then to withdraw commitment to the mathematical part. Similarly, Rosen holds that it is rational to regard mathematized science as 'nominalistically adequate', that is, to regard it as saying that things are 'in all concrete respects *as if* [the theory] were true' (Rosen [2001], p. 75).

reals has been conclusively established. But we also noted that the existence of the real numbers is not the only assumption taken for granted in the use of continuous mathematics to model motion: the modeling also carries with it deep structural assumptions about space and time, in particular, that both are continuous, and this physical assumption is made despite uncertainties over whether or not time and space are truly continuous. Thus, it seems that the Fictionalist's proposal for a new metaphysics naturalized—believe the physical but not the mathematical portion of our best scientific theory—delivers the wrong answer in this case.[7] It would (perhaps properly)[8] discount the existential assumptions about real numbers, but it would go on to endorse the continuity of time and space as part of the physical content of the theory, thus begging a serious question that practitioners regard as open (see IV.2.i).[9]

The stage for this less-than-satisfying ontological conclusion was set when the reaction to Quine focused so narrowly on the status of abstracta. The original project of metaphysics naturalized—how can we use our best science to determine what there is?—was transformed into the project of defending nominalism—how can we embrace our best science without believing in abstracta? Let's return here to the larger and deeper question: what does our best science tell us about the furniture of the world? How can Quine's method of assessing ontology be reconfigured?

Another recent Fictionalist, Stephen Yablo, takes up the broader question. In particular, Yablo suggests that

[7] This is the second way Balaguer distances himself from his Fictionalist: Balaguer recognizes that we may not want to commit ourselves to the full 'nominalistic content' of our theories; his point is that, even if we do so commit ourselves, we aren't thereby committed to abstracta. (This is implicit in some of the hedged formulations in his [1998] and explicit in personal correspondence.) As he emphasizes, this portion of the debate over ontological commitment is addressed exclusively to the problem of abstracta.

[8] As we've seen (in IV.2.i), the Second Philosopher takes the question of mathematical ontology not to be settled by natural science, but to be left to mathematics; wearing her mathematician's hat (in IV.4) she views both Thin Realism and Arealism as viable and in fact as essentially the same. So she would have no quarrel with the Fictionalist's claim that mathematical abstracta don't exist—this is to speak with the Arealist—though she would disagree if he were to reject Thin Realism as an equally acceptable way of speaking and she regards the Fictionalist metaphor as an unnecessary, potentially distorting addition.

[9] Field's nominalization project (Field [1980]) has a similar failing when regarded as an attempt at post-Quinean metaphysics naturalized: it assumes the full continuity of space and time. Let me add that though here and elsewhere I express this as a concern over the continuity of space or time or spacetime, those who doubt that these items are physically real can easily rephrase the concern in terms of a mathematically closely related structure whose physical status is less problematic, like the electromagnetic field. I take up the physical reality of the electromagnetic field below.

of the things that regularly crop up in people's *apparently* descriptive utterances, not all really exist, or are even believed to exist by the speaker.... the practice of associating oneself with sentences that don't, as literally understood, express one's true meaning is extraordinarily familiar and common. ... the sentences' literal content (if any) is not what the speaker believes, or what she is trying to get across. (Yablo [2000], p. 212)

In such cases, the speaker engages in a sort of make-believe; by uttering the metaphorical claim, she doesn't assert its literal meaning, she 'portray[s] the world as holding up its end of the bargain, by being in a condition to make a pretense like that appropriate' (Yablo [1998], p. 248).[10] To take a simple example, 'he goes by the book' doesn't mean that he walks past a salient bound volume: it takes place against the backdrop of a make-believe that there is an actual book of rules for the situations in which he generally finds himself, and in that make-believe, it asserts that he guides his actions by following each rule in turn, perhaps even running his finger along the lettering as he goes; it actually asserts that he so behaves in reality as to make this make-believe assertion 'sayable' in the make-believe context; in other words, it asserts that he is very attentive to formal procedures, or something like that. This 'or something like that' brings out that it is often difficult or even impossible to state literally what the metaphor enables us to state figuratively.

For ontological purposes, Yablo concentrates on *existential* metaphor: a metaphor making play with a special sort of object to which the speaker is not committed ... and to which she adverts only for the light it sheds on other matters. (Yablo [1998], p. 250)

Abstract entities are among these, according to Yablo, but he also holds that they include scientific idealizations and more. For this Metaphorical Fictionalist, then, Quine's implementation of metaphysics naturalized—our best science tells us that what exists are those things it asserts to exist—is to be replaced by a new criterion: our best science tells us that what exists are those things it asserts to exist in its literal (= non-metaphorical) claims. Here we might hope to find our bearings in the post-Quinean world, but alas, Yablo continues 'one unbreakable rule in the world of metaphor is that there is no consensus on ... what should be counted as a metaphor and what

[10] This way of describing the world's contribution goes back to Balaguer [1996], p. 312: 'the physical world ... holds up *its end* of the ... bargain'.

should not' (Yablo [2000], p. 214). By Yablo's reckoning, determining whether a delicate case is metaphorical or not requires us to answer the ontological question first, so the replacement criterion can only be applied if it is in fact unnecessary![11]

Another recent examination of the problem of assessing ontology comes from Jody Azzouni, but sadly he also concludes that the problem can't be solved:

> A [criterion for what a discourse commits us to] is so fundamental that there's no hope of slipping a rationale *under* it ... there is no bedrock *below* one's [criterion for what a discourse commits us to], no place to get a foothold to apply pressure against an opponent ... the question of *what there is* ... is ... philosophically indeterminate (Azzouni [1998], pp. 10–11; see also Azzouni [1997a], p. 208, Azzouni [2004], chapters 3 and 4)

The position here is even more despairing than Yablo's: it isn't that we have a criterion we can't apply in difficult cases; it's that there is no defensible criterion in the first place! What inspires this ontological nihilism?[12]

Azzouni begins by distinguishing what he calls a 'criterion for what exists' from a 'criterion for recognizing what a discourse commits us to'; let's call these ur-criteria and discourse criteria, respectively.[13] For example, recall (from IV.2.i) that the early Quine once advocated a 'renunciation' of abstract objects 'based on a philosophical intuition that cannot be justified by appeal to anything more ultimate' (Goodman and Quine [1947], p. 105), thus proposing a nominalistic ur-criterion. Combined with Quine's well-known discourse criterion—if a discourse includes the claim

[11] This last is the moral of Yablo [1998]. Even if this problem could be set aside, the Second Philosopher might well wonder why our literal or non-literal intentions are a reliable guide to what exists, unless they're tied to something else, like confirmation.

[12] Unlike the Fictionalist, Azzouni retains Quine's holism (though see Azzouni [2000], part I, for some caveats on the isolation of the special sciences) and instead rejects his criterion of ontological commitment (i.e., the claim that if 'there are so-and-sos' is true, then so-and-sos exists). For Azzouni, the moral of the atoms case isn't that a theory isn't a homogeneous whole confirmed as one, but that not all existential claims have the same status, that the truth of existential assertions must be separated from ontology (the 'separation thesis' of Azzouni [1998], p. 2, [2004], chapters 3 and 4), that 'there is a so-and-so' can be true without a so-and-so existing. Much of this springs from Azzouni's version of deflationism about truth, particularly his treatment of blind ascription (see Azzouni [2004], chapters 1 and 2). Fortunately, the relevant aspects of Azzouni's take on how to evaluate scientific ontology are easily translated into the Second Philosopher's terms: where Azzouni asks whether or not a given true existential claim carries ontological commitment, we ask whether or not that same existential claim is confirmed.

[13] See Azzouni [1998], p. 2, [2004], chapters 3 and 4. Azzouni uses the acronyms CWE and CRD in place of my 'ur-criterion' and 'discourse criterion'.

that 'so-and-sos exist', then it's committed to so-and-sos—this leads to trouble, as our favored theories happily assert the existence of abstracta. Quine's solution, of course, was to give up on the nominalistic ur-criterion, but Azzouni suggests that he might just as well have revised his discourse criterion by denying ontological force to the existential quantifier, adding a predicate 'x is concrete', and counting only things in the extension of that predicate as existing. The point is general:

> one can mix and match [discourse criteria] with [ur-criteria] ... all the competing pairs of matching [ur-criteria] and [discourse criteria] are entirely on a par with regard to pertinent evidence (namely, the scientific theories we accept). (Azzouni [1998], p. 10)

Thus Azzouni arrives at the hopelessness of metaphysics naturalized.

There is much to say about these moves, but let's concentrate here on the naturalist's perspective: are ur-criteria and discourse criteria really so freely adjustable? The naturalist engages in scientific inquiry to determine what the world is like, including the question of what there is; no ur-criterion that precedes science will be appropriate from this point of view. Some naturalists might adopt the ur-criterion 'science tells us what there is' but as we've seen throughout, the Second Philosopher doesn't rely on any meta-principle about the reliability of so-called 'science', on any official demarcation between 'science' and 'non-science'; her reason for believing in atoms isn't that science says so, but the range of evidence provided by the likes of Einstein and Perrin. So it's hard to see how the Second Philosopher can offer any contentful ur-criterion at all.

What about discourse criteria? Given a successful bit of science, the naturalist, with the scientist, is perfectly free to ask, 'what does this tell us about how the world is?' Much of the time, the answer is straightforward: the presence of medium-sized physical objects in our theories of macro-events is due to their presence in the world, combined with our evolutionarily and experientially generated ability to detect them, visually and otherwise; this aspect of our science tells us that there are medium-sized physical objects.[14] But, often enough—as in the case of atomic theory and wherever idealizations and other mathematizations abound—we've seen that we need a more nuanced evaluation of what our theories are telling

[14] See III.4 and III.5.

us. This may be difficult, but so far we've been given no reason to think it impossible.

Azzouni considers various alternatives to Quine's discourse criterion—like the proposal that we're only committed to the causally efficacious posits of our physical theories[15]—but his most suggestive ideas come in connection with his distinction between thick and thin evidence.[16] We have thick evidence for what we observe, but also for what we detect, in the sense that Perrin showed us how to detect atoms. In fact, before the shift to detection in the atoms case, practitioners seemed inclined to restrict thick evidence to observation;[17] it was the specifics of evidence like the Perrin experiments that motivated departure from the stricter requirement. As this case demonstrates, one type of scientific progress lies in developing new and previously unimagined ways of detecting; ultimately, the task of adjudicating what counts as detection and why will rest on scientific considerations as particular and detailed as those that motivated the acceptance of atoms. Thin evidence, in contrast, is simply what accrues to the indispensable posits of a theory that enjoys the Quinean virtues.

Might a defensible discourse criterion be developed from this seed?[18] The Second Philosopher's anti-holistic point (from I.6) is that thin evidence isn't enough to confirm the existence of a posit. It might be tempting to conclude that we have thin evidence for our mathematical posits—they do play a role in our holistically evaluated best current theory—but that we don't have thick evidence, and thus that their existence isn't confirmed. The trouble with this suggestion is that the epistemic status of the functions, numbers, etc., of our science is not the same as that of atoms before 1900:

[15] If Azzouni actually endorsed this proposal, I'd express second-philosophical concerns about relying so heavily on the notion of causation. See, e.g., Skyrms's analysis ([1984], p. 254): 'our ordinary, everyday conception of causation is an amiably confused jumble ... one with real heuristic value ... in the noisy macro-world of everyday life ... [but] the game has been altered by the progress of science. ... qualified by relativity ... in the quantum domain, the old cluster concept loses its heuristic value and becomes positively misleading'. Not to mention that the causal criterion seems to deliver the wrong answer for atoms in 1900 (see below).

[16] Azzouni actually speaks of thick and thin epistemic access; I switch to 'evidence' to avoid appearing to beg the ontological question.

[17] Though observation through a microscope was allowed, *pace* van Fraaseen [1980].

[18] For the sake of discussion, I take a stab at devising a new discourse criterion consistent with second-philosophical thinking as so far presented. I don't try to sort out the many points of agreement and disagreement with Azzouni's more elaborate and evolving thoughts on these matters (in Azzouni [1997a], [1997b], [1998], [2004]) because my primary concern here is with his nihilism about the possibility of any discourse criterion whatsoever.

scientists bemoaned the lack of thick evidence for atoms—they demanded it, they set out to secure it despite the difficulties—while they happily adopt mathematical machinery exclusively for its effectiveness and convenience, without incurring any further epistemic burden. The attendant existential claims, both for abstracta and for related physical structures, are accepted without a demand for thick evidence, and even the thin evidence gained by the success of the theory isn't regarded as confirmatory (see IV.2.i).

So it seems that there are three rough categories of posits: those for which we have thick evidence (e.g., atoms now), those for which we have only thin evidence (e.g., atoms in 1900), and those for which evidence is irrelevant, for which only effectiveness and convenience are required (e.g., real numbers, functions, etc.). A tentative candidate for a new discourse criterion begins to emerge: we're committed to the posits of our best science in the first category, but not to those in the other two categories; in other words, we've established the existence (within the limitations of ordinary fallibilism) of what we detect, but not of abstracta or of those posits for which our evidence is exclusively thin, that is, of those posits that merely play an indispensable role in a theory with the Quinean virtues.

I don't mean to recommend this proposal, for reasons that will come out in a moment,[19] but our topic for the present is Azzouni's ontological nihilism: our preference for one criterion over another is mere philosophical prejudice, supported by mere philosophical intuitions, not by good reasons. Despite his interest in thick and thin evidence and related matters, Azzouni isn't actually engaged in the project of figuring out what there is—this project he takes to be impossible. Rather, the discourse criterion he ends up proposing is designed to capture the conditions under which '*our* linguistic and epistemic community takes something to exist' (Azzouni [2004], p. 10). Now the Second Philosopher regards the investigation of such 'folk ontologizing' (Azzouni's term) as a legitimate, indeed a fascinating study, in psychology or sociology or anthropology, but when the question is ontology, her focus is on what there is, not on what various people are inclined to think or say there is.[20]

[19] Among others, for example, it makes no separate provision for the physical structural assumptions that accompany mathematization, whose status isn't the same as atoms in 1900 or abstracta.

[20] Cf. footnote 11 on Yablo's Fictionalism.

So before attending to the viability of the proposed criterion, we need to ask why Azzouni thinks there can be no good arguments on this matter. What sort of argument does Azzouni have in mind? Approaching the matter naturalistically, it seems we certainly do have good reasons, for example, for believing in what we observe: given our understanding of what the world is like and of how our sensory organs (and eyewear and microscopes) work, we conclude that under certain conditions our observational beliefs are reliable; let's assume for now that we can also make a persuasive case for the reliability of Perrin's detection. On the other hand, perhaps it's not unreasonable to abstain from belief in real numbers or functions, given that we add them without concern for ontology.[21] I'm inclined to think that there are good arguments in the vicinity of this candidate discourse criterion, whatever its ultimate shortcomings.

The odd thing is that Azzouni seems to agree:

> The required epistemic stories, about how instrumental thick [evidence] is possible, are ones told by scientists (in various subdisciplines) about how their instruments, and their other means of interactions with the world, function. So, for example, there's the story physicists tell about how the instruments they've developed to measure acoustical phenomena operate. There's also the story that psychologists borrow from physiologists about how our visual systems work. These stories are invariably incomplete, as stories often are in the sciences, but ... they're clearly designed to indicate how we know what we know about something—how, as far as the science of a particular time can tell us, the means of access we have to things operates ... Neither observation nor thick epistemic [evidence] (which generalizes observation) is a metaphysical black box that achieves its epistemic deeds by magic. They're both as transparent to scientific inquiry as any other aspect of our universe. (Azzouni [2004], pp. 135–136)

So we're faced with this question: what is the source of Azzouni's conviction that there are no grounds on which to argue for or against a discourse criterion? Why isn't it reasonable, for example, to argue that a criterion requiring observability is too restrictive (because of the status of atoms now) or that a criterion admitting anything for which there is thin evidence is too lenient (because of the status of atoms in 1900)? By rejecting these arguments a priori, we seem to be ignoring the force of ordinary scientific evidence like that obtained by Einstein and Perrin.

[21] See footnote 8.

One simple answer to this question is that Azzouni isn't at heart a naturalist. If this wasn't entirely clear in his earlier writings, it's explicit in his recent book:

> There are no philosophically conclusive ways to argue for *our* criterion for what exists. That is, we can imagine alternative communities with the same science we have but with different beliefs about what exists because (and solely because) they have a different criterion for what exists; they're otherwise unaffected by their choice. In particular, there is nothing we can point to, either practically speaking or in terms of some implicit incoherence in their practices or theories, that shows they've got the *wrong* criterion for what exists. (Azzouni [2004], pp. 5–6)[22]

Now imagine my Second Philosopher, believing in atoms on the Einstein/Perrin evidence, confronted with someone from 'another community' who has exactly 'the same science' but denies that atoms exist. Her first thought is to explain the details of the calculations and experiments, to imagine that her opponent has ordinary scientific objections to the evidence cited; such a debate would be settled on standard scientific grounds, but the assumption that the other has 'the same science' suggests that his objections are not of this sort. Thus it seems his reasons for resisting the force of her evidence must be extra-scientific, perhaps a philosophical commitment to some strong version of empiricism.[23] The Second Philosopher will then ask that those reasons be explained, so as to weigh them against the strong evidence she has in hand, but given that the new reasons aren't addressed to her ordinary scientific grounds and that the evidence she has in hand has already persuaded her to go beyond observation, she's unlikely to be convinced. In other words, the Second Philosopher, unlike Azzouni, would argue that the other fellow is making a mistake, that he's failing to recognize good evidence when he sees it.

But beyond its brute unnaturalism, I suspect there is a second factor underlying Azzouni's position that's worth digging out. To see what this is, return for a moment to Quine's original discourse criterion: we are committed to whatever posits our best science asserts to exist. This criterion is intended to have normative force: if a scientist adopts a mathematized theory and denies the existence of mathematical objects, she

[22] This appears in a short passage addressed in part to an earlier (2002) draft of the present discussion.

[23] As in van Fraaseen [1980] (see IV.1).

stands convicted of 'philosophical double think' (Quine [1960], p. 242) or 'intellectual dishonesty' (Putnam [1971], p. 347). Now consider the present proposal: we're committed to those things for which we have thick evidence, that is, to those things we've detected. The question is: does this criterion have the normative force of Quine's original?

This brings us to the first of two concerns about the proposed replacement criterion. The suggestion is that we're committed to the existence of a particle if and only if we're able to detect it, but given the poor prospects for a strict definition of 'detection', the proposed criterion must either decline to specify precisely what counts as detected or leave open the possibility of shifting its current specification in light of scientific progress. Now imagine that we of the scientific community claim to have detected a so-and-so by some new method (as Perrin did), that we give good scientific grounds for the reliability of this method (as we've been assuming Perrin did), and that we conclude there are so-and-sos. Does the proposed criterion have any independent judgment to pass on this matter? It seems not: at issue is whether the new method falls under the old, vague notion of 'detection' (if the criterion is indeterminate) or whether the old notion of 'detection' should be replaced with one that admits the new method (if the criterion is precise), and the only grounds available on which to base either decision are the ordinary scientific grounds we gave for so-and-sos in the first place. In other words, the ordinary science, not the criterion, is what's doing the work; the criterion provides no independent ground for charging the scientist with 'double-think' or 'intellectual dishonesty'. This isn't to say that the scientist's judgment can't be questioned—all her evidence and reasonings are open to perfectly ordinary scrutiny and critique—but the criterion alone isn't providing any leverage. So I wonder: perhaps the sort of toothless criterion under consideration here—with its vague and/or easily disposable notion of 'detection'—isn't a full-fledged discourse criterion in Azzouni's eyes.

Look at it this way. The Second Philosopher simply sees herself as inquiring into the way the world is, including its ontology. Along the way, she continually analyzes and assesses the methods she takes to establish the existence of previously debated entities. This inquiry may or may not produce a general characterization of what's needed in such cases, but even if it does, that characterization will be open to modification as new methods are added. We could understand this process as the search for a discourse criterion, but it's not clear that phrasing the matter in these terms is particularly helpful. In any case, an attempt to formulate general criteria must begin by asking when the existence of particular posits is confirmed, by what sort of evidence, and why. What's needed isn't overarching philosophical theorizing; it's individual case studies.[24]

So perhaps in the end what drives Yablo and Azzouni to despair of metaphysics naturalized is the unspoken assumption that it must proceed by producing a general, normative criterion, coupled with the belief that no such thing is within our grasp. On this last point perhaps I share their pessimism—where I differ is in my conviction that this in no way undermines the original project of metaphysics naturalized, or perhaps I should say, of Second Metaphysics, as characterized here. Second Metaphysics is emphatically not a purely descriptive enterprise: the Second Philosopher holds that the existence of atoms had not been established in 1900 and that it had been established by 1910, not merely that scientists thought this or that at various times. Where it differs from the old normative project is not in renouncing normativity, but in its piecemeal approach: it doesn't begin with the demand for a general criterion.

So, how *does* the Second Metaphysician proceed? For her, the answer to 'what is there?' takes the form of a list; what she actually confronts are a series of particular existence questions. What reason do we have, for example, to believe in the existence of medium-sized physical objects? Leaving radical skepticism aside (see I.2), the answer is by now familiar: our chemical and physical story of such things as the apple on the table supports the commonsense view that the stuff making up the apple is importantly different from the stuff making up its surroundings; our electromagnetic and subatomic story explains how it holds together and resists penetration; our optics, physiology, cognitive science, biology and evolutionary theory describe how the underlying structures of our brain and nervous system react to light and other inputs from the apple to produce our belief in it. On these grounds, we hold that there is an apple there, that our

[24] Some readers of my [2001b] have imagined that I take the rejection of general analyses of notions like confirmation, explanation, and so on to be part of naturalism per se. I don't. What I do think is that (i) we shouldn't assume that general analyses are the only acceptable outcome of our scientific study of science, (ii) the desire for general analyses has often distorted our scientific study of science, and (iii) that study itself may well lead us away from the general to an appreciation of the variety and significance of the particular.

belief that there is an apple there is veridical; more generally, we trust the evidence of our senses under certain circumstances and not under others.

But what of atoms, the case that's so exercised us in these pages? Throughout, we've assumed that in light of Einstein's work, the Second Philosopher holds that Perrin's experiments established the existence of atoms (within reasonable limits of scientific fallibility), but to say this is only to point in the general direction of the grounds for a proper second-philosophical defense of the atomic hypothesis. Philosophers of science have reconstructed the case for atoms in a bewildering variety of ways: atoms are indispensable posits of a theory with simplicity, familiarity of principle, scope, fecundity, conformity with experimental tests (Quine); we infer the existence of atoms as the best explanation of the various phenomena (Harman);[25] the existence of atoms is part of the only plausible explanation of the success of our scientific theories (Boyd);[26] we know atoms exist because they cause observable effects (Cartwright).[27] Unfortunately all these general accounts founder on the same rock as Quine's: the atomic hypothesis already enjoyed the preferred features by 1900; the Einstein/Perrin evidence 'should have been an anti-climax ... simply more of the same' (Miller [1987], p. 470). Wesley Salmon ([1984], pp. 213–229) and more recently Peter Achinstein ([2002]) attend more carefully to the particulars of Perrin's experiments, but in both cases, the type of evidence highlighted was already available: Salmon, following the historian Mary Jo Nye [1972], emphasizes the surprising convergence of thirteen different methods for computing Avogadro's number, including several of Perrin's, but a similar convergence of different methods for computing atomic weights emerged by 1860; Achinstein sees Perrin's experiments as increasing the probability of the atomic hypothesis, but again this would appear to be 'simply more of the same'.

[25] See, e.g., Harman [1965], p. 89: 'When a scientist infers the existence of atoms and subatomic particles, he is inferring the truth of an explanation for various data which he wishes to account for'.

[26] See, e.g., Boyd [1983], p. 207: 'the reliability of theory-dependent judgments of projectibility and degrees of confirmation can only be satisfactorily explained on the assumption that the theoretical claims embodied in the background theories which determine those judgments are relevantly approximately true, and that scientific methodology acts dialectically so as to produce in the long run an increasingly accurate theoretical picture of the world'.

[27] See Cartwright [1983], p. 85: 'In ... Perrin's [experiments] we infer a concrete cause from a concrete effect. [This case does] not vindicate a general method for inferring the truth of explanatory laws [but] a far more restrictive kind of inference: inference to the best cause.'

Now I've been suggesting all along, on the Second Philosopher's behalf, that the evidence involved in establishing the atomic hypothesis wasn't just more of the same, but a new type of evidence altogether, what we've been calling 'detection'. So far, I've set aside the challenge of explaining why detection is reliable, or better, why Perrin's particular way of detecting molecules was reliable, but for purposes of Second Metaphysics, that is the crucial issue. This is no place to attempt a full analysis of the Einstein/Perrin case, even if I were expert enough to provide one, but let me venture a few observations.

There were, at the time, two general schools of thought.[28] The first began with those chemists who favored the atomic hypothesis as early as 1860 and now embraced the kinetic theory. The second took classical thermodynamics as the paradigm of an empirically supported science, applauded Maxwell's gradual replacement of mechanical vortices with his purely mathematical treatment of electromagnetism, and generally rejected hypotheses of discontinuous underlying structures in favor of differential equations describing continuously varying observed phenomena (transmission of heat, propagation of electricity, movement of fluids, ...). In contrast, members of the first school took those same differential equations to result from the application of limiting operations to discrete systems.

Members of the second group naturally regarded the atomic hypothesis as a flight of fancy, an unnecessary hypothesis reaching beyond the realm of possible experience and into metaphysics, but from their point of view, kinetic theory also had another, more specific flaw: it conflicts with the classical understanding of the second law of thermodynamics. (Classically, entropy never decreases in a closed system; in the kinetic theory, it is only highly unlikely to decrease.[29]) Thus the atomic hypothesis, an unattractive theory based on poor methodology, claims to correct thermodynamics, the epitome of a well-confirmed, purely empirical description of the world!

In this context, Gouy first proposed Brownian motion as potential confirmation of the kinetic theory.[30] In Perrin's words:

Every granule suspended in a fluid is being struck continually by the molecules in its neighborhood and receives impulses from them that do not in general

[28] See Nye [1972], pp. 13–20. Also my [1996], pp. 330–331, Wilson [2006], pp. 154–7, 654–659.

[29] See Perrin [1913], pp. 86–88, for discussion.

[30] See my [1997], pp. 138–143, for discussion and references, or better, the classic Perrin [1913].

exactly counterbalance each other; consequently it is tossed hither and thither in an irregular fashion. (Perrin [1913], p. 86)

Einstein converted this qualitative explanation into a series of precise quantitative predictions for the behavior of a truly random walk. Perrin writes:

Such, in broad outlines, is the remarkable theory we owe to Einstein. It is well adapted to experimental verification, *provided we are able to prepare spherules of measureable radius.* (Perrin [1913], p. 114, emphasis in the original)

This Perrin was able to do: he traced 'the complicated nature of the trajectory of a granule', verified its 'complete irregularity', and confirmed Einstein's predictions.[31]

Before Perrin's successes, the case for the existence of atoms had hinged on aggregate behavior: the hypothesized activities of large groups of molecules were used to explain phenomena from combining volumes to heat. Underlying these accounts was the picture of a vast number of identical bodies separated by distances large in comparison with their size, moving entirely at random, meeting in perfectly elastic collisions. Perrin's accomplishment was to establish a link to this behavior of individual molecules. Though we don't observe the atoms themselves, we do observe the pattern of their motion:

Molecular movement has not been made visible. The Brownian movement is a faithful reflection of it, or, better, it is molecular movement in itself, in the same sense that the infra-red is still light. From the point of view of agitation, there is no distinction between nitrogen molecules and the visible ... grains of an emulsion. (Perrin [1913], p. 105)

Thus we are privy not only to a striking aspect of individual behavior, but in fact to its most counterintuitive aspect—the random walk—the very source of the conflict with classical thermodynamics. I think these considerations give some sense of how Perrin's evidence was qualitatively different from what came before, and why it was rightly regarded as a watershed.

Around the same time, C. T. R. Wilson hoped to discover how rain droplets form in clouds by recreating the phenomenon in the laboratory.[32]

[31] See Perrin [1913], §72, §73, and chapter IV, respectively.

[32] See Galison [1997], chapter 2, for an account of Wilson's work.

In the course of this pursuit, he developed his ground-breaking cloud chamber and recorded streaks of condensed droplets that marked the trajectories of individual subatomic particles. The historian Peter Galison writes:

> the importance of the Wilson chamber lay in its ability to display individual processes, directly ... Unlike instruments that measured molecular properties in the aggregate, the cloud chamber singled them out. ... the cloud chamber, perhaps better than any instrument, instantiates the production of direct evidence for the subvisible world of microphysics. (Galison [1997], pp. 66, 73)

Here Wilson moved past Perrin's reflection of individual behavior to track the behavior itself. These were both early instances of 'detection', a notion that subsequently grew to embrace the outputs of nuclear emulsions, bubble chambers, and a growing, bewildering array of increasingly elaborate detectors.[33]

The job of the Second Metaphysician, as I see it, is to flesh out this story—perhaps a corrected version of this story!—to reveal and explicate the rationality of existence claims based on each of these particular varieties of evidence. It is possible, of course, that such an examination will instead uncover shortcomings, but they will be shortcomings couched in ordinary scientific terms, not complaints based on philosophical preconceptions. Those trained in philosophy have sometimes discerned methodological errors that scientists themselves have missed—for example, in some feministic critiques of primatology[34]—but I venture that the likelihood of such a development in the present case is low. Still, those trained in philosophy are well situated to articulate the structure and strength of the argumentation involved, aided immeasurably by such invaluable histories as Nye [1972] and Galison [1997].

What we've seen so far illustrates my first concern about the Azzouni-like criterion floated earlier: that calling a particular process 'detection' only masks the difficult task of explaining what makes it good evidence for existence. My second concern is that the notion of 'detection' derives from the investigation of unobservable particles, but these alone don't seem to exhaust the furniture of the world: 'The electromagnetic field is, for the modern physicist, as real as the chair in which he sits' (Einstein and Infeld [1938], p. 151). Leaving the chair aside, the electromagnetic field is

[33] Galison [1997] gives a monumental and fascinating history.

[34] See Hrdy [1999], pp. xviii–xix; quoted in III.4, footnote 3.

most often taken to be as real as the particles that have been our focus so far: 'We have as good reason to believe in the *fields* of the electron theory as we have to believe in the *electrons*' (Stein [1970], p. 284). Here the problem doesn't hinge on any kind of inaccessibility—we experience electromagnetic force as directly as anything when we sit down in that chair of Einstein's![35]—rather, the question is whether or not that force is the manifestation of a field that 'exists, acts and changes according to Maxwell's laws' (Einstein and Infeld [1938], p. 146).

As it happens, Marc Lange has organized his recent introduction to the philosophy of physics as a compelling book-length explication of the case for the reality of the electromagnetic field, an argument that draws on the reflections of a pantheon of great physicists: Faraday, Maxwell, Heaviside, Einstein, and others. The final lap begins from Einstein's discomfort with asymmetries in classical electromagnetism:

> Recall, for example, the electrodynamic interaction between a magnet and a conductor. The observable phenomenon depends here only on the relative motion of conductor and magnet, while according to the customary conception the two cases, in which, respectively, either the one or the other of the two bodies is the one in motion, are to be strictly differentiated from each other. For if the magnet is in motion and the conductor is at rest, there arises in the surroundings of the magnet an electric field endowed with a certain energy value that produces a current in the places where parts of the conductor are located. But if the magnet is at rest and the conductor is in motion, no electric field arises in the surroundings of the magnet, while in the conductor an electromotive force will arise, to which in itself there does not correspond any energy, but which, provided that the relative motion in the two cases considered is the same, gives rise to electrical currents that have the same magnitude and the same course as those produced by the electric forces in the first-mentioned case. (Einstein 1905, quoted by Lange [2002], pp. 190–191)

[35] Cf. Wilson [2000b], p. 232: 'any of Sonny Liston's opponents would have been surprised to learn that *force* is a non-observable'. Wilson describes how the historical idea that force is unobservable traces to nineteenth-century difficulties in the foundations of continuum mechanics: 'what did nineteenth century scientists do, given that they didn't see how a straightforward mathematical framework for continua might be assembled? Answer: they filled in with *anti-realist philosophy*' (p. 237). But the relevant foundational problems have subsequently been solved—'greater mathematical patience was wanted, not antirealist philosophizing' (p. 238)—and our 'conceptually unproblematic [contemporary] physics becomes garnished with quotations from venerable authorities [e.g., Duhem]' (p. 238) whose worries no longer exist. See Wilson [2006] (as cited in I.6, footnote 17, II.6, footnote 16, IV.4, footnote 48) for more on these issues.

Lange argues that Einstein's distaste for this distinction without a difference is not 'a mere aesthetic preference' (Lange [2002], p. 200), that it forms part of the justification for Einstein's application of his principle of relativity:[36] the motion in such cases, and thus the nature of the force involved (electrical or magnetic), depends on the choice of reference frame. In the resulting relativistic context, spatial and temporal separations are frame-dependent manifestations of the underlying spacetime, electricity and magnetism are frame-dependent manifestations of the underlying electromagnetic field, and energy and momentum are frame-dependent manifestations of the invariant mass.[37] Lange concludes by arguing that 'The electromagnetic field is a full-fledged constituent of a system, as real as the system's bodies since, like them, it possesses mass' (Lange [2002], p. 243). In sum: mass is real because it's invariant; the electromagnetic field is real because it has mass.

Obviously I can't do justice to Lange's case in the course of a single paragraph, but I hope at least to have given a sense of its purely intra-scientific flavor. Lange explicitly challenges the idea that there is a

> sharp, permanent line between a scientific theory (whose business it is to predict our observations) and its philosophical interpretation (which specifies what reality would or could be like if the theory succeeds in predicting our observations). (Lange [2002], p. 250)

Lange's 'philosophical interpretations' fall within the purview of our Second Metaphysician, and his point is that they are not idle. The Einsteinian argument cited above provides one example of how a matter of interpretation can cast doubt on a theory; Lange observes that

> This moral was also implicit throughout [the book] where we witnessed Faraday, Maxwell, Heaviside, Lodge, and Poynting arriving at electromagnetic field theory partly as a result of their struggles with interpretive ... questions about how charged bodies manage to act on one another despite being separated in space and time. (Lange [2002], p. 201)

[36] Like his distaste for the Lorentz–FitzGerald contraction hypothesis (see Lange [2002], p. 199), which posits a spatial contraction of bodies in the direction of motion to explain away the apparent constancy of the speed of light (e.g., in the Michelson–Morley experiments). This puts Lange at odds with, e.g., DiSalle [2002], who sees the move to special relativity as a Kuhnian revolution, mediated not by scientific methods, but by philosophy. Cf. IV.1, footnote 19.

[37] See Lange [2002], chapters 7 and 8.

Indeed with sensitive enough instrumentation, the mass distribution in a region of the electromagnetic field could be detected by its gravitational effects, and

> if new empirical predictions can be derived from a theory's interpretation [then] the interpretation ... is just more physical theory! ... [Thus] the interpretation of scientific theories, although traditionally 'philosophical' in its concern with rigorous conceptual foundations, is also an essential component of doing science. (Lange [2002], p. 250)

Lange's conclusion here dovetails with my description of the Second Philosopher: though she is motivated by purely scientific concerns and employs purely scientific methods, she ends up deliberating effectively on traditional metaphysical questions of what there is.

Of course, what's been suggested so far on behalf of certain particles and fields must ultimately face the challenge of quantum mechanics, where it's unclear what—if anything!—our fantastically successful theories tell us about the structure of the world. Having traced the historical development of field theories from Newton to Maxwell to Einstein, Howard Stein remarks:

> Despite these profound revisions of the fundamental physics, the basic conceptual structure remains the same. That cannot be said, however, in the case of the quantum theory: here, I think, even the metaphysics fails, and has to be replaced by a conceptual structure of a thoroughly new order. One of the points about quantum field theory is that, in its domain, the necessary conceptual structure has in fact not yet been found; I am tempted to say that the quantum theory of fields is the contemporary locus of metaphysical research. (Stein [1970], pp. 284–285)

Here Stein's 'metaphysics' is Lange's 'interpretation' and my 'Second Metaphysics'.[38] Once again the suggestion is that matters of ontology are internal to science.[39]

Finally, even a complete investigation of the ontology of physical theory leaves a vast realm of questions untouched. I trust we will continue to admit plants and planets and people, regardless of how we come to regard

[38] Or perhaps not quite in Stein's case, as a trace of lingering positivism threads through his [1970]. For present purposes I ignore this.

[39] Quantum field theory as now understood is an uncongenial home for particles: see Malament [1996]. Lange [2002], chapter 6, credits Faraday as first to suggest that fields may be all there is. Of course, some version of 'particles' and particle talk would almost certainly survive even this revision.

their composition as physical objects, but are there tadpoles?—that is, is being a tadpole an objective feature of some physical things (as assumed in III.4)?—are there acids, species, conscious experiences? Though we no doubt require some understanding of how these items relate to the underlying physics, it seems to me that the sort of explanatory role a given posit is supposed to play will dictate the particulars of what's required in each case, and that no decision can be made ahead of time, a priori, for all such things. In other words, I don't think it's helpful to try to formulate a doctrine of 'physicalism' that specifies the appropriate notion of 'reduction' in advance (see II.3).

Let me stop here, with this sketch of what I take to be involved in Second Metaphysics, my proposal as a post-Quinean successor to metaphysics naturalized. The Second Philosopher conducts her metaphysical inquiry as she does every other inquiry, beginning with observation, experimentation, theory formation and testing, revising and refining as she goes, but without recourse to any official notion of what constitutes 'science', without any means of justification beyond her tried and true methods. Nevertheless, I hope to have shown that she has plenty to do, and that—despite the poignant concerns of some philosophers—she has all the necessary tools with which to do it!

References

Achinstein, Peter [2002] 'Is there a valid experimental argument for scientific realism?', *Journal of Philosophy* 99, pp. 470–495.

Ahlfors, Lars [1979] *Complex Analysis*, third edition (New York: McGraw Hill).

Akiba, Ken [2000] 'Indefiniteness of mathematical objects', *Philosophia Mathematica* 8, pp. 26–46.

Allison, Henry [1973] *The Kant-Eberhard Controversy* (Baltimore: Johns Hopkins University Press).

——[1976] 'The non-spatiotemporality of things in themselves for Kant', *Journal of the History of Philosophy* 14, pp. 313–321.

——[1983] *Kant's Transcendental Idealism* (New Haven: Yale University Press).

——[1996] 'Transcendental idealism: a retrospective', in his *Idealism and Freedom* (Cambridge: Cambridge University Press), pp. 3–26.

——[2004] *Kant's Transcendental Idealism*, revised and enlarged edition (New Haven: Yale University Press).

Anderson, R. Lanier [2005] 'Neo-Kantianism and the roots of anti-psychologism', *British Journal for the History of Philosophy* 13, pp. 287–323.

Andrews, Paul, Gangestad, Steven, and Matthews, Dan [2002] 'Adaptionism—how to carry out an exaptationist program', *Behavioral and Brain Sciences* 25, pp. 489–504.

Antell, S., and Keating, D. [1983] 'Perception of numerical invariance in neonates', *Child Development* 54, pp. 695–701.

Apostol, Tom [1967] *Calculus*, vol. i, 2nd edn. (New York: John Wiley & Sons).

Avigad, Jeremy [2006] 'Methodology and metaphysics in the development of Dedekind's theory of ideals', in J. Ferreirós and J. Gray, eds., *The Architecture of Modern Mathematics* (Oxford: Oxford University Press), pp. 159–286.

Awodey, Steve, and Klein, Carsten, eds. [2004] *Carnap Brought Home: The View from Jena* (Chicago: Open Court).

Ayer, A. J. [1936] *Language, Truth and Logic* (New York: Dover, 1952).

Ayers, Michael [1997] 'Is *Physical Object* a sortal concept? A reply to Xu', *Mind and Language* 12, pp. 393–405.

Azzouni, Jody [1994] *Mathematical Myths, Mathematical Practice* (Cambridge: Cambridge University Press).

——[1997a] 'Applied mathematics, existential commitment, and the Quine-Putnam indispensability thesis', *Philosophia Mathematica* 5, pp. 193–209.

——[1997b] 'Thick epistemic access: distinguishing the mathematical from the empirical', *Journal of Philosophy* 94, pp. 472–484.

——[1998] 'On "On what there is"', *Pacific Philosophical Quarterly* 79, pp. 1–18.

——[2000] 'Applying mathematics: an attempt to design a philosophical problem', *Monist* 83, pp. 209–227.

——[2004] *Deflating Ontological Consequence* (New York: Oxford University Press).

Baars, Bernard, Banks, William, and Newman, James, eds. [2003] *Essential Sources in the Scientific Study of Consciousness* (Cambridge, Mass.: MIT Press).

Baillargeon, Renée [1993] 'The object concept revisited: new directions in the investigation of infants' physical knowledge', in Granrud [1993], pp. 265–315.

——Spelke, Elizabeth, and Wasserman, Stanley [1985] 'Object permanence in five-month-olds', *Cognition* 20, pp. 191–208.

Bak, Joseph, and Newman, Donald [1997] *Complex Analysis*, second edition (New York: Springer-Verlag).

Balaguer, Mark [1996] 'A fictionalist account of the indispensable applications of mathematics', *Philosophical Studies* 83, pp. 291–314.

——[1998] *Platonism and Anti-Platonism in Mathematics* (New York: Oxford University Press).

Baldwin, Thomas [2003] 'From knowledge by acquaintance to knowledge by causation', in N. Griffin, ed., *The Cambridge Companion to Bertrand Russell* (Cambridge: Cambridge University Press), pp. 420–448.

Bangu, Sorin [2006] 'Steiner on the applicability of mathematics', *Philosophia Mathematica* 14, pp. 26–43.

Barwise, Jon, ed. [1977] *Handbook of Mathematical Logic* (Amsterdam: North Holland).

Beck, Lewis White [1956a] 'Kant's theory of definition', reprinted in Wolff [1967], pp. 23–36, and in Beck [1965b], pp. 61–73. (Page references to the 1967 version.)

——[1956b] 'Can Kant's synthetic judgments be made analytic?', reprinted in Wolff [1967], pp. 3–22, and in Beck [1965b], pp. 74–91. (Page reference to the 1967 version.)

——[1965a] 'Kant's letter to Marcus Hertz', reprinted in his [1965b], pp. 54–60.

——[1965b] *Studies in the Philosophy of Kant* (Indianapolis: Bobbs-Merrill).

——[1976] 'Toward a meta-critique of pure reason', reprinted in his *Essays on Kant and Hume* (New Haven: Yale University Press, 1978), pp. 20–37.

——[1992] 'Foreword' to Reich [1932], pp. xi–xx.

Bell, E. T. [1937] *Men of Mathematics* (New York: Simon and Schuster).

Bell, John [1998] *A Primer of Infinitesimal Analysis* (Cambridge: Cambridge University Press).

Benacerraf, Paul [1973] 'Mathematical truth', reprinted in Benacerraf and Putnam [1983], pp. 403–420.

——and Putnam, Hilary, eds. [1983] *Philosophy of Mathematics*, second edition (Cambridge: Cambridge University Press).

Berkeley, George [1710] *A Treatise Concerning the Principles of Human Understanding*, ed. J. Dancy (Oxford: Oxford University Press, 1998).

——[1734] *The Analyst*, in *The Works of George Berkeley, Bishop of Cloyne*, ed. A. A. Luce and T. E. Jessop (Edinburgh: Nelson, 1948–57), volume 4, pp. 65–102.

Bernays, Paul [1935] 'On platonism in mathematics', reprinted in Benacerraf and Putnam [1983], pp. 258–271.

Bird, Graham [1962] *Kant's Theory of Knowledge* (London: Routledge and Kegan Paul).

Blackburn, Simon [1994] *The Oxford Dictionary of Philosophy* (Oxford: Oxford University Press).

——and Simmons, Keith, eds. [1999] *Truth* (Oxford: Oxford University Press).

Block, Ned [1995] 'On a confusion about the function of consciousness', *Behavioral and Brain Sciences* 18, pp. 227–247.

Bloom, Paul [1998] 'Some issues in the evolution of language and thought', in Cummins and Allen [1998], pp. 204–223.

——[2000] 'Object names and other common nouns', chapter 4 of *How Children Learn the Meanings of Words* (Cambridge, Mass.: MIT Press), pp. 89–120.

Boghossian, Paul, and Peacocke, Christopher, eds. [2000] *New Essays on the A Priori* (Oxford: Oxford University Press).

Bolzano, Bernard [1837] *Theory of Science: Attempt at a Detailed and in the Main Novel Exposition of Logic with Constant Attention to Earlier Authors*, ed. and trans. R. George (Berkeley: University of California Press, 1972).

Borstein, Marc H. [1985] 'Habituation of attention as a measure of visual information processing in human infants: summary, systematization, and synthesis', in Gottlieb and Krasnegor [1985], pp. 253–300.

Boyd, Richard [1983] 'On the current status of scientific realism', reprinted in R. Boyd, P. Gasper, and J. D. Trout, eds., *The Philosophy of Science* (Cambridge, Mass.: MIT Press, 1991), pp. 195–222.

——[1990] 'Realism, approximate truth, and philosophical method', reprinted in Papineau [1996], pp. 215–255.

Boyer, Carl [1949] *The History of the Calculus and its Conceptual Development* (New York: Dover, 1959).

Braine, Martin, and Rumain, Barbara [1983] 'Logical reasoning', in Mussen [1983], pp. 263–340.

Brandt, Reinhardt [1995] *The Table of Judgments: Critique of Pure Reason A67–76/B92–101*, trans. E. Watkins (Atascadero, Calif.: Ridgeview Publishing Company).

Broughton, Janet [1983] 'Hume's skepticism about causal inferences', *Pacific Philosophical Quarterly* 64, pp. 3–18.

——[1992] 'What does the scientist of man observe?', *Hume Studies* 18, pp. 155–168.

——[2002] *Descartes's Method of Doubt* (Princeton, NJ: Princeton University Press).

——[2003] 'Hume's naturalism about cognitive norms', *Philosophical Topics* 31, pp. 1–19.

Burgess, J. A. [1990a] 'Vague objects and indefinite identity', *Philosophical Studies* 59, pp. 263–287.

——[1990b] 'The sorites paradox and higher-order vagueness', *Synthese* 85, pp. 417–474.

Burgess, John [1990] 'Epistemology and nominalism', in Irvine [1990], pp. 1–15.

——[2004] 'Mathematics and *Bleak House*', *Philosophia Mathematica* 12, pp. 18–36.

——and Rosen, Gideon [1997] *A Subject with no Object* (Oxford: Oxford University Press).

——[2005] 'Nominalism reconsidered', in Shapiro [2005], pp. 515–535.

Carey, Susan [1995] 'Continuity and discontinuity in cognitive development', in D. Osherson and E. Smith, eds., *Thinking: An Invitation to Cognitive Science*, second edition, volume 3 (Cambridge, Mass.: MIT Press), pp. 101–129.

——and Xu, Fei [2001] 'Infant's knowledge of objects: beyond object files and object tracking', *Cognition* 80, pp. 179–213.

Carnap, Rudolph [1928] *Logical Structure of the World*, trans. R. George (Berkeley: University of California Press, 1967).

——[1934] *Logical Syntax of Language*, trans. A. Smeaton (London: Routledge and Kegan Paul, 1937).

——[1950a] 'Empiciricism, semantics and ontology', reprinted in Benacerraf and Putnam [1983], pp. 241–257.

——[1950b] *Logical Foundations of Probability* (Chicago: University of Chicago Press).

——[1963] 'Intellectual autobiography' and 'Replies and systematic expositions', in P. A. Schilpp, ed., *The Philosophy of Rudolf Carnap* (La Salle, Ill.: Open Court), pp. 3–84, 859–1013.

——[1966] *An Introduction to the Philosophy of Science*, ed. M. Gardner (New York: Dover, 1995).

Carroll, Lewis [1895] 'What the tortoise said to Achilles', reprinted in his *Symbolic Logic*, ed. W. W. Bartley (New York: Clarkson N. Potter, Inc., Publishers, 1977), pp. 431–434.

Cartwright, Nancy [1983] *How the Laws of Physics Lie* (Oxford: Oxford University Press).

Case, Robbie [1999] 'Conceptual development in the child and in the field: a personal view of the Piagetian legacy', in Scholnick et al. [1999], pp. 23–51.

Chihara, Charles [1973] *Ontology and the Vicious Circle Principle* (Ithaca, NY: Cornell University Press).

——[1990] *Constructibility and Mathematical Existence* (Oxford: Oxford University Press).

Churchland, Paul, and Hooker, Clifford, eds. [1985] *Images of Science* (Chicago: University of Chicago Press).

Clark, Andy [1989] *Microcognition* (Cambridge, Mass.: MIT Press).

——[1997] *Being There: Putting Brain, Body, and World Together Again* (Cambridge, Mass.: MIT Press).

Coffa, Alberto [1991] *The Semantic Tradition from Kant to Carnap: To the Vienna Station* (Cambridge: Cambridge University Press).

Cohen, Leslie [1988] 'An information-processing approach to infant cognitive development', in Weiskrantz [1988], pp. 211–228.

——and Younger, Barbara [1983] 'Perceptual categorization in the infant', in Scholnick [1983], pp. 197–220.

Colyvan, Mark [2001] *The Indispensability of Mathematics* (New York: Oxford University Press).

——[2005] 'Ontological independence as the mark of the real', review of Azzouni [2004], *Philosophia Mathematica* 13, pp. 216–225.

Cummins, Denise, and Allen, Colin [1998] *The Evolution of Mind* (New York: Oxford University Press).

dalla Chiara, Maria Luisa, and Guintini, Roberto [2002] 'Quantum logics', in Gabbay and Guenthner [2002], pp. 129–228.

Dedekind, Richard [1872] *Essays on the Theory of Numbers* (La Salle, Ill.: Open Court, 1948).

Dehaene, Stanislaus [1997] *The Number Sense* (New York: Oxford University Press).

——Dehaene-Lambertz, Ghislaine, and Cohen, Laurent [1998] 'Abstract representations of numbers in the animal and human brain', *Trends in Neurosciences* 21, pp. 355–361.

——Spelke, Elizabeth, Pinel, P., Stanescu, R., and Tsivkin, S. [1999] 'Sources of mathematical thinking: behavioral and brain-imaging evidence', *Science* 284, pp. 970–974.

DeRose, Keith, and Warfield, Ted, eds. [1999] *Skepticism* (New York: Oxford University Press).

Descartes, René [1641a] *Meditations on First Philosophy*, in *The Philosophical Writings of Descartes*, trans. J. Cottingham et al., volume 2 (Cambridge: Cambridge University Press, 1984), pp. 3–62.

——[1641b] 'Letter to Mersenne', in *The Philosophical Writings of Descartes*, trans. J. Cottingham et al., volume 3 (Cambridge: Cambridge University Press, 1991), pp. 171–173.

——[1642] *Objections and Replies*, in *The Philosophical Writings of Descartes*, trans. J. Cottingham et al., volume 2 (Cambridge: Cambridge University Press, 1984), pp. 66–383.

——[1644] *Principles of Philosophy*, in *The Philosophical Writings of Descartes*, trans. J. Cottingham et al., volume 1 (Cambridge: Cambridge University Press, 1985), pp. 179–291.

Detlefsen, Michael [1986] *Hilbert's Program* (Dordrecht: Kluwer).

Devitt, Michael [1981] *Designation* (New York: Columbia University Press).

Diaconis, Persi [1978] 'Statistical problems in ESP research', *Science* 201, pp. 131–136.

——and Mosteller, Frederick [1989] 'Methods for studying coincidences', *Journal of the American Statistical Association* 84, pp. 853–861.

Dieks, Dennis [1990] 'Quantum statistics, identical particles, and correlations', *Synthese* 82, pp. 127–155.

Dieterle, J. M. [1999] 'Mathematical, astrological, and theological naturalism', *Philosophia Mathematica* 7, pp. 129–135.

DiSalle, Robert [2002] 'Reconsidering Kant, Friedman, Logical Positivism, and the Exact Sciences', *Philosophy of Science* 69, pp. 191–211.

——[2006] *Understanding Spacetime: The Philosophical Development of Physics from Newton to Einstein* (Cambridge: Cambridge University Press).

Duhem, Pierre [1906] *The Aim and Structure of Physical Theory*, trans. P. Weiner (Princeton: Princeton University Press, 1954).

Dummett, Michael [1963] 'Realism', reprinted in his [1978], pp. 145–165.

——[1973a] 'The philosophical basis of intuitionistic logic', reprinted in his [1978], pp. 215–247. (Also reprinted in Benacerraf and Putnam [1983], pp. 97–129.)

——[1973b] 'The justification of deduction', reprinted in his [1978], pp. 290–318.

——[1975] 'Wang's paradox', reprinted in his [1978], pp. 248–268, and in Keefe and Smith [1997a], pp. 99–118. (Page references to the 1997 version.)

——[1978] *Truth and Other Enigmas* (Cambridge, Mass.: Harvard University Press).

——[1981] *The Interpretation of Frege's Philosophy* (Cambridge, Mass.: Harvard University Press).

Dunn, J. Michael [1986] 'Relevance logic and entailment', in Gabbay and Guenthner [1986], pp. 117–224.

Dunn, J. Michael and Restall, Greg [2002] 'Relevance logic', in Gabbay and Guenthner [2002], pp. 1–128.

Dyson, Freeman [1964] 'Mathematics in the physical sciences', reprinted in Kline [1968b], pp. 249–257.

Earman, John, and Fine, Arthur [1977] 'Against indeterminacy', *Journal of Philosophy* 87, pp. 535–538.

Ebbs, Gary [1997] *Rule-Following and Realism* (Cambridge, Mass.: Harvard University Press).

Edgington, Dorothy [2001] 'Conditionals', in Goble [2001], pp. 385–414.

Einstein, Albert [1949] 'Autobiographical notes' and 'Reply to criticisms' in Schilpp [1949], pp. 1–94, 663–688.

—— and Infeld, Leopold [1938] *The Evolution of Physics* (New York: Simon and Schuster, 1966).

Enderton, Herbert [1972] *A Mathematical Introduction to Logic* (San Diego: Academic Press).

—— [1977] *Elements of Set Theory* (New York: Academic Press).

Etchemendy, John [1988] 'Tarski on truth and logical consequence', *Journal of Symbolic Logic* 53, pp. 51–79.

Evans, Gareth [1978] 'Can there be vague objects?', reprinted in Keefe and Smith [1997], p. 317.

Evans, Jonathan, Newstead, Stephen, and Byrne, Ruth [1993] *Human Reasoning: The Psychology of Deduction* (Hillsdale, NJ: Lawrence Erlbaum Associates).

Feferman, Solomon [1964] 'Systems of predicative analysis', *Journal of Symbolic Logic* 29, pp. 1–30.

—— [1987] 'Infinity in mathematics: is Cantor necessary?', reprinted in his *In the Light of Logic* (New York: Oxford University Press, 1998), pp. 28–73.

Feigenson, Lisa, Carey, Susan, and Spelke, Elizabeth [2002] 'Infants' discrimination of number vs. continuous extent', *Cognitive Psychology* 44, pp. 33–66.

Feynman, Richard [1965] *The Character of Physical Law* (Cambridge, Mass.: MIT Press, 1967).

—— [1998] *The Meaning of it All: Thoughts of a Citizen-Scientist* (Reading, Mass.: Perseus Books).

—— Leighton, Robert, and Sands, Matthew [1964] *The Feynman Lectures on Physics: Mainly Electromagnetism and Matter* (Reading, Mass.: Addison-Wesley Publishing Company).

—— [1965] *The Feynman Lectures on Physics: Quantum Mechanics* (Reading, Mass.: Addison-Wesley Publishing Company).

Field, Hartry [1972] 'Tarski's theory of truth', reprinted with a new postscript in his [2001], pp. 3–29.

——[1973] 'Theory change and indeterminacy of reference', reprinted with a new postscript in his [2001], pp. 177–198.

——[1980] *Science without Numbers* (Oxford: Basil Blackwell).

——[1986a] 'The deflationary conception of truth', in G. McDonald and C. Wright, eds., *Fact, Science and Morality* (Oxford: Blackwell Publishers), pp. 55–117.

——[1986b] 'Correspondence truth, disquotational truth, and deflationism', excerpted with changes from [1986a], in Lynch [2001], pp. 483–503.

——[1989] *Realism, Mathematics and Modality* (Oxford: Basil Blackwell).

——[1992] 'Physicalism', in J. Earman, ed., *Inference, Explanation, and Other Frustrations* (Los Angeles: University of California Press), pp. 271–291.

——[1994a] 'Deflationist views of meaning and content', reprinted with a new postscript in his [2001], pp. 104–156.

——[1994b] 'Disquotational truth and factually defective discourse', in his [2001], pp. 222–258.

——[1998a] 'Some thoughts on radical indeterminacy', reprinted with a new postscript in his [2001], pp. 259–277.

——[1998b] 'Which undecidable mathematical sentences have determinate truth values?', reprinted with new postscript in his [2001], pp. 332–360.

——[1998c] 'Mathematical objectivity and mathematical objects', in S. Laurence and C. MacDonald, eds., *Contemporary Readings in the Foundations of Metaphysics* (Oxford: Blackwell), pp. 387–403.

——[2000] 'Indeterminacy, degree of belief, and excluded middle', reprinted with a new postscript in his [2001], pp. 278–311.

——[2001] *Truth and the Absence of Fact* (Oxford: Oxford University Press).

Fine, Arthur [1986] *The Shaky Game* (Chicago: University of Chicago Press).

Fine, Kit [1975] 'Vagueness, truth and logic', reprinted in Keefe and Smith [1997a], pp. 119–150.

Fodor, Jerry [1975] *The Language of Thought* (New York: Thomas Crowell).

——and McLaughlin, Brian [1990] 'Connectionism and the problem of systematicity; why Smolensky's solution doesn't work', *Cognition* 35, pp. 183–204.

——and Pylyshyn, Zenon [1988] 'Connectionism and cognitive architecture', *Cognition* 28, pp. 2–71.

Fogelin, Robert [1997] 'Quine's limited naturalism', *Journal of Philosophy* 94, pp. 543–563.

FOM Foundations of Mathematics e-mail list: www.cs.nyu.edu/mailman/listinfo/fom/.

Fourier, Joseph [1822] *The Analytic Theory of Heat*, trans. A. Freeman (Mineola, NY: Dover, 1955).

Fourman, Michael [1977] 'The logic of topoi', in Barwise [1977], pp. 1053–1090.
Fraenkel, Abraham, Bar-Hillel, Yehoshua, and Levy, Azriel [1973] *Foundations of Set Theory*, second edition (Amsterdam: North Holland).
Frege, Gottlob [1879] *Begriffsschrift*, trans. S. Bauer-Mengelberg, in van Heijenoort [1967], pp. 5–82.
——[1880/1] 'Boole's logical calculus and the concept-script', in his [1979], pp. 9–52.
——[1884] *Foundations of Arithmetic*, second edition, trans. J. L. Austin (Evanston, Ill.: Northwestern University Press, 1968).
——[1891] 'Function and concept', reprinted in his [1997], pp. 130–148.
——[1892] 'On *Sinn* and *Bedeutung*', reprinted in his [1997], pp. 151–171.
——[1893] *Grundgesetze der Arithmetik*, volume 1 (Jena: H. Pohle), selections translated as his [1964] and by M. Beaney in Frege [1997], pp. 194–223. Page references are to the 1964 version.
——[1897] 'Logic', in his [1979], pp. 126–151.
——[1903] *Grundgesetze der Arithmetic*, volume 2 (Jena: H. Pohle), selections translated by P. Geach in Frege [1964], pp. 127–143, and by M. Beaney in Frege [1997], pp. 258–289. Page references are to the 1964 version.
——[1906] 'A brief survey of my logical doctrines', in his [1979], pp. 197–202.
——[1918] 'The thought', reprinted in his [1997], pp. 325–345.
——[1924/5a] 'Sources of knowledge of mathematics and the mathematical natural sciences', in his [1979], pp. 267–274.
——[1924/5b] 'Numbers and arithmetic', in his [1979], pp. 275–277.
——[1924/5c] 'A new attempt at a foundation of arithmetic', in his [1979], pp. 278–281.
——[1964] *Basic Laws of Arithmetic*, trans. and ed. M. Furth (Berkeley: University of California Press).
——[1979] *Posthumous Writings*, ed. H. Hermes, F. Kambartel, and F. Kaulbach, trans. P. Long and R. White (Chicago: University of Chicago Press).
——[1997] *The Frege Reader*, ed. M. Beaney (Oxford: Blackwell).
Friedman, Michael [1979] 'Truth and confirmation', *Journal of Philosophy* 76, pp. 361–382.
——[1988] 'Analytic truth in Carnap's *Logical Syntax of Language*', reprinted in his [1999b], pp. 165–176.
——[1995] 'Carnap and Weyl on the foundations of geometry and relativity theory', reprinted in his [1999b], pp. 44–58.
——[1997] 'Carnap and Wittgenstein's *Tractatus*', reprinted in his [1999b], pp. 177–197.
——[1999a] 'Tolerance and analyticity in Carnap's philosophy of mathematics', reprinted in his [1999b], pp. 198–233.

——[1999b] *Reconsidering Logical Positivism* (Cambridge: Cambridge University Press).

——[2001] *Dynamics of Reason* (Stanford, Calif.: CSLI Publications).

——[2002] 'Kant, Kuhn, and the rationality of science', *Philosophy of Science* 69, pp. 171–190.

Gabbay, D. M., and Guenthner, F., eds. [1986] *Handbook of Philosophical Logic*, volume 3 (Dordrecht: Kluwer Academic Publishers).

——[2001] *Handbook of Philosophical Logic*, second edition, volume 2 (Dordrecht: Kluwer Academic Publishers).

——[2002] *Handbook of Philosophical Logic*, second edition, volumes 5 and 6 (Dordrecht: Kluwer Academic Publishers).

Galison, Peter [1997] *Imagine and Logic: A Material Culture of Microphysics* (Chicago: University of Chicago Press).

Garber, Daniel [1986] '*Semel in vita*: the scientific background to Descartes' *Meditations*', in A. Rorty, ed., *Essays on Descartes Meditations* (Berkeley: University of California Press).

Gardner, Michael [1972] 'Two deviant logics for quantum theory: Bohr and Reichenbach', *British Journal for the Philosophy of Science* 23, pp. 89–109.

Gelman, R., and Gallistel, C. R. [1978] *The Child's Understanding of Number* (Cambridge, Mass.: Harvard University Press).

George, Rolf [1972] 'Editor's introduction' to Bolzano [1837], pp. xxiii–xlvi.

Gibbins, Peter [1987] *Particles and Paradoxes: The Limits of Quantum Logic* (Cambridge: Cambridge Univeristy Press).

Giere, Ronald, and Richardson, Alan, eds. [1996] *Origins of Logical Empiricism*, Minnesota Studies in the Philosophy of Science 16 (Minneapolis: University of Minnesota Press).

Glymour, Clark [1980] *Theory and Evidence* (Princeton: Princeton University Press).

Goble, Lou, ed. [2001] *The Blackwell Guide to Philosophical Logic* (Malden, Mass.: Blackwell).

Gödel, Kurt [1944] 'Russell's mathematical logic', reprinted in his [1990], pp. 119–141, and in Benacerraf and Putnam [1983], pp. 447–469. (Page references are to the 1990 version.)

——[1964] 'What is Cantor's continuum problem?', reprinted in his [1990], pp. 254–270, and in Benacerraf and Putnam [1983], pp. 470–485. (Page references are to the 1990 version.)

——[1990] *Collected Works*, volume 2, S. Feferman et al., eds. (New York: Oxford University Press).

Goldfarb, Warren [1997] 'Semantics in Carnap: a rejoinder to Alberto Coffa', *Philosophical Topics* 25, pp. 51–66.

Goldfarb, Warren, and Ricketts, Thomas [1992] 'Carnap and the philosophy of mathematics', in D. Bell and W. Vossenkuhl, eds., *Science and Subjectivity* (Berlin: Akademie), pp. 61–78.

Goodman, Nelson, and Quine, W. V. [1947] 'Steps toward a constructive nominalism', *Journal of Symbolic Logic* 12, pp. 105–122.

Gopnik, Alison, and Meltzoff, Andrew [1997] *Words, Thoughts, and Theories* (Cambridge, Mass.: MIT Press).

Gottlieb, Gilbert, and Krasnegor, Norman, eds. [1985] *Measurement of Audition and Vision in the First Year of Postnatal Life* (Norwood, NJ: Ablex).

Grabiner, Judith [1981] *The Origins of Cauchy's Rigorous Calculus* (Cambridge, Mass.: MIT Press).

Graff, Delia, and Williamson, Timothy [2002] *Vagueness* (Brookfield, Vt.: Dartmouth Publishing Company).

Granrud, Carl, ed. [1993] *Visual Perception and Cognition in Infancy* (Hillsdale, NJ: Lawrence Erlbaum).

Gregory, Frederick [1977] *Scientific Materialism in Nineteenth Century Germany* (Dordrecht: Reidel Publishing Company).

Guyer, Paul [1987] *Kant and the Claims of Knowledge* (Cambridge: Cambridge University Press).

Haack, Susan [1976] 'The justification of deduction', reprinted in her [1996], pp. 183–191.

——[1982] 'Dummett's justification of deduction', reprinted in her [1996], pp. 192–213.

——[1996] *Deviant Logic, Fuzzy Logic: Beyond the Formalism* (Chicago: University of Chicago Press).

Hale, Bob, and Wright, Crispin [2001] *Reason's Proper Study: Essays towards a Neo-Fregean Philosophy of Mathematics* (Oxford: Oxford University Press).

Halliday, David, Resnick, Robert, and Walker, Jearl [2005] *Fundamentals of Physics*, extended seventh edition (Hoboken, NJ: John Wiley and Sons).

Hamming, R. W. [1980] 'The unreasonable effectiveness of mathematics', *American Mathematical Monthly* 87, pp. 81–90.

Harman, Gilbert [1965] 'Inference to the best explanation', *Philosophical Review* 74, pp. 88–95.

Harman, P. M. [1998] *The Natural Philosophy of James Clerk Maxwell* (Cambridge: Cambridge University Press).

Hauser, Marc, and Carey, Susan [1998] 'Building a cognitive creature from a set of primitives', in Cummins and Allen [1998], pp. 51–100.

Helmholtz, Hermann [1867] *Treatise on Physiological Optics*, trans. J. P. C. Southall (New York: Dover, 1962).

Horwich, Paul [1990/8] *Truth*, first and second editions (Oxford: Oxford University Press).

Hrdy, Sarah Blaffer [1999] *The Woman that Never Evolved*, updated edition (Cambridge, Mass.: Harvard University Press).

Hughes, R. I. G. [1989] *The Structure and Interpretation of Quantum Mechanics* (Cambridge, Mass.: Harvard University Press).

Hume, David [1739] *A Treatise of Human Nature*, ed. D. F. Norton and M. J. Norton (Oxford: Oxford University Press, 2000).

____[1748] *An Enquiry Concerning Human Understanding*, ed. T. L. Beauchamp (Oxford: Oxford University Press, 1999).

Huntley-Fenner, Gavin, Carey, Susan, and Solimando, Andrea [2002] 'Objects are individuals but stuff doesn't count: perceived rigidity and cohesiveness influence infants' representations of small groups of discrete entities', *Cognition* 85, pp. 203–221.

Ihde, Aaron J. [1964] *The Development of Modern Chemistry* (New York: Dover, 1984).

Irvine, Andrew [1990] *Physicalism in Mathematics* (Dordrecht: Kluwer).

Jackson, Reginald [1941] *An Examination of the Deductive Logic of John Stuart Mill* (London: Oxford University Press).

James, William [1890] *The Principles of Psychology*, vol. 1 (New York: Dover, 1950).

Jeffrey, Richard [2002] 'Logicism lite', *Philosophy of Science* 69, pp. 447–451.

Johnsen, Bredo [2005] 'How to read "Epistemology naturalized"', *Journal of Philosophy* 102, pp. 78–93.

Johnson, Kent [2004] 'On the systematicity of language and thought', *Journal of Philosophy* 101, pp. 111–139.

Kahneman, Daniel, Treisman, Anne, and Gibbs, Brian [1992] 'The reviewing of object files: object-specific integration of information', *Cognitive Psychology* 24, pp. 175–219.

Kant, Immanuel [1772] 'Letter to Marcus Hertz', in *Kant: Philosophical Correspondence 1759–99*, ed. and trans. A. Zweig (Chicago: University of Chicago Press, 1967), pp. 70–76.

____[1781/7] *The Critique of Pure Reason*, trans. P. Guyer and A. Wood (Cambridge: Cambridge University Press, 1997).

____[1783] *Prolegomena to Any Future Metaphysics*, trans. G. Hatfield, revised edition (Cambridge: Cambridge University Press, 2004).

____[1790a] *Critique of the Power of Judgment*, ed. P. Guyer, trans. P. Guyer and E. Matthews (Cambridge: Cambridge University Press, 2000).

____[1790b] 'On a discovery according to which any new critique of pure reason has been made superfluous by an earlier one', trans. and repr. in Allison [1973], pp. 105–60.

Kant, Immanuel [1792] 'The Dohna-Wundlacken Logic', in Kant [1992], pp. 425–516.

—— [1800] 'The Jäsche Logic', in Kant [1992], pp. 517–640.

—— [1992] *Lectures on Logic*, ed. and trans. J. M. Young (Cambridge: Cambridge University Press).

Keefe, Rosanna, and Smith, Peter, eds. [1997a] *Vagueness* (Cambridge, Mass.: MIT Press).

—— [1997b] 'Introduction', in Keefe and Smith [1997a], pp. 1–57.

Keller, Joseph [1986] 'The probability of heads', *American Mathematical Monthly* 93, pp. 191–197.

Kellman, Philip [1993] 'Kinematic foundations of infant visual perception', in Granrud [1993], pp. 121–173.

—— and Spelke, Elizabeth [1983] 'Perception of partly occluded objects in infancy', *Cognitive Psychology* 15, pp. 483–524.

Kitcher, Philip [1978] 'Theories, theorists, and theoretical change', *Philosophical Review* 87, pp. 519–547.

—— [1982] *Abusing Science: The Case Against Creationism* (Cambridge, Mass.: MIT Press).

—— [1993] *The Advancement of Science* (New York: Oxford University Press).

Klahr, David [1999] 'The conceptual habitat: in what kind of system can concepts develop?', in Scholnick et al. [1999], pp. 131–161.

Kline, Morris [1968a] 'The import of mathematics', in Kline [1968b], pp. 232–237.

—— ed. [1968b] *Mathematics in the Modern World* (San Francisco: W. H. Freeman and Company).

—— [1972] *Mathematical Thought from Ancient to Modern Times* (New York: Oxford University Press).

—— [1980] *Mathematics: The Loss of Certainty* (New York: Oxford University Press).

Koellner, Peter [2006] 'On the question of absolute undecidability', *Philosophia Mathematica* 14, 153–188.

Koperski, Jeffrey [2001] 'Has chaos been explained?', *British Journal for the Philosophy of Science* 52, pp. 683–700.

Körner, Stephan [1969] *Fundamental Questions of Philosophy: One Philosopher's Answers* (Harmondsworth: Penguin Books).

Kripke, Saul [1972] 'Naming and necessity', in D. Davidson and G. Harman, eds., *Semantics of Natural Language*, second edition (Dordrecht: Reidel Publishing Company), pp. 253–355.

—— [1982] *Wittgenstein on Rules and Private Language* (Cambridge, Mass.: Harvard University Press).

Kuhn, Thomas [1962] *The Structure of Scientific Revolutions*, third edition (Chicago: University of Chicago Press, 1996).

Lambert, Karel [2001] 'Free logics', in Goble [2001], pp. 258–279.

Lange, Marc [2002] *An Introduction to the Philosophy of Physics* (Malden, Mass.: Blackwell).

Leeds, Stephen [1978] 'Theories of truth and reference', *Erkenntnis* 13, pp. 111–129.

—— [1995] 'Truth, correspondence, and success', *Philosophical Studies* 79, pp. 1–36.

—— [1997] 'Incommensurability and vagueness', *Nous* 31, pp. 385–407.

Leslie, Alan [1988] 'The necessity of illusion: perception and thought in infancy', in Weiskrantz [1988], pp. 185–210.

—— and Keeble, Stephanie [1987] 'Do six-month-old infants perceive causality?', *Cognition* 25, pp. 265–288.

Levy, Azriel, and Solovay, Robert [1967] 'Measurable cardinals and the continuum hypothesis', *Israel Journal of Mathematics* 5, pp. 234–248.

Lewis, David [1969] *Convention* (Cambridge, Mass.: Harvard University Press).

—— [1996] 'Elusive knowledge', reprinted in Derose and Warfield [1999], pp. 220–239.

Liston, Michael [2000] 'Review of Steiner [1998]', *Philosophia Mathematica* 8, pp. 190–207.

Locke, John [1690] *An Essay Concerning Human Understanding*, ed. A. C. Fraser (New York: Dover, 1959).

Longuenesse, Beatrice [1993] *Kant and the Capacity to Judge*, trans. C. T. Wolfe (Princeton: Princeton University Press, 1998).

Löwe, Benedikt [2001] 'A first glance at non-restrictiveness', *Philosophia Mathematica* 9, pp. 347–354.

—— [2003] 'A second glance at non-restrictiveness', *Philosophica Mathematica* 11, pp. 323–331.

Lynch, Michael, ed. [2001] *The Nature of Truth* (Cambridge, Mass.: MIT Press).

MacFarlane, John [2002] 'Frege, Kant, and the logic in logicism', *Philosophical Review* 111, pp. 25–65.

Macnamara, John [1986] *A Border Dispute: The Place of Logic in Psychology* (Cambridge, Mass.: MIT Press).

Maddy, Penelope [1990] *Realism in Mathematics* (Oxford: Oxford University Press).

—— [1996] 'Ontological commitment: between Quine and Duhem', in J. Tomberlin, ed., *Philosophical Perspectives* 10, *Metaphysics 1996* (Cambridge, Mass.: Blackwell), pp. 317–341.

—— [1997] *Naturalism in Mathematics* (Oxford: Oxford University Press).

Maddy, Penelope [1999] 'Logic and the discursive intellect', *Notre Dame Journal of Formal Logic* 40, pp. 94–115.

——[2000] 'Naturalism and the a priori', in Boghossian and Peacocke [2000], pp. 92–116.

——[2001a] 'Some naturalistic reflections on set theoretic method', *Topoi* 20 (2001), pp. 17–27.

——[2001b] 'Naturalism: friends and foes', in J. Tomberlin, ed., *Philosophical Perspectives* 15, *Metaphysics 2001* (Malden, Mass.: Blackwell, 2001), pp. 37–67.

——[2002] 'A naturalistic look at logic', *Proceedings of the American Philosophical Association* 76, pp. 61–90.

——[2003] 'Second philosophy', *Journal of the Indian Council of Philosophical Research* 20 (2003), pp. 73–106.

——[2005a] 'Three forms of naturalism', in S. Shapiro, ed., *Oxford Handbook of Philosophy of Mathematics and Logic* (Oxford: Oxford University Press), pp. 437–459.

——[2005b] 'Mathematical existence', *Bulletin of Symbolic Logic* 11, pp. 351–376.

Malament, David [1996] 'In defense of a dogma: why there cannot be a relativistic quantum mechanics of (localizable) particles', in R. Clifton, ed., *Perspectives on Quantum Reality* (Dordrecht: Kluwer Academic Publishers), pp. 1–10.

Malinowski, Grzegorz [2001] 'Many-valued logics', in Goble [2001], pp. 309–335.

Mares, Edwin [2006] 'Relevance logic', in *The Stanford Encyclopedia of Philosophy (Spring 2006 Edition)*, ed. Edward N. Zalta, <http://plato.stanford.edu/archives/spr2006/entries/logic-relevance/>.

——and Meyer, Robert [2001] 'Relevant logics', in Goble [2001], pp. 280–308.

Marino, Patricia [2002] '*Language and the world: correspondence vs. deflationary theories of truth*', Ph.D. dissertation, University of California, Irvine.

——[2005] 'Expressivism, deflationism, and correspondence', *Journal of Moral Philosophy* 2, pp. 171–191.

——[2006] 'What should a correspondence theory be and do?', *Philosophical Studies* 127, pp. 415–457.

——[forthcoming] 'Minimalism, anti-realism and quasi-realism', in preparation.

Markman, Ellen [1983] 'Two different kinds of hierarchical organization', in Scholnick [1983], pp. 165–184.

Martin, D. A. [1998] 'Mathematical evidence', in H. G. Dales and G. Oliveri, eds., *Truth in Mathematics* (Oxford: Oxford University Press), pp. 215–231.

——[2005] 'Gödel's conceptual realism', *Bulletin of Symbolic Logic* 11, pp. 207–224.

Mates, Benson [1972] *Elementary Logic*, second edition (New York: Oxford University Press).

Mathias, A. R. D. [2001] 'Slim models of Zermelo set theory', *Journal of Symbolic Logic* 66, pp. 487–496.

Matthews, H. E. [1969] 'Strawson on transcendental idealism', *Philosophical Quarterly* 19, pp. 204–220.

Meck, W. H., and Church, R. M. [1983] 'A mode control model of counting and timing processes', *Journal of Experimental Psychology: Animal Behavior Processes* 9, pp. 320–334.

Melia, Joseph [2000] 'Weaseling away the indispensability argument', *Mind* 109, pp. 455–479.

Meltzoff, Andrew, and Moore, M. Keith [1999] 'A new foundation for cognitive development in infancy: the birth of the representational infant', in Scholnick et al. [1999], pp. 53–78.

Michens, Ronald, ed. [1990] *Mathematics and Science* (Singapore: World Scientific).

Mill, John Stuart [1865] *An Examination of Sir William Hamilton's Philosophy*, selections in *John Stuart Mill's Philosophy of Scientific Method*, ed. E. Nagel (New York: Hafner, 1950), pp. 359–404.

Miller, Richard [1987] *Fact and Method* (Princeton: Princeton University Press).

Moore, G. E. [1925] 'A defence of common sense', in his [1959], pp. 32–59.

—— [1939] 'Proof of an external world', in his [1959], pp. 127–150.

—— [1959] *Philosophical Papers* (London: George Allen and Unwin Ltd.).

Musgrave, Alan [1972] 'George Boole and psychologism', *Scientia* 107, pp. 593–608.

Mussen, Paul, ed. [1983] *Handbook of Child Psychology*, fourth edition, volume 3 (New York: John Wiley and Sons).

Narens, Louis, and Luce, Duncan [1990] 'Three aspects of the effectiveness of mathematics in science', in Michens [1990], pp. 122–135.

Norton, John [2003] 'A material theory of induction', *Philosophy of Science* 70, pp. 647–670.

Nye, Mary Jo [1972] *Molecular Reality* (New York: Elsevier).

Oakes, Lisa, and Cohen, Leslie [1990] 'Infant perception of a causal event', *Cognitive Development* 5, pp. 193–207.

Oldershaw, Robert [1990] 'Mathematics and natural philosophy', in Michens [1990], pp. 136–153.

Pais, Abraham [1982] *Subtle is the Lord: The Science and the Life of Albert Einstein* (Oxford: Oxford University Press).

—— [1986] *Inward Bound: Of Matter and Forces in the Physical World* (Oxford: Oxford University Press).

—— [1991] *Niels Bohr's Times, in Physics, Philosophy and Polity* (Oxford: Oxford University Press).

—— [1998] 'Paul Dirac: aspects of his life and work', in Pais et al. [1998], pp. 1–45.

Pais, Abraham Jacob, Maurice, Olive, David, and Atiyah, Michael [1998] *Paul Dirac: The Man and his Work* (Cambridge: Cambridge University Press).

Papineau, David, ed. [1996] *The Philosophy of Science* (Oxford: Oxford University Press).

Parsons, Charles [1990] 'The structuralist view of mathematics', *Synthese* 84, pp. 303–346.

—— [1995] 'Structuralism and the concept of set', in W. Sinnott-Armstrong et al., eds., *Modality, Morality and Belief* (Cambridge: Cambridge University Press), pp. 74–92.

—— [2004] 'Structuralism and metaphysics', *Philosophical Quarterly* 54, pp. 56–77.

Parsons, Terence [2000] *Indeterminate Identity* (Oxford: Oxford University Press).

—— and Woodruff, Peter [1995] 'Worldly indeterminacy of identity', reprinted in Keefe and Smith [1997a], pp. 321–337.

Partington, J. R. [1957] *A Short History of Chemistry*, third edition (New York: Dover, 1989).

Pearce, John [1988] 'Stimulus generalization and the acquisition of categories in pigeons', in Weiskrantz [1988], pp. 132–155.

Perrin, Jean [1913] *Atoms*, trans. D. Hammick (Woodbridge, Conn.: Ox Bow Press).

Piaget, Jean [1937] *The Construction of Reality in the Child*, trans. M. Cook (New York: Basic Books, 1954).

Pitcher, George [1977] *Berkeley* (London: Routledge and Kegan Paul).

Prichard, H. A. [1909] *Kant's Theory of Knowledge* (Oxford: Oxford University Press).

Priest, Graham [2001] *An Introduction to Non-Classical Logic* (Cambridge: Cambridge University Press).

—— [2002] 'Paraconsistent logic', in Gabbay and Guenthner [2002], volume 6, pp. 287–393.

—— [2004] 'Dialetheism', in *The Stanford Encyclopedia of Philosophy (Summer 2004 Edition)*, ed. Edward N. Zalta, <http://plato.stanford.edu/archives/sum2004/entries/dialetheism/>.

—— and Tanaka, Koji [2004] 'Paraconsistent Logic', in *The Stanford Encyclopedia of Philosophy (Winter 2004 Edition)*, ed. Edward N. Zalta, <http://plato.stanford.edu/archives/win2004/entries/logic-paraconsistent/>.

Putnam, Hilary [1957] 'Three-valued logic', reprinted in his [1979], pp. 166–173.

—— [1962] 'It ain't necessarily so', reprinted in his [1979], pp. 237–249.

—— [1965] 'Philosophy of physics', reprinted in his [1979], pp. 79–92.

—— [1968] 'The logic of quantum mechanics', reprinted in his [1979], pp. 174–197.

——[1971] 'Philosophy of logic', reprinted in his [1979], pp. 323–357.
——[1975a] 'The meaning of "meaning"', reprinted in his [1975c], pp. 215–271.
——[1975b] 'What is mathematical truth', reprinted in his [1979], pp. 60–78.
——[1975c] *Mind, Language and Reality, Philosophical Papers*, volume 2 (Cambridge: Cambridge University Press).
——[1976a] 'Meaning and knowledge: the John Locke Lectures', in his [1978b], pp. 7–80.
——[1976b] 'Reference and understanding', reprinted in his [1978b], pp. 95–119.
——[1976c] 'Realism and reason', reprinted in his [1978b], pp. 121–140.
——[1977] 'Models and reality', reprinted in his [1983], pp. 1–25.
——[1978a] 'There is at least one *a priori* truth', reprinted in his [1983], pp. 98–114.
——[1978b] *Meaning and the Moral Sciences* (London: Routledge and Kegan Paul).
——[1979] *Mathematics, Matter and Method, Philosophical Papers*, volume 1, second edition (Cambridge: Cambridge University Press).
——[1981a] *Reason, Truth and History* (Cambridge: Cambridge University Press).
——[1981b] 'Why there isn't a ready-made world', reprinted in his [1983], pp. 205–228.
——[1981c] 'Why reason can't be naturalized', reprinted in his [1983], pp. 229–247.
——[1983] *Realism and Reason, Philosophical Papers*, volume 3 (Cambridge: Cambridge University Press).
——[1995] 'Mathematical necessity reconsidered', in P. Leonardi and M. Santambrogio, eds., *On Quine* (Cambridge: Cambridge University Press), pp. 267–282.
Quine, W. V. [1936] 'Truth by convention', reprinted in his [1976], pp. 77–106.
——[1948] 'On what there is', reprinted in his [1980], pp. 1–19.
——[1951] 'Two dogmas of empiricism', reprinted in his [1980], pp. 20–46.
——[1954] 'Carnap and logical truth', reprinted in his [1976], pp. 107–132.
——[1955] 'Posits and reality', reprinted in his [1976], pp. 246–254.
——[1960] *Word and Object* (Cambridge, Mass.: MIT Press).
——[1968] 'Ontological relativity', reprinted in his [1969b], pp. 26–68.
——[1969a] 'Epistemology naturalized', in his [1969b], pp. 69–90.
——[1969b] *Ontological Relativity and Other Essays* (New York: Columbia University Press).
——[1970a] 'Homage to Rudolf Carnap', reprinted in his [1976], pp. 40–43.
——[1970b] *Philosophy of Logic*, first edition (Englewood Cliffs, NJ: Prentice Hall).
——[1974] *The Roots of Reference* (La Salle, Ill.: Open Court).
——[1975] 'Five milestones of empiricism', reprinted in his [1981b], pp. 67–72.
——[1976] *The Ways of Paradox*, revised and enlarged edition (Cambridge, Mass.: Harvard University Press).

Quine, W. V. [1980] *From a Logical Point of View*, second edition (Cambridge, Mass.: Harvard University Press).

——[1981a] 'What price bivalence?', reprinted in his [1981b], pp. 31–37.

——[1981b] *Theories and Things* (Cambridge, Mass.: Harvard University Press).

——[1981c] 'Reply to Stroud', *Midwest Studies in Philosophy* 6, pp. 473–475.

——[1986] *Philosophy of Logic*, second edition (Cambridge, Mass.: Harvard University Press).

——[1990] *Pursuit of Truth* (Cambridge, Mass.: Harvard University Press).

——[1995] *From Stimulus to Science* (Cambridge, Mass.: Harvard University Press).

——and Ullian, J. S. [1970] *The Web of Belief* (New York: Random House).

Quinn, Paul [1994] 'The categorization of above and below spatial relations by young infants', *Child Development* 65, pp. 58–69.

——[2003] 'Concepts are not just for objects: categorization of spatial relation information by infants', in Rakinson and Oakes [2003], pp. 50–76.

——Eimas, Peter, and Rosenkrantz, Stacey [1993] 'Evidence for representations of perceptually similar natural categories by 3-month-old and 4-month-old infants', *Perception* 22, pp. 463–475.

Rakison, David, and Oakes, Lisa, eds. [2003] *Early Category and Concept Development: Making Sense of the Blooming, Buzzing Confusion* (New York: Oxford University Press).

Reich, Klaus [1932] *The Completeness of Kant's Table of Judgments*, trans. J. Kneller and M. Losonsky (Stanford, Calif.: Stanford University Press, 1992).

Reichenbach, Hans [1920] *The Theory of Relativity and A Priori Knowledge*, trans. M. Reichenbach (Berkeley: University of California Press, 1965).

——[1928] *The Philosophy of Space and Time*, trans. M. Reichenbach and J. Freund (New York: Dover, 1957).

——[1931] 'Aims and methods of the modern philosophy of nature', reprinted in M. Reichenbach and R. S. Cohen, eds., *Hans Reichenbach, Selected Writings: 1909–1953, volume 1* (Dordrecht: Reidel, 1978), pp. 359–388.

——[1936] 'Logical empiricism in Germany and the present state of its problems', *Journal of Philosophy* 33, pp. 141–160.

——[1938] *Experience and Prediction* (Chicago: University of Chicago Press).

——[1944] *Philosophical Foundations of Quantum Mechanics* (Berkeley: University of California Press).

——[1949] 'The philosophical significance of the theory of relativity', in Schilpp [1949], pp. 287–311.

Resnik, Michael [1980] *Frege and the Philosophy of Mathematics* (Ithaca, NY: Cornell University Press).

——[1997] *Mathematics as a Science of Patterns* (Oxford: Oxford University Press).

Richardson, Alan [1996] 'From epistemology to the logic of science: Carnap's philosophy of empirical knowledge in the 1930s', in Giere and Richardson [1996], pp. 309–332.

——[1997] 'Two dogmas about logical empiricism: Carnap and Quine on logic, epistemology, and empiricism', *Philosophical Topics* 25, pp. 145–168.

——[1998] *Carnap's Construction of the World* (Cambridge: Cambridge University Press).

——[2004] 'Tolerating semantics: Carnap's philosophical point of view', in Awodey and Klein [2004], pp. 63–78.

Ricketts, Thomas [1994] 'Carnap's principle of tolerance, empiricism, and conventionalism', in P. Clark and B. Hale, eds., *Reading Putnam* (Oxford: Blackwell), pp. 176–200.

——[1996] 'Carnap: from logical syntax to semantics', in Giere and Richardson [1996], pp. 231–250.

——[2003] 'Languages and calculi', in G. Hardcastle and A. Richardson, eds., *Logical Empiricism in North America*, Minnesota Studies in the Philosophy of Science 18 (Minneapolis: University of Minnesota Press), pp. 257–280.

Rosen, Gideon [1999] 'Review of Maddy [1997]', *British Journal for the Philosophy of Science* 50, pp. 467–474.

——[2001] 'Nominalism, naturalism, epistemic relativism', *Philosophical Perspectives* 15, *Metaphysics 2001*, pp. 69–91.

Roush, Sherrilyn [2006] *Tracking Truth* (Oxford: Oxford University Press).

Rowlands, Mark [1994] 'Connectionism and the language of thought', *British Journal of the Philosophy of Science* 45, pp. 485–503.

Russell, Bertrand [1914] *Our Knowledge of the External World* (London: Routledge, 1993).

——[1923] 'Vagueness', reprinted in Keefe and Smith [1997a], pp. 61–68.

Sainsbury, R. M. [1990] 'Concepts without boundaries', in Keefe and Smith [1997a], pp. 251–264.

Salmon, Wesley [1984] *Scientific Explanation and the Causal Structure of the World* (Princeton: Princeton University Press).

Schilpp, P. A. [1949] *Albert Einstein: Philosopher-Scientist* (La Salle, Ill.: Open Court).

Schlick, Moritz [1921] 'Critical or empiricist interpretation of modern physics', reprinted in his *Philosophical Papers*, vol. 1, ed. H. L. Mulder and B. F. B. van de Velde-Schlick, trans. P. Heath (Dordrecht: Reidel), pp. 322–334.

Scholnick, Ellin Kofsky [1983] *New Trends in Conceptual Representation: Challenges to Piaget's Theory?* (Hillsdale, NJ: Lawrence Erlbaum).

——Nelson, Katherine, Gelman, Susan, and Miller, Patricia, eds. [1999] *Conceptual Development: Piaget's Theory* (Mahwah, NJ: Lawrence Erlbaum).

Shapiro, Stewart [1997] *Philosophy of Mathematics: Structure and Ontology* (New York: Oxford University Press).

——[2000] *Thinking about Mathematics* (Oxford: Oxford University Press).

——, ed. [2005] *Oxford Handbook of Philosophy of Mathematics and Logic* (Oxford: Oxford University Press).

Siegel, Daniel [1991] *Innovation in Maxwell's Electromagnetic Theory* (Cambridge: Cambridge University Press).

Simons, Peter [2001] 'Review of Steiner [1998]', *British Journal for the Philosophy of Science* 52, pp. 181–184.

Sklar, Lawrence [1993] *Physics and Chance: Philosophical Issues in the Foundations of Statistical Mechanics* (Cambridge: Cambridge University Press).

Skyrms, Brian [1984] 'EPR: lessons for metaphysics', *Midwest Studies in Philosophy* 9, *Causation and Causal Theories* (Minneapolis: University of Minnesota Press), pp. 245–255.

——[1999] 'Evolution of inference', in T. Kohler and G. Gumerman, eds., *Dynamics in Human and Primate Societies* (New York: Oxford University Press), pp. 77–88.

Sluga, Hans [1980] *Gottlob Frege* (London: Routledge and Kegan Paul).

Smolensky, Paul [1988] 'On the proper treatment of connectionism', *Behavioral and Brain Sciences* 11, pp. 1–23.

Soames, Scott [1984] 'What is a theory of truth?', reprinted in Lynch [2001], pp. 397–418.

Sober, Elliott [1981] 'The evolution of rationality', *Synthese* 46, pp. 95–120.

Sorenson, Roy [2002] 'Vagueness', in *The Stanford Encyclopedia of Philosophy (Fall 2003 Edition)*, ed. Edward N. Zalta, <http://plato.stanford.edu/archives/fall2003/entries/vagueness/>.

Spelke, Elizabeth [1985] 'Preferential-looking methods as tools for the study of cognition in infancy', in Gottlieb and Krasnegor [1985], pp. 323–363.

——[1988] 'The origins of physical knowledge', in Weiskrantz [1988], pp. 168–184.

——[2000] 'Core knowledge', *American Psychologist* 55, pp. 1233–1243.

——and Newport, Elissa [1998] 'Nativism, empiricism, and the development of knowledge', in W. Damon and R. M. Lerner, eds., *Handbook of Child Psychology*, volume 1 (New York: Wiley), pp. 275–340.

——Breinlinger, Karen, Macomber, Janet, and Jacobson, Kristen [1992] 'Origins of knowledge', *Psychological Review* 99, pp. 605–632.

——Gutheil, Grant, and Van de Walle, Gretchen [1995a] 'The development of object perception', in S. M. Kosslyn and D. N. Osherson, eds., *Visual Cognition: An Invitation to Cognitive Science*, volume 2 (Cambridge, Mass.: MIT), pp. 297–330.

——Kestenbaum, Roberta, Simons, Daniel, and Wein, Debra [1995b] 'Spatiotemporal continuity, smoothness of motion and object identity in infancy', *British Journal of Developmental Psychology* 13, pp. 113–142.

——Vishton, Peter, and Von Hofsten, Claes [1995c] 'Object perception, object-directed action, and physical knowledge in infancy', in M. S. Gazzaniga, ed., *The Cognitive Neurosciences* (Cambridge, Mass.: MIT Press), pp. 165–179.

Stanford, P. Kyle, and Kitcher, Philip [2000] 'Refining the causal theory of reference for natural kind terms', *Philosophical Studies* 97, pp. 99–129.

Steel, John [2000] 'Mathematics needs new axioms', *Bulletin of Symbolic Logic* 6, pp. 422–433.

——[2004] 'Generic absoluteness and the continuum problem', www.lps.uci.edu/home/conferences/Laguna-Workshops/LagunaBeach2004/laguna1.pdf.

Stein, Howard [1970] 'On the notion of field in Newton, Maxwell, and beyond', in R. Stuewer, ed., *Historical and Philosophical Perspectives of Science, Minnesota Studies in the Philosophy of Science*, volume 5 (Minneapolis: University of Minnesota Press), pp. 264–287.

——[2004] 'The enterprise of understanding and the enterprise of knowledge', *Synthese* 140, pp. 135–176.

Steiner, Mark [1989] 'The application of mathematics in natural science', *Journal of Philosophy* 86, pp. 449–480.

——[1998] *The Applicability of Mathematics as a Philosophical Problem* (Cambridge, Mass.: Harvard University Press).

Sterelny, Kim [1990] *The Representational Theory of Mind* (Oxford: Basil Blackwell).

Stich, Stephen [1990] 'Evolution and rationality', chapter 3 of his *Fragmentation of Reason: Preface to a Pragmatic Theory of Cognitive Evaluation* (Cambridge, Mass.: MIT Press), pp. 55–74.

Stillwell, John [2002] *Mathematics and its History*, second edition (New York: Springer-Verlag).

Strawson, P. F. [1966] *The Bounds of Sense* (London: Routledge, 1975).

——[1985] *Skepticism and Naturalism* (New York: Columbia University Press).

Stroud, Barry [1977] *Hume* (London: Routledge and Kegan Paul).

——[1984] *The Significance of Philosophical Skepticism* (Oxford: Oxford University Press).

——[1989] 'Understanding human knowledge in general', reprinted in his [2000], pp. 99–121.

——[1994] 'Scepticism, "externalism", and the goal of epistemology', reprinted in his [2000], pp. 139–154, and in DeRose and Warfield [1999], pp. 292–304.

——[1996] 'Epistemological reflection on knowledge of the external world', reprinted in his [2000], pp. 122–138.

Stroud, Barry [2000] *Understanding Human Knowledge* (Oxford: Oxford University Press).

Tappenden, James [1994] 'Some remarks on vagueness and a dynamic conception of language', *Southern Journal of Philosophy* 33 (Supplement), pp. 193–201.

—— [2001] 'Recent work in philosophy of mathematics', *Journal of Philosophy* 98, pp. 488–497.

Tarski, Alfred [1933] 'The concept of truth in formalized languages', in his [1956], pp. 152–278.

—— [1936] 'The establishment of scientific semantics', in his [1956], pp. 401–408.

—— [1944] 'The semantic conception of truth and the foundations of semantics', reprinted in many places, including Blackburn and Simmons [1999], pp. 115–143.

—— [1956] *Logic, Semantics and Meta-Mathematics*, second edition, trans. J. H. Woodger, ed. J. Cocoran (Indianapolis: Hackett Publishing Company, 1983).

Tennant, Neil [2000] 'What is naturalism in mathematics, really?' (review of Maddy [1997]), *Philosophia Mathematica* 8, pp. 316–338.

Thagard, Paul [2005] *Mind: Introduction to Cognitive Science* (Cambridge, Mass.: MIT Press).

Tritton, D. J. [1988] *Physical Fluid Dynamics*, second edition (Oxford: Oxford University Press).

Troelstra, A. S., and van Dalen, Dirk [1988] *Constructivism in Mathematics* (Amsterdam: North Holland).

Tye, Michael [1990] 'Vague Objects', *Mind* 99, pp. 535–557.

—— [1994] 'Sorites paradox and the semantics of vagueness', reprinted in Keefe and Smith [1997a], pp. 281–293.

Urquhart, Alasdair [2001] 'Basic many-valued logics', in Gabbay and Guenthner [2001], pp. 249–295.

van Dalen, Dirk [1999] *Mystic, Geometer, and Intuitionist: The Life of L. E. J. Brouwer* (Oxford: Oxford University Press).

—— [2001] 'Intuitionistic logic', in Goble [2001], pp. 224–257.

—— [2002] 'Intuitionistic logic', in Gabbay and Guenthner [2002], volume 5, pp. 1–114.

van Fraassen, Bas C. [1980] *The Scientific Image* (Oxford: Oxford University Press).

—— [1985] 'Empiricism in the philosophy of science', in Churchland and Hooker [1985], pp. 245–308.

van Gelder, Tim [1990] 'Compositionality: a connectionist variation on a classical theme', *Cognitive Science* 14, pp. 355–384.

—— [1995] 'What might cognition be, if not computation?', *Journal of Philosophy* 91, pp. 345–381.

van Heijenoort, Jean, ed. [1967] *From Frege to Gödel: A Sourcebook in Mathematical Logic, 1879–1931* (Cambridge, Mass.: Harvard University Press).

Wald, Robert [1992] *Space, Time and Gravity: The Theory of the Big Bang and Black Holes*, second edition (Chicago: University of Chicago Press).

Weiskrantz, Lawrence [1986] *Blindsight* (Oxford: Oxford University Press).

—— ed. [1988] *Thought without Language* (Oxford: Oxford University Press).

—— [1997] *Consciousness Lost and Found: A Neuropsychological Exploration* (Oxford: Oxford University Press).

Weyl, Hermann [1918] *The Continuum*, trans. S. Pollard and T. Bole (New York: Dover, 1994).

Wiese, Heike [2003] *Numbers, Language and the Human Mind* (Cambridge: Cambridge University Press).

Wiggins, David [1980] *Sameness and Substance* (Oxford: Basil Blackwell).

—— [1997] 'Sortal concepts: a reply to Xu', *Mind and Language* 12, pp. 413–421.

Wigner, Eugene [1960] 'The unreasonable effectiveness of mathematics in the natural sciences', reprinted in his *Symmetries and Reflections* (Indianapolis: Indiana University Press, 1967), pp. 222–237.

Williams, Michael [1988] 'Epistemological realism and the basis for skepticism', *Mind* 97, pp. 415–439.

—— [1996] *Unnatural Doubts* (Princeton: Princeton University Press).

Williamson, Timothy [1992] 'Vagueness and ignorance', reprinted in Keefe and Smith [1997a], pp. 265–280.

—— [1994] *Vagueness* (London: Routledge).

Wilson, Mark [1982] 'Predicate meets property', *Philosophical Review* 91, pp. 549–589.

—— [1994] 'Can we trust logical form?', *Journal of Philosophy* 91, pp. 519–544.

—— [2000a] 'The unreasonable uncooperativeness of mathematics in the natural sciences', *Monist* 83, pp. 296–314.

—— [2000b] 'Where did the notion that forces are unobservable come from?', in E. Agazzi and M. Pauri, eds., *The Reality of the Unobservable* (Dordrecht: Kluwer), pp. 231–239.

—— [2006] *Wandering Significance* (Oxford: Oxford University Press).

Wittgenstein, Ludwig [1921] *Tractatus Logico-Philosophicus*, trans. C. K. Odgen (London: Routledge and Kegan Paul, 1922).

Wolff, Robert Paul, ed. [1967] *Kant: A Collection of Critical Essays* (Notre Dame, Ind.: University of Notre Dame Press).

Woodin, Hugh [2001] 'The continuum hypothesis, parts I and II', *Notices of the American Mathematical Society* 48, pp. 567–576, 681–690.

Worrall, John [1999] 'Two cheers for naturalised philosophy of science', *Science and Education* 8, pp. 339–361.

Wright, Crispin [1975] 'On the coherence of vague predicates', *Synthese* 30, pp. 325–365, excerpted as 'Language-mastery and the sorites paradox' in Keefe and Smith [1997a], pp. 151–173.

—— [1983] *Frege's Conception of Numbers as Objects* (Aberdeen: Aberdeen University Press).

—— [1992] *Truth and Objectivity* (Cambridge, Mass.: Harvard University Press).

—— [1996] 'Précis of *Truth and Objectivity*', reprinted in Wright [2003], pp. 3–10.

—— [1999] 'Truth: a traditional debate reviewed', reprinted in Blackburn and Simmons [1999], pp. 203–238.

—— [2001] 'Minimalism, deflationism, pragmatism, pluralism', in Lynch [2001], pp. 751–787.

—— [2003] *Saving the Differences* (Cambridge, Mass.: Harvard University Press).

Wussing, Hans [1969] *The Genesis of the Abstract Group Concept*, trans. A. Shenitzer, (Cambridge, Mass.: MIT Press, 1984).

Wynn, Karen [1990] 'Children's understanding of counting', *Cognition* 36, pp. 155–193.

—— [1992] 'Addition and subtraction by human infants', *Nature* 358, pp. 749–750.

—— [1998a] 'Psychological foundations of number: numerical competence in human infants', *Trends in Cognitive Sciences* 2, pp. 296–303.

—— [1998b] 'An evolved capacity for number', in Cummins and Allen [1998], pp. 107–126.

Xu, Fei [1997] 'From Lot's wife to a pillar of salt: evidence that *physical object* is a sortal concept', *Mind and Language* 12, pp. 365–392.

—— and Carey, Susan [1996] 'Infants' metaphysics: the case of numerical identity', *Cognitive Psychology* 30, pp. 111–153.

—— and Spelke, Elizabeth [2000] 'Large number discrimination in 6-month-old infants', *Cognition* 74, pp. B1–B11.

Yablo, Stephen [1998] 'Does ontology rest on a mistake?', *Proceedings of the Aristotelian Society*, supplementary volume 72, pp. 229–262.

—— [2000] 'A priority and existence', in Boghossian and Peacocke [2000], pp. 197–228.

Yi, Byeong-uk [2005/6] 'The logic of plurals', *Journal of Philosophical Logic* 34, pp. 459–506, 35, pp. 239–288.

Younger, Barbara [1985] 'The segregation of objects into categories by ten-month-old infants', *Child Development* 56, pp. 1574–1583.

—— [1986] 'Developmental change in infants' perception of correlations among attributes', *Child Development* 57, pp. 803–815.

——— [2003] 'Parsing objects into categories: infants' perception and use of correlated attributes', in Rakinson and Oakes [2003], pp. 77–102.

——— and Cohen, Leslie [1983] 'Infant perception of correlations among attributes', *Child Development* 54, pp. 858–867.

——— and Fearing, D. [1999] 'Parsing items into separate categories: developmental change in infant categorization', *Child Development* 70, pp. 291–303.

Zemanian, A. H. [1965] *Distribution Theory and Transform Analysis: An Introduction to Generalized Functions, with Applications* (New York: Dover, 1987).

Index